Evolving Issues Surrounding Technoethics and Society in the Digital Age

Rocci Luppicini
University of Ottawa, Canada

A volume in the Advances in Human and Social Aspects of Technology (AHSAT) Book Series

Managing Director: Lindsay Johnston
Production Editor: Jennifer Yoder
Development Editor: Erin O'Dea
Acquisitions Editor: Kayla Wolfe
Typesetter: James Knapp
Cover Design: Jason Mull

Published in the United States of America by
Information Science Reference (an imprint of IGI Global)
701 E. Chocolate Avenue
Hershey PA 17033
Tel: 717-533-8845
Fax: 717-533-8661
E-mail: cust@igi-global.com
Web site: http://www.igi-global.com

Library of Congress Cataloging-in-Publication Data

Evolving Issues Surrounding Technoethics and Society in the Digital Age.
Rocci Luppicini, editor
Includes bibliographical references and index. ISBN 978-1-4666-6122-6 (hardcover) -- ISBN 978-1-4666-6123-3 (ebook) --ISBN 978-1-4666-6125-7 (print & perpetual access) 1. Technology--Moral and ethical aspects. 2. Technology--Social-aspects. 3. Technology and civilization. I. Luppicini, Rocci. BJ59.E96 2014
170--dc23

2014012883

This book is published in the IGI Global book series Advances in Human and Social Aspects of Technology (AHSAT) (ISSN: 2328-1316; eISSN: 2328-1324)

British Cataloguing in Publication Data
A Cataloguing in Publication record for this book is available from the British Library.

Advances in Human and Social Aspects of Technology (AHSAT) Book Series

Ashish Dwivedi
The University of Hull, UK

ISSN: 2328-1316
EISSN: 2328-1324

Mission

In recent years, the societal impact of technology has been noted as we become increasingly more connected and are presented with more digital tools and devices. With the popularity of digital devices such as cell phones and tablets, it is crucial to consider the implications of our digital dependence and the presence of technology in our everyday lives.

The **Advances in Human and Social Aspects of Technology (AHSAT) Book Series** seeks to explore the ways in which society and human beings have been affected by technology and how the technological revolution has changed the way we conduct our lives as well as our behavior. The AHSAT book series aims to publish the most cutting-edge research on human behavior and interaction with technology and the ways in which the digital age is changing society.

Coverage

- Activism and ICTs
- Digital Identity
- Philosophy of Technology
- Human-Computer Interaction
- Human Development and Technology
- End-User Computing
- Cyber Behavior
- Cultural Influence of ICTs
- Technology and Social Change
- Cyber Bullying

IGI Global is currently accepting manuscripts for publication within this series. To submit a proposal for a volume in this series, please contact our Acquisition Editors at Acquisitions@igi-global.com or visit: http://www.igi-global.com/publish/.

The Advances in Human and Social Aspects of Technology (AHSAT) Book Series (ISSN 2328-1316) is published by IGI Global, 701 E. Chocolate Avenue, Hershey, PA 17033-1240, USA, www.igi-global.com. This series is composed of titles available for purchase individually; each title is edited to be contextually exclusive from any other title within the series. For pricing and ordering information please visit http://www.igi-global.com/book-series/advances-human-social-aspects-technology/37145. Postmaster: Send all address changes to above address.

Titles in this Series

For a list of additional titles in this series, please visit: www.igi-global.com

Technological Advancements and the Impact of Actor-Network Theory
Arthur Tatnall (Victoria University, Australia)
Information Science Reference • copyright 2014 • 320pp • H/C (ISBN: 9781466661264) • US $195.00 (our price)

Gender Considerations and Influence in the Digital Media and Gaming Industry
Julie Prescott (University of Bolton, UK) and Julie Elizabeth McGurren (Codemasters, UK)
Information Science Reference • copyright 2014 • 313pp • H/C (ISBN: 9781466661424) • US $195.00 (our price)

Human-Computer Interfaces and Interactivity Emergent Research and Applications
Pedro Isaías (Universidade Aberta (Portuguese Open University), Portugal) and Katherine Blashki (Noroff University College, Norway)
Information Science Reference • copyright 2014 • 325pp • H/C (ISBN: 9781466662285) • US $200.00 (our price)

Political Campaigning in the Information Age
Ashu M. G. Solo (Maverick Technologies America Inc., USA)
Information Science Reference • copyright 2014 • 359pp • H/C (ISBN: 9781466660625) • US $210.00 (our price)

Handbook of Research on Political Activism in the Information Age
Ashu M. G. Solo (Maverick Technologies America Inc., USA)
Information Science Reference • copyright 2014 • 498pp • H/C (ISBN: 9781466660663) • US $275.00 (our price)

Interdisciplinary Applications of Agent-Based Social Simulation and Modeling
Diana Francisca Adamatti (Universidade Federal do Rio Grande, Brasil) Graçaliz Pereira Dimuro (Universidade Federal do Rio Grande, Brasil) and Helder Coelho (Universidade de Lisboa, Portugal)
Information Science Reference • copyright 2014 • 376pp • H/C (ISBN: 9781466659544) • US $225.00 (our price)

Examining Paratextual Theory and its Applications in Digital Culture
Nadine Desrochers (Université de Montréal, Canada) and Daniel Apollon (University of Bergen, Norway)
Information Science Reference • copyright 2014 • 419pp • H/C (ISBN: 9781466660021) • US $215.00 (our price)

Global Issues and Ethical Considerations in Human Enhancement Technologies
Steven John Thompson (Johns Hopkins University, USA)
Medical Information Science Reference • copyright 2014 • 322pp • H/C (ISBN: 9781466660106) • US $215.00 (our price)

www.igi-global.com

701 E. Chocolate Ave., Hershey, PA 17033
Order online at www.igi-global.com or call 717-533-8845 x100
To place a standing order for titles released in this series, contact: cust@igi-global.com
Mon-Fri 8:00 am - 5:00 pm (est) or fax 24 hours a day 717-533-8661

Table of Contents

Section 2
Un/Ethical Waves in Cyberspace

Section 3
New Waves in Cyberspace

Detailed Table of Contents

Internet company ethics encountered when attempting to navigate differing social, cultural, and political landscapes.

Chapter 9 provides a critical account of the impact of the information revolution on data protection. The chapter contributes by drawing on current information system cases (video surveillance networks, smart cards for biometric identifiers) to unpack the challenges of ensuring privacy.

Chapter 10 delves into the murky waters of law and ethics in cyberspace. This chapter draws on selected legal and ethical issues that occur at the intersection of online activity and offline life to show how greater awareness of legal and ethical complexities is required to ensure a safeguarded online experience for all and to identify appropriate interventions to deal with negative online experiences such as the development of online gambling addictions.

Chapter 11 focuses on cybercrime in India and examines the motives underlying cybercrime perpetration. This chapter contributes by documenting the state of cybercrime in one part of the world and offers useful advice on measures for crime prevention in India.

Chapter 12 investigates the connection between ethics, reasoning, and the media as it bears on the role of ethical reasoning education for communication and media professionals. This chapter looks at the current state of globalization and the ever-increasing media system connectivity to illustrate the intricacies involved in how different media systems operate around the world.

Chapter 13 explores changing human values with the current information age. The author provides an insightful discussion on how information technologies can, at the same time, empower individuals and organizations, while creating new challenges to sustaining a sense of individuality and freedom. The

Preface

INTRODUCTION

The Human Face of Technology (And Vice Versa)

Selves are the ultimate negentropic technologies, through which information temporarily overcomes its own entropy, becomes conscious, and is finally able to recount the story of its own emergence in terms of a progressive detachment from external reality. There are still only informational structures. But some are things, some are organisms, and some are minds, intelligent and self-aware beings. Only minds are able to interpret other informational structures as things or organisms or selves. And this is part of their special position in the universe. (Floridi, 2011)

This provocative statement from Floridi illustrates the interconnectedness of humans, science, and technology within our evolving society. The advancement of technologies in the 20th century has radically altered life and society in substantial ways by offering new tools and techniques to leverage core areas of human activity including health, education, transportation, construction, communication, medicine, and economics. These transformations have been both outward and inward, challenging the very nature of human existence and what it means to live harmoniously with one another in a society that revolves around technological progress and innovation.

Much of the current research in technoethics mirrors key areas of technological development where society is divided. For instance, one emerging area of technoethical inquiry and debate revolves around the very nature of our human condition as technoselves (Luppicini, 2012a) and what it means to be human amidst a myriad of human enhancement technologies that allow us to augment human bodies and minds in substantive ways. There are a diversity of human enhancing technologies currently available or in development which have spurred public debate and concern (e.g., plastic surgery, prosthetic limbs, exoskeletons, performance and mind-enhancing drugs, biosensors, neural implants, wearable computers, etc.). These technologies (and others to come) are being scrutinized as an (un)ethical means of altering or transforming the human condition itself. The ethical debate over human nature and changing boundaries of humanity and the human body is a core concern within this technoethical area. Another area of technoethical inquiry examines the moral dilemmas revolving around the boundaries of the social world itself as we struggle to navigate the juxtaposition of humans and autonomous social robots, which for some warrant social rights and responsibilities. This type of ethical inquiry brings the debate concerning human enhancement to a new level of importance. Are human and artificial moral agents equivalent on some level that should be acknowledged? Beyond the realm of human embodiment and agency, there are also technological applications revolving around the Internet and social networking technologies that invoke public concern (digital democracy, media ethics, cybercrime, and online plagiarism). What are the ethical problems and dilemmas that individuals face in cyberspace? What are the privacy concerns with posting personal information online, and how much is too much? What are the trends in cyber-criminal activity and other areas of ethical misconduct that occur using the Internet? How does the use of sensationalized writing techniques used by mainstream journalists influence the credibility of online news? Grounded in the research domain of technoethics which places human ethics and values at the centre of technological inquiry, these questions and others are taken up in this volume.

Objectives of this Book

This edited book continues to expand the growing body of work on the existing intellectual platform within the field of technoethics. As a rapidly expanding area of inquiry, technoethics draws on and goes beyond traditional "ethics of technology" and "philosophy of technology" approaches, which highlight longstanding ethical theories and controversies for intellectual analysis. Technoethics also deals with current (and future) problems in science and technology innovation at the intersection of human life and society. As such, it brings into play an interdisciplinary base of scholarly contributions from pure philosophy, the social sciences, humanities, engineering, computing, applied sciences, and other areas of scholarly inquiry into technology and ethics. This interdisciplinary focus is helping to leverage ethical analysis, risk analysis, technology evaluation, and the combination of ethical and technological analyses within a variety of real life decision-making contexts and future planning situations faced by 21st century society.

The significance of the selected chapters appearing in this volume can be attributed to a variety of factors including high peer review standards, the timeliness of the topics covered, author attention to research protocols and methodological procedures, contribution to research knowledge on ethics and technology, and contribution to practice. Both empirical and theoretically oriented chapters are offered to provide a current snapshot of new developments in the field today. The remaining sections in this preface outline essential background knowledge, key areas of current research in the field, and brief sketches of chapters covered in this volume.

BACKGROUND

General Background

As the unity of the modern world becomes increasingly a technological rather than a social affair, the techniques of the arts provide the most valuable means of insight into the real direction of our own collective purposes. (McLuhan, 1951)

The modern world, as depicted by McLuhan over 60 years ago is continuing to evolve (technologically, individually, and socially) as humans attempt to weave the benefits of technological progress with the need to further collective purposes of humans living together in a society that embraces progress and innovation. In response, the field of Technoethics has developed to deal with the mounting ethical debates and dilemmas that accompany the co-evolution of technology and human society.

There are a number of seminal publications that can be cited as key drivers in the formalization of Technoethics as an interdisciplinary research field. These publications brought together leading technology and ethics scholars from around the world and provided a solid intellectual platform upon which to grow. First, the two volume *Handbook of Research on Technoethics* (Luppicini & Adel, 2009) drew on the contributions of over 100 experts from around the world working on a diversity of areas where technoethical inquiry. Next, the first reader in *Technoethics, Technoethics and the Evolving Knowledge Society: Ethical Issues in Technological Design, Research, Development, and Innovation* (Luppicini, 2010) was published for use at the undergraduate and graduate level in a variety of courses that focus on technology and ethics in society. The text focused on a broad base of human activities connected to technology interwoven within social, political, and moral spheres of life. It engaged readers in the study of key ethical dimensions of a technological society and helped reinforce work found in the handbook. Together, these publications helped set the stage for the creation of the *International Journal of Technoethics* in 2010.

In 2012, the *International Journal of Technoethics* (Rocci Luppicini, Founding Editor-in-Chief) completed its third year of publication. This international peer-reviewed journal provides a forum for scholarly exchange among philosophers, researchers, students, social theorists, ethicists, historians, practitioners, and technologists working in areas of human activity affected by technological advancements and applications. With the strong support of IGI Global, the journal retains its founding Editor-in-Chief and its 12 Associate Editors, namely Allison Anderson (University of Plymouth), Keith Bauer (Marquette University), Josep Esquirol (University of Barcelona), Deb Gearthart (Ohio University), Pablo Iannone (Central Connecticut State University), Mathias Klang (University of Lund), Andy Miah (University of the West of Scotland), Lynne Roberts (Curtin University of Technology), Neil Rowe (U.S. Naval Postgraduate School), Martin Ryder (Sun Microsystems), John Sullins III (Sonoma State University), and Mary Thorseth (NTNU). The mission of the journal is as follows:

The mission of the International Journal of Technoethics (IJT) is to evolve technological relationships of humans with a focus on ethical implications for human life, social norms and values, education, work, politics, law, and ecological impact. This journal provides cutting edge analysis of technological innovations, research, developments policies, theories, and methodologies related to ethical aspects of technology in society. IJT publishes empirical research, theoretical studies, innovative methodologies, practical applications, case studies, and book reviews. IJT encourages submissions from philosophers, researchers, social theorists, ethicists, historians, practitioners, and technologists from all areas of human activity affected by advancing technology. (International Journal of Technoethics, n.d.).

Technoethics developed through the coming together of technology experts (philosophers, technicians, administrators, instructors, students, and researchers) struggling with the many dilemmas arising from public controversies and ethical debate created by technological advancement in society. This interdisciplinary research field provided a means to transcend traditional approaches in the study of ethics and technology driven by existing philosophical approaches, intellectual analyses of pervasive problems, and logical reasoning. It provided a multifaceted intellectual platform for experts working at the nexus of applied work in technology and ethics (e.g., bioethics, engineering ethics, computer ethics, etc.). The types of scholars attracted to this field are scholars and technology experts working in new areas of technology research where social and ethical issues emerge (i.e., genetic research, nanotechnology, human enhancement, neurotechnology, robotics, reproductive technologies, etc). The current state of Technoethics is marked by an openness to multiple forms of scholarly inquiry and practical real world value. As stated in Luppicini (2012b):

As pioneering breakthroughs are made in technological advancements and applications, novel questions arise regarding human values and ethical implications for society, many of which give rise to ethical dilemmas where conflicting viewpoints cannot be solved by relying on any one ethical theory or set of moral principles. Accordingly, the field of Technoethics takes a practical focus on the actual impacts (and potential impacts) of technology on human beings struggling to navigate the "real world" of technology. In many cases, this leads to the creation of more questions than answers in an effort to discern the underlying ethical complexities connected to the application of technology within real-life situations.

The topics covered in this volume expand on scholarship covered in the *International Journal of Technoethics* and supporting publications. It provides up to date coverage of cutting edge work from a variety of areas where technoethical inquiry is currently being applied.

KEY AREAS OF TECHNOETHICS IN FOCUS

Caution: Humans under Construction

It may seem odd to some to bring up such an entrenched notion as the homo sapiens (Latin for wise man) and its claim to fame as representing the most sentient beings on Earth. This popular nomenclature aligned with evolutionary theory and the morphological change that separated homo sapiens from earlier primates who lacked the precision grip offered by the opposable thumb, along with lowered larynx and hyoid bone, which made speech possible (Brues & Snow, 1965). Although these evolutionary developments were significant, equally significant were the historical and behavioural changes that accompanied the development of material culture resulting from the industrial revolution. It was at this time that the human construction of objects became intimately intertwined with human life and society in a way that raised public fear and awareness about the close connection between technology, human life, and society. Under the industrial revolution, the nature of human work, community structures, and the structure of society became reorganized under an industrial model. It was at this point in history where scientific and technological advancement became core drivers of social interaction and human development. In this way, the industrial and post-industrial age became the most significant developments in differentiating humans from all other living entities on Earth since the time of Darwin.

In the 21st century, the priority position given to science and technology in driving human and social evolution has not declined. If anything, it has become more pronounced in the rise of the digital age according to academic work, which acknowledges the close link and importance of technology in the evolution of society and human development within it (Richerson & Boyd, 2005). Within this digital age, the material culture is not only concerned with external objects (digital devices, tablets, laptops, cell phones, etc.), which humans can use, but it also includes transformative technologies (objects), which are more intimately connected to human beings (neural implants, prosthetics, performance-enhancing substances). It is the shift from external object focus to internal object focus that marks an important shift in human culture and values, which defines this stage of the digital society. This shift has also created some of the most challenging ethical dilemmas that society must face.

Will enhanced humans become the new norm, and if so, what will this mean for those that are not enhanced? How do we navigate the changing conceptualization of what constitutes human nature and what are/should be the safe and ethical limits of human enhancement to preserve the integrity and meaning of what it means to be human? As a young academic, my supervisor used to remind me that we have always been technical beings (what I later expanded on and developed with the notion of "homotechnicus" (with Jose Galvin) and, more recently, the notion of the "technoself"). This helped pull together the amassing conceptual and empirical investigations from established scholars around the world who, like me, recognized the intrinsic technological nature of human beings and the need to update the existing conceptualization of what it means to be human. The notion of "technoself" was created by pulling together the core aspects of technical and non-technical features of human nature expressed in polar notions of a technical human (cyborg, posthuman, transhuman, beman, etc.). At its core, this provides a more accurate representation of human nature to advance knowledge.

One major driving force for the reconceptualization of humans aligns with the recent development of new human enhancement and therapeutic technologies that create neurological changes in humans to augment or restore brain functions. The use of technology in neurological treatments is just one example of the plethora of new tools available to the technoself. For example, an electrode placed in the brain has been found to be effective in stopping limb tremors in Parkinson's patients. However, there are other technological advances, such as the use of neurochips and other types of brain-machine interfaces, which may have unintended consequences. If many alterations are made to the brain, it must be questioned as to whether the issue of personal liability could be affected. Who is responsible for actions in humans subjected to such technological alterations at the neurological level? It is no surprise that there is a growing body of work in neuroethics focusing on technological alterations to brain chemistry and function. This is just one of the topic areas covered in this volume.

Extending Human Ethics into our Creations

Does it make sense to attribute moral agency and ethical responsibility to technological creations such as social robots and other types of autonomous agents? Given the vivid images imparted to us from science fiction literature and films (and extreme proponents of the transhumanist movement like Ray Kurweil), it becomes easy to dream of such a reality at this point in history or in the near future. Certainly the computational dream of downloading one's consciousness into a second vessel (human, android, or digital hologram) appears to be completely unfounded (and absurd). However, there is a larger public concern about the ethical limits of human enhancement and the augmentation of status for social robotic applications becoming increasingly entrenched in human life and society. For example, through her scholarly writings, MIT professor Sherry Turkle has explored the connection of children and the elderly to sociable robots designed for personal companionship and care. There is the very real and growing fear that the increased personal investment and time in the relationships built with artificial agents/devices will detract from the quality of personal investment and time in building relationships with human beings. This represents one major area of focus in this volume.

Riding the Waves in Cyberspace

There is the old saying, "the old grey mare, she ain't what she used to be," and this is definitely true of the Internet and how it is being used today. In the early days of the Internet, beginning in 1969 (called ARPANET), it was used as a measure to safeguard information and communications in the case of war. With the advent of personal computing and popularization of the Internet in the 1990s, the digital "wild wild west" was open for business, and it gradually became known that the Internet was not always a positive driver of culture and society. The lives of too many people have been negatively impacted by the unethical waves of exploitation, harassment, and abuse, which have allowed cybercrime to develop and flourish.

What is cybercrime? Cybercrime is a fairly new type of crime that has only become possible with the advent of the Internet and advanced digital technologies. Cybercrime refers to any crime that involves a computer and a network. According to the Britannica Online Encyclopedia (Cybercrime, 2012):

Cybercrime, also called computer crime, the use of a computer as an instrument to further illegal ends, such as committing fraud, trafficking in child pornography and intellectual property, stealing identities, or violating privacy. Cybercrime, especially through the Internet, has grown in importance as the computer has become central to commerce, entertainment, and government.

Given the widespread growth of the Internet and networking technologies within the global economy and social life, efforts to locate and eliminate cybercrime represents a serious challenge for law enforcement agencies around the world. It is trivially true that we live at a time when computing is at the heart of the knowledge economy and social life itself (Luppicini, 2009). Unfortunately, there are myriad cybercrime varieties that exist, including cyber espionage, cyber terrorism, cyber stalking, cyberbullying, identity theft, and information theft (to name just a few). In a more subtle, yet equally important area of public concern, there is continual criticism against sensationalistic media, which distorts the truth and biases the content of online news media to highlight what is exciting and controversial rather than trying to provide fair and comprehensive coverage of actual events. It raises the question as to what online news should cater to, what is actually happening or what broadcasters believe people want to see for entertainment value.

Within the sea of information online are the ingredients for a paradise of calm ethical waves and a dangerous tsunami with the power to crush everyone it its path. In an effort to mitigate the unethical waves crashing up against the shores of global cyberspace, there are a number of countermeasures aimed at curbing negative consequences faced by digital natives while highlighting the positive. For instance, the formulation of rules and guidelines for netiquette along with the promotion of broader programs of e-learning ethics provide a means of setting visible boundaries for what should and what should not take place with respect to student use of the Internet and digital tools for academic purposes. These types of measures play an important role in defining social expectations to help steer away from the turbulent seas of online plagiarism and other academic offences. Building on this framework of appropriate standards are broader societal aims geared towards increasing democratic participation in society through the use of the Internet and social networking. This volume provides coverage of new ethical and unethical waves found on the Internet.

ORGANIZATION OF THE BOOK

This volume pulls together current scholarship pertaining to recent developments within the field of Technoethics. In terms of book organization, this book contains 16 chapters divided into 3 sections to highlight a logical flow of writing organized into key thematic areas of technoethical inquiry. Section 1, "Ethical Boundaries of Humans and Robots," contains 6 chapters: Chapter 1, "Redefining the Boundaries of Humanity and the Human Body: From Homo Sapiens to Homo Technicus" (Jose Galván and Rocci Luppicini); Chapter 2, "The Relevance of Value Theory for the Ethical Discussion of Human Enhancement" (Tobias Hainz); Chapter 3, "Personal Liability and Human Free Will in the Background of Emerging Neuroethical Issues: Some Remarks Arising from Recent Case Law up to 2013" (Angela Di Carlo and Elettra Stradella); Chapter 4, "Artificial Ethics: A Common Way for Human and Artificial Moral Agents and an Emergent Technoethical Field" (Laura Pană); Chapter 5, "Of Robots and Simulacra: The Dark Side of Social Robots" (Pericle Salvini); and Chapter 6, "Nature and Cases (Peter Heller). Section 2, "Unethical Waves in Cyberspace," contains 5 chapters: Chapter 7, "Sacrificing Credibility for Sleaze: Mainstream Media's Use of Tabloidization" (Jenn Mackay and Erica Bailey); Chapter 8, "Internet Companies and the Great Firewall of China: Google's Choices" (Richard A. Spinello); Chapter 9, "The Legal Challenges of the Information Revolution and the Principle of 'Privacy by Design'" (Ugo Pagallo); Chapter 10, "Gambling with Laws and Ethics in Cyberspace" (Lee Gillam and Anna Vartapetiance); and Chapter 11, "A Study of Cyber Crime and Perpetration of Cyber Crime in India (Saurabh Mittal and Ashu Singh). Finally, Section 3, "New Waves in Cyberspace," contains 5 chapters: Chapter 12, "Ethics, Media, and Reasoning: Systems and Applications" (Mahmoud Eid); Chapter 13, "Antinomies of Values under Conditions of Information Age" (Liudmila Baeva); Chapter 14, "Shaping Digital Democracy in the United States: my.barackobama.com and Participatory Democracy" (Rachel Baarda and Rocci Luppicini); Chapter 15, "Ethics for eLearning: Two Sides of the Ethical Coin" (Deb Gearhart); and Chapter 16, "Privacy vs. Security: Smart Dust and Human Extinction" (Mark Walker). More detailed descriptions of the chapters are as follows:

- Chapter 1, "Redefining the Boundaries of Humanity and the Human Body: From Homo Sapiens to Homo Technicus," by Jose Galván and Rocci Luppicini, takes a fresh look at the recent debate concerning what it means to be human within a technological society and what attributes are core to human beings with respect to human enhancement technologies. This chapter explores key concepts that are helping to redefine how human beings, as homotechnicus, provide a more accurate vision of human beings and the priority of ethics over technics within the evolving technological society. A technoethical perspective of the human being is presented in order to highlight defining characteristics of humans entrenched within a technological society. Under this framework, symbolic capacity and technical ability are assumed to be grounded within the free and ethical nature of human beings. Ideas derived from Modernity and Postmodernity are drawn upon in an effort to find a more encompassing view of humans, one which accommodates both its technical and ethical dimensions.
- Chapter 2, "The Relevance of Value Theory for the Ethical Discussion of Human Enhancement," by Tobias Hainz, explores core aspects of value theory to show its utility in leveraging ethical debates concerning human enhancement technologies. Key dimensions of value theory discussed include value lexicality, the monism-pluralism dichotomy, and incommensurability. In addition to providing new concepts to help navigate debates concerning human enhancement technologies,

practical examples are drawn upon to highlight key arguments for and against human enhancement. The chapter makes a solid contribution by demonstrating how value theory can leverage the ethical discussion of human enhancement in an effort to raise awareness of value-theoretical issues revolving around human enhancement technologies.

- Why are people concerned about the possible legal and ethical issues connected to the uptake of new neurotechnologies? Chapter 3, "Personal Liability and Human Free Will in the Background of Emerging Neuroethical Issues: Some Remarks Arising from Recent Case Law up to 2013" (Angela Di Carlo and Elettra Stradella), examines ethical and legal issues connected to emerging neurotechnologies in relation to relevant legal concepts, including capacity, liability, testimony, evidence, fundamental constitutional rights and freedoms, the principle of human dignity, etc. More specifically, the authors question how the nature of personal liability is changing due to new scientific developments. This chapter contributes by highlighting the opportunities and challenges of neurolaw and neurethics within the traditional legal framework.
- Chapter 4, "Artificial Ethics: A Common Way for Human and Artificial Moral Agents and an Emergent Technoethical Field" (Laura Pană), envisions a new type of morality arising from public debates concerning current scientific and technical developments. In this provocative chapter, the author appeals to the need for a new approach to ethics, one in which both individual and social morality are key priorities with respect to controversial scientific and technical developments. To this end, the author attempts to unpack key scientific, technical, and philosophical premises to help ground a program of artificial ethics. Pană's innovative approach to artificial ethics contributes by establishing the place of artificial ethics within the group of new and emergent ethical fields of research connected to our technological society.
- Chapter 5, "Of Robots and Simulacra: The Dark Side of Social Robots" (Pericle Salvini), elaborates on a promising theoretical framework for assessing the ethical acceptability of robotic technologies (including social robots). In this chapter, the author conceptualizes robots as a form of mediation between human actions and their ethical acceptance based on the notion of human presence. The author makes a convincing case to show that human presence is characterised by a system of mutual relations among human beings and the environment shaped through technological mediation. The chapter makes a valuable contribution by drawing attention to the increasing ethical impact of social robots and simulated forms of human presence in the lives of human beings.
- Chapter 6, "Nature and Cases" (Peter Heller), examines the complexity and typical ethical trade-offs that occur when new technologies are applied in core areas of life and society (health, medical, military, engineering, etc.). The chapter is grounded in the belief that new technological advances often lead to ethical trade-offs in an effort to balance the desirable versus undesirable ethical aspects of a new technology or technological application. The discussion contributes by drawing on empirical cases to address the difficult challenges faced by decision makers struggling to decide on overall moral merit based on numerous variables examined.
- Chapter 7, "Sacrificing Credibility for Sleaze: Mainstream Media's Use of Tabloidization" (Jenn Mackay and Erica Bailey), draws on an empirical study to help explain how sensationalized (tabloid-style) writing techniques employed by mainstream journalists influence the credibility of online news. The chapter documents a study where participants read a series of news stories and rated perceived credibility of writing using McCroskey's Source Credibility Scale. The results showed that tabloid style articles were rated less credible than traditional stories and that

online news media using tabloidized writing techniques may limit perceived article credibility. Furthermore, participants were less likely to enjoy stories written in a tabloidized style.

- Chapter 8, "Internet Companies and the Great Firewall of China: Google's Choices" (Richard A. Spinello), addresses a recent technoethical dilemma in the world of the Internet. The chapter focuses on the search engine company Google and the types of ethical challenges faced when dealing with current censorship policies in China. This case study makes an important contribution to the discussion of Internet company ethics encountered when attempting to navigate differing social, cultural, and political landscapes. Based on the authors' analysis and an appeal to the ideal of universal human rights, including the natural right of free expression, the case is made that a socially responsible company should resist the implementation of censorship regulations within authoritarian political regimes.
- Chapter 9, "The Legal Challenges of the Information Revolution and the Principle of 'Privacy by Design'" (Ugo Pagallo), provides a critical account of the impact of the information revolution on data protection in a twofold way. First, the chapter addresses the cross-border interaction taking place in cyberspace and how it challenges current legal frameworks. Second, it draws on the principle of privacy to show how normative limits can be established to safeguard data protection and guard against liability issues. The chapter contributes by drawing on current information system cases (video surveillance networks, smart cards for biometric identifiers) to unpack the challenges of ensuring privacy.
- Chapter 10, "Gambling with Laws and Ethics in Cyberspace" (Lee Gillam and Anna Vartapetiance), delves into the murky waters of law and ethics in cyberspace. The author makes the case that, while many Internet users ignore policies and laws that govern online activities, there are valid reasons and existing laws in the physical world, which apply to online conduct that Internet users should follow. This chapter draws on selected legal and ethical issues that occur at the intersection of online activity and offline life to show how greater awareness of legal and ethical complexities is required to safeguard the online experience for all and to identify appropriate interventions to deal with negative online experiences such as the development of online gambling addictions.
- Chapter 11, "A Study of Cyber Crime and Perpetration of Cyber Crime in India" (Saurabh Mittal and Ashu Singh), focuses on cybercrime in India and examines the motives underlying cybercrime perpetration. This chapter contributes by documenting the state of cybercrime activity in one part of the world and offers useful advice on measures for crime prevention in India.
- Chapter 12, "Ethics, Media, and Reasoning: Systems and Applications" (Mahmoud Eid), investigates the connection between ethics, reasoning, and the media as it bears on the role of ethical reasoning education for communication and media professionals. This thought-provoking chapter looks at the current state of globalization and the ever-increasing media system connectivity to illustrate the intricacies involved in how different media systems operate around the world. The author makes a strong case for the need to pay greater attention to cultivate rational and ethical skills to cope with the huge impact that media has on society.
- Chapter 13, "Antinomies of Values under Conditions of Information Age" (Liudmila Baeva), explores changing human values in the current information age. The author provides an insightful discussion on how information technologies can, at the same time, empower individuals and organizations and create new challenges to sustaining a sense of individuality and freedom. The chapter discusses how the virtualization of core areas of life and society (communication, education, leisure, art) has led to a substitution of virtual versions and simulacra for traditional relations and

amenities. The chapter is devoted to studying value antinomies of the modern age: information and knowledge, virtuality and reality, feelings and game, friendship and contacts, etc. The chapter contributes by providing a rich conceptual understanding of key ethical challenges faced by individuals. Since values are the projections of the future in the present, this chapter helps to elicit to some extent the main trend of the social-cultural dynamics of the modern high-tech society.

- Chapter 14, "Shaping Digital Democracy in the United States: my.barackobama.com and Participatory Democracy" (Rachel Baarda and Rocci Luppicini), puts ideas about participatory democracy to the test by examining the ethical implications of new digital technologies and social media within the context of e-politics. This chapter reviews an empirical study conducted on Barack Obama's campaign social networking site, my.barackobama.com. It explores the ways in which the site was used to create participatory democracy within an online community. A content analysis of the Website and interviews with members of groups on the site provide insight into the uses and abuses of the Internet.
- Chapter 15, "Ethics for eLearning: Two Sides of the Ethical Coin" (Deb Gearhart), reveals that there are multiple ethical issues an eLearning program administrator faces that merit serious consideration. This chapter discusses core ethical concerns facing eLearning administrators, including internal ethical issues (eLearning program quality control) and external ethical issues (un/ethical behaviors from eLearning students and counter measures). This chapter provides helpful insight into the ethical challenges faced by eLearning program administrators in their profession.
- Chapter 16, "Privacy vs. Security: Smart Dust and Human Extinction" (Mark Walker), investigates the dilemma created by intrusive surveillance technologies required to safeguard people's security. At the same time, it explores the potential negative consequences such technologies might have on individual privacy. The chapter contributes valuable policy recommendations to help guide surveillance organizations such as the National Security Agency.

In summary, this collection provides an excellent collection of the latest developments in technoethics in the hopes of helping readers better navigate the murky ethical and social waters created by a broad range of "new" technological advances within society today. As the editor of this volume, it is a privilege to present the following 16 chapters, which delve into the current state of technoethics and new dilemmas in the age of technology.

Rocci Luppicini
University of Ottawa, Canada

REFERENCES

Brues, A. M., & Snow, C. C. (1965). Physical anthropology. *Biennial Review of Anthropology*, *4*, 1–39.

Cybercrime. (2012). *Britannica online encyclopedia*. Retrieved, October 15, 2012, from www.britannica.com/EBchecked/topic/130595/cybercrime

Floridi, L. (2011). The informational nature of personal identity. *Minds and Machines*, *21*(4), 549–566. doi:10.1007/s11023-011-9259-6

International Journal of Technoethics. (n.d.). Retrieved from http://www.igi-global.com/journal/international-journal-technoethics-ijt/1156

Luppicini, R. (2009). Technoethical inquiry: From technological systems to society. *Global Media Journal*, *2*(1), 5–21.

Luppicini, R. (2010). *Technoethics and the evolving knowledge society*. Hershey, PA: Idea Group Publishing. doi:10.4018/978-1-60566-952-6

Luppicini, R. (Ed.). (2012a). *Handbook of research on technoself: Identity in a technological society*. Hershey, PA: Idea Group Publishing. doi:10.4018/978-1-4666-2211-1

Luppicini, R. (Ed.). (2012b). *Ethical impact of technological advancements and applications in society*. Hershey, PA: Idea Group Publishing. doi:10.4018/978-1-4666-1773-5

Luppicini, R., & Adell, R. (Eds.). (2009). *Handbook of research on technoethics*. Hershey, PA: Idea Group Publishing.

McLuhan, M. (1951). *Mechanical bride: Folklore of industrial man*. New York: Vanguard Press.

Richerson, R., & Boyd, R. (2005). *Not by genes alone: How culture transformed human evolution*. Chicago: University of Chicago Press.

Section 1

Ethical Boundaries of Humans and Robots

Chapter 1
Redefining the Boundaries of Humanity and the Human Body: From Homo Sapiens to Homo Technicus

José M. Galván
Pontificia Università della Santa Croce, Italy

Rocci Luppicini
University of Ottawa, Canada

ABSTRACT

What are the boundaries of humanity and the human body within our evolving technological society? Within the field of technoethics, inquiry into the origins of the species is both a biological and ethical question as scholars attempt to grapple with conflicting views of what it means to be human and what attributes are core to human beings within the era of human enhancement technologies. Based on a historical and conceptual analysis, this chapter uses a technoethical lens to discuss defining characteristics of the human species as homo technicus. Under this framework, both symbolic capacity and technical ability are assumed to be grounded within the free and ethical nature of human beings. Ideas derived from Modernity and Postmodernity are drawn upon to provide a more encompassing view of humans that accommodates both its technical and ethical dimensions as homo technicus.

OLD WINE NEW BOTTLES? TRANSHUMANISM AND THE POST-HUMAN DREAM

In recent years, new advances in human enhancement technologies have stimulated academic interest about the future of human-technology relationships. Transhumanist scholarship is one major area where the transformation of the human condition through technological enhancement is being discussed. However, the transformation of humanity proposed within popular transhumanist writing is not a new notion and harkens back to the story of the Great Flood found in the Book of Genesis. In this story God provides Noah with the tools to survive a great flood that will wipe out everything and provide opportunity for a re-birth (or transformation) of humanity:

DOI: 10.4018/978-1-4666-6122-6.ch001

And God said to Noah, " have decided to put an end to all mortals on earth; the earth is full of lawlessness because of them. So I will destroy them and all life on earth. Make yourself an ark of gopherwood, put various compartments in it, and cover it inside and out with pitch. This is how you shall build it: the length of the ark shall be three hundred cubits, its width fifty cubits, and its height thirty cubits. Make an opening for daylight in the ark, and finish the ark a cubit above it. Put an entrance in the side of the ark, which you shall make with bottom, second and third decks. I, on my part, am about to bring the flood (waters) on the earth, to destroy everywhere all creatures in which there is the breath of life; everything on earth shall perish. But with you I will establish my covenant; you and your sons, your wife and your sons' wives, shall go into the ark."(Genesis 6:13-18)

Common to both transhumanist speculations and the story of the Great Flood was the idea of human longevity beyond normal human parameters. In the story of the Great Flood, human lifespan prior to the Great Flood was much longer than within contemporary society. Noah was noted to have lived nine hundred and fifty years, an unfathomable lifespan for humans today.

What is unique about the current context of human transformation focused on by the transhumanists it that the means of this transformation are not viewed in terms of divine intervention or concrete historical terms of post-industrial advances in technology that have allowed humans to survive and harness control over the environment. Rather, within the current context of human transformation there is a focus on the advancement of human enhancement technologies that directly affect the human body and mind. This "nward turn of technology"(Luppicini, 2010) presents new ethical challenges to address that humans have never had to deal with in real life circumstances at any other point in human history. Prior to this inward turn, such considerations were restricted to the arts and entertainment domain where science fiction writing and film sparked the imagination without the social responsibility to worry about any real life implications. However, the inward turn of technology has closed the gap between the imagination and the real as we struggle to grapple with new technologies that could potentially threaten how human beings are conceptualized and (de)valued. This is the problematic with which this article is concerned. To this end, a technoethical framing is offered which attempts to provide a conceptualization of human beings that accommodates the complexity of human, technological, and ethical relations present within a technological society.

CONCEPTUALIZING THE BODY AS TECHNOETHICAL MATTER

In the past decade, the discussion of the nature of the human body has been the basis of bioethical debate, conditioning the form of judging many of the emerging biotechnologies, both in medical and surgical fields. Bioethics has lost the body (Meilaender, 1995). One of the most significant examples is a book by Campbell (2009), in which the need "o re-establish the importance of the human body in bioethics"(p. 1) is indicated. The author shows the need for such a rediscovery in the biomedical sciences and the humanities and social sciences to combat possible risks of reducing the body of the person to a mere instrument, " branded body"p. 75). The risk does not concern the alternation or elimination of the human body but rather a substantial loss of its anthropological significance as a result of inappropriately applying new human altering technologies. But what does this mean and how could this alter the anthropological significance of the human body?

One must recognize that within the technological society we live, the human body is not only subject to alteration through one type of technology that alters one aspect of the human condition. There are many other human altering technologies to consider, including prosthetic appendages, neurotechnologies, and nanotechnologies. Thanks to the development of new techno-sciences, scientists are able to integrate the organic and inorganic in new and powerful ways. Thus ever-increasingly, a plethora of new technologies are becoming available to integrate into our physiological make-up, opening new topic areas not covered in traditional bioethics.

In our current world where technological enhancement of humans is possible, technoethics treats the human body as a technical being and ethical, largely because we need to question the, more or less, intrusive presence in the organism of devices with different degrees of autonomous functioning. This includes simple tools as well as cybernetic machines driven by powerful AI systems which show a very high level of autonomous ability which could interfere with human agency and how human beings are defined. This could affect how much we do for ourselves versus how much is programmed to be done for us at work and even in everyday life interactions.

Transhumanist writings are one particularly popular areas of contemporary scholarly work that addresses the transforming role of technology in defining the future of humanity. One tenacious transhumanist position holds that the human body acts as a necessary wrapping, but is not essential to the constitution of a human being. For example, Kevin Warwick affirmed without hesitation that the physical dimension is not important for the ethical evaluation of cyborg question: you have to take into account only the "uman and machine mental functioning"(2003, p. 131). There is the assumption among many transhumanists that if one could find a better way to contain a human, a human could get along without a biological body, perhaps by downloading oneself from a decaying body to a computer or some type of robotic body. This type of stance highlights the evolving interaction between living organisms and cybermachines in terms of a new species (a tertium quid) in transition- the posthuman.

Unfortunately, this type of redefinition of human nature can run the risk of removing much of the value and meaning we attribute to being human. In our view, regardless of the human enhancement technology that comes along, we believe it is unwise to accept visions which push the human body into a background, allowing a simple instrumental consideration of the body as, for example, in the paper of Thomas Schramme, in which this author asserts: "n conclusion, I see no convincing support for prohibition of voluntary mutilations"(2008, p. 15). We assume that humans take care of their bodies, not just as tools, but as expressions of themselves, alive or dead. Even in the case of death, humans do not easily dispose of the bodies of their dearly departed. Despite all the human efforts to get rid of the body, it seems not possible to have experience of being without a body. In the discussion that follows, key concepts are drawn upon to help highlight important aspects of what it means to be a human being living within a technological society.

WHAT IS A HUMAN BODY AND WHAT IS A HUMAN BEING?

First of all we would like to clarify this section title "hat is the human body,"in order to avoid a dualistic assumption that it is possible to question the human body without including a consideration about the human being; the question, in fact, has a preconception. No one among us discovering a never seen animal would question "hat is the body of that animal?"but rather one would say "hich animal is this?"In the case of a human being, "iped without feathers acting autonomously, quite weird,"a Martian would ask "hat is this?"instead of "hat is the body of this?"So the title of this section could be better expressed into "hat is the human being,"considering this question as a result of a simple phenomenological contemplation and of corporeality.

At the beginning of one of his works Robert Spaemann (1991) dialectically expressed that the question "hat is a sparrow?"is a different kind of question from "hat is a human being?"In line with this provocation, we would like to specify that the expression "uman body"includes a meaning of the word "ody"which is not shareable with a concept of inanimate bodies or with the body of plants and animals. We realize that some contemporary authors, such as Peter Singer, do not accept this distinction (Cavalieri & Singer, 1993); but it seems to us that not to accept it is not simply phenomenological, but comes from an rational pre-understanding of human and animal nature (Torralba Roselló, 2005, pp. 97-193; Schaler, 2009).

The meanings of the word "ody"are at least threefold. First, there is the "ody"of the matter of which is made whatever animated or inanimate being. Second, there is the concept of "iological body"(alive) that designates any living being able to develop on its own accord as a self-producing system (different compared virus that depends on a host body to spread). Third, there is the idea of a specific type of living body, the "uman body" The human body shares with other living bodies the processes of living as a being like getting sick and dying. But, as we will see, how humans experience life and derive meaning from it is different and much more complex compared to other living beings.

The human body utilizes a unique symbol system that allows it to express externally an internal reality. Lain Entralgo (1991, 1995) has proposed an analytic scheme of the human body based on external perception of the human behaviour which could be resumed in the following points:

- **Free will:** Through observation human seems to be able to act or not, make decisions or not, and follow principles which are not deducible from external elements.
- **Symbolic capacity:** Humans communicate freely using signs which become symbols through a conventional decision taken by individuals or groups;
- **Inconclusiveness:** The human being extends his or her actions to the future, he/she may desire to improve them, and he/she is able to promise;
- **Self-reflection:** A human is able to self-reflect on in his/her own personal subjectivity;
- **Awareness of reality:** To be in the cosmos for animals is to be related to external stimuli, but for the human person part of what makes up reality depends on the actions of human beings within it.

In other words, external perception of the human behaviour reveals the human being as constituted with a unique capacity to decide his/her own fate in a free way independently from following the laws of "ature."Beyond natural actions, there are actions specific to each human's personal condition that are considered to be culturally shaped (Polo, 1993) and beyond any rules governing nature. That is, a natural action is the same and can be generalized for all living beings of the same nature, whereas a cultural act is unique and non-transferable between different groups of individuals of the same nature. This change in core attributes linked to the definition of what it means to be human paved the way for a number of contenders for a better definition of humans as a species. As will be discussed, recent views of humans as homo technicus (technical) along with other related views of humans (homotechnologicus, technoself) have helped public opinion move away from traditional anthropomorphic views of humans as natural beings separate from the technological world. Instead, new conceptualizations of human beings with a natural and artificial nature.

LONG LIVE HOMO TECHNICUS!

A number of recent terms have been posited to provide a richer conceptualization of humans beings as technical beings living in a technological society. Human reality is constituted, in part, by an individual's ability to modify his/her environment and adapt it to himself/herself in life changing ways. This kind of relation with the environment is specific to the human species. Although animals do adapt themselves to the environment where they live, through mutagenesis; they are linked to their environment and the adaptation to other environments requires a long period of biological evolution. Humans, instead, are not related to the environment in the same way. For instance, when food of a specific habitat is too hard to eat, an animal has two options: starving to death or developing a more functional mastication system. Humans instead learn to cook (Polo, 1996). This is one of the first characteristics which define humans, and what we refer to when describing their technical capacity (see description of technical processes in Luppicini, 2010).

Animals in themselves do not have such technical capacities. Sometimes their capacity to create tools is superior to the human beings, if we consider the complexity of a bee hive or some bird nests. However, these are operational capacities coming from nature (biological evolution) which may contribute to survival, but these capacities cannot be transferred to other individuals, unless through a very long evolution process. In the case of the human beings such capacities could be transferred and integrated in the common experience of the specific species by virtue of being derived through their technical capacities. In other words, technical human beings (homo technicus) generate a technological culture.

It is interesting to note that a closely related term, homo technological (technological being) has also been used in the field to define our species from a technological standpoint. Gringas (2005) posited the term homo technologicus (technological person) to describe homo sapiens (wise person) as homo faber (craftsperson or toolmaker) surrounding themselves with artificial creations (counter-nature) and becoming homo technologicus. The term was intended to capture the essence of humans as technological and intelligent beings capable of creating their own artificial nature or "anti-nature" (Gringas, 2005, p.12). The same term was also employed by the late cultural historian Walter Ong (2002). Another closely related term from philosopher of technology, Michel Puech, posits that human beings are best described as 'homo sapien technologicus' (Puech, 2008). Both terms helped capture the richness of human co-evolution that takes place within a technological society.

Although the multiple terms discussed have strong merit, we use the term, 'homo technicus' which highlights humans as technical beings (Galvan, 2003; Galvan and Luppicini, 2012). The term 'homo technicus' is employed for multiple reasons, First, 'homo technicus' avoids possible criticisms against speculating on the evolutionary changes which transformed homo faber into homo technologicus. Instead, homo technicus assumes simply that humans are and have always been technical beings. Second, 'homo technicus' is a simpler (shorter term) and aligns more closer to the longstanding definition of technology as 'techne' (art or craft) which includes technologies and techniques used by technical humans (homo technicus).

As technical humans, we are part of a species which owns a body which is not only material and alive, but which allow us to be free and symbolic, and cultural. The inclusion of the three corporal levels (physical, biological, human), required by the uniqueness of being (Wojtyła, 1982) gives rise to a unique nature and personal condition for humans. There is not an opposition between being natural and being personal, because humans are naturally personal. Therefore, from the link we previously expressed between one's personal condition and human technical capacity we can conclude that the humans are naturally artificial since technical capacities are an important component of the human personal condition.

The term 'homo technicus' is also consistent with early biblical writings and stands in opposition to the dualist tradition of splitting the uniqueness of the human being into biological bodies and minds (soul). It is worth noting that long before dualist views became popular, non-dualist views of the human condition were available. As Guardini said (1994), the Platonic splitting between physical and psyche does not correspond to the Jewish–Christian tradition: the story of the Adam's creation in the second chapter of the Genesis doesn't report a human biological body in which a soul is instilled by God, but an inactive reality which God moulds through clay. Of course, the biblical story is compatible with any theory of evolution regarding the origin of the human species, since it is stated that the human being comes from a pre-existing matter, and the clay could be an image of an animated matter; what is implied is that this is not a human being as long as God has not blow in it his living spirit. The divine blow of Genesis is what the Christian tradition intends as "oul" but the starting act of a new being is all body and all soul from now on. Biblical tradition does not contemplate a dualistic vision as the Platonic one; the sculpture modelled by God was not a human body, but a representation of the communion of humans with the rest of the material world (Galvan, 2008).

It is, hence, useful to appreciate the importance of the body entrenched in the Jewish-Christian tradition where Christianity overcame any need to rely on any kind of dualism and the view that what in human is spiritual is good and what is material is evil. The relation between body and soul was considered as substantial unity (Fabro, 1955). Pulling together the ideas drawn from reflections on early human civilization, homo technicus appears to provide an adequate explanation of the human body and mind with technical capacities entrenched within the human condition as a substantial unity. Moreover, it is the non-dualist relational view which comes closer to providing a cohesive conceptualization of the human condition that aligns with the position espoused in this article (see Luppicini, 2008, 2010, for a more detailed description of the relational perspective). The next section introduces key terms to further articulate the relationship between human bodies and minds entrenched within a technological society where human values (ethics) and interests are at play.

Post-Modernity and the Ethical Body of Homo Technicus

Besides being known as a modern proponent of dualism, Descartes can also be attributed a precursor to the notion of "umanoid"for his treatment of the truth problem when he engages in methodological doubt calling into question held assumptions concerning truth and certainty of human understanding. The truth problem can be summed up as an attitude where subjective knowledge is reduced to a matter of human consciousness, and where objective knowledge can be treated as a mathematical topic and scientific activity. Under this view, the physical body should be treated as something to which can be applied physical rules (and thus mathematics) of the res extensa; Its origins are knowable from the res cogitans, which will be able one day to understand and reproduce that which is derived from the res extensa. Descartes argues in his Discours de la méthode that humans (thanks to their technical capacities), will be able one day to reproduce their res extensa, generating in this way the first mechanical human being. In this respect, Descartes was a prophet of the humanoid!

In the last two centuries under Modernity, the entrenchment of Cartesian model has helped ground a scientific consideration of the body and the dominion-paradigm, typical of modern mentality (Galvan, 2004; Irrgang, 2005, pp. 28-46); it must be said that a defense of the Cartesian position has been attempted in order to ensure a role of the divinity in affirming the non-material dimension of mankind (Foster, 2001), but we consider that this point of view is only possible in the realm of theological reflection. In Modernity, the body is different from the person. The immediate consequence is the oblivion of the person. That is, if it is necessary to distinguish between body and person, it is also clear that the person defies any scientific definition; the realm of the person is dominated by rationality, or, in a very recent expression of Bibeau (2011), it is a genomythology. The dominion-paradigm becomes a technological imperative: tutto ciò che si può fare è buono (all you can do is good).

In response to Modernist views of the body, postmodern writings opposed the dominion-paradigm and advanced views that align with an emergent relational-paradigm that highlights the body, not as an object to be dominated, but rather as the destination of a positive interweaving of the person and the material reality which are taken together. Under this view, the body manifests the relational nature of the person. This view grounds the ethical significance of the human body and the need to address the personal human condition (including human values and technical capacities) in considerations of the human body and human reality.

This paradigm-shift includes the reversal of priorities attributed to Science and Technics. During the era of the dominion-paradigm, experimental sciences had the leading role of providing knowledge about humanity and cosmos, thus allowing the dominant but mere instrumental role of Technics (Science over Technics); whereas in the relational-paradigm, science still remains of paramount importance as a source of knowledge about reality, but this is only a first step, which requires to be perfected by the technical capability: when Science becomes Technics, becomes more human (Galvan, 2004).

To take this discussion on step further, we assume that in postmodernity the body is the person. But is it possible to say that the person is the body? A positive answer (including the total reversibility of the proposition) entails the non-existence of a formalizing action beyond the material dimension in the human person. Under this condition, the relational model would be related only to the technical capacity of human beings. As in the dominion-paradigm, the philosophical study of technology and ethics was reduced to Technics over Ethics and there was no other way to human perfection other than the technological enhancement of the body. In our opinion, this is the way in which many of the transhumanist theories understand the changing nature of human beings through technological enhancement (Galimberti, 2000; Signore, 2006). They apply the concept of transcendence (transhuman) to describe the end product of this coupling.

Alternatively, if the answer to the question is that the person is the body, but not only the body, a principle (not separated) of formalizing activity is assumed to be present in the human, and this principle grounds the relational value of the whole person. The body is always seen as the subject of technological enhancement, but the framework of ethical reference, and therefore the key for human perfection, is given by the (so-called) spiritual dimension of the human being (Elliot, 2011) or at least by the consideration of the human being as a hierarchical complex system with two sub-systems (body and mind) in which the presence of emergent properties irreducible to the sub-systems permits technical human enhancement that maintains, improves or repairs those emergent properties, but not the substantial alteration of those properties (Deplazes, 2011).

To sum up, the conceptualization of the human species as homo technicus, mirrors the shifting values in society with respect to the relationship between science, technology, and ethics. In other words, the role of Science is to provide an adequate description of humans which requires an inquiry into human values and technical capacities (Technics over Science!). We are homo technicus!

Discussion

Much of the modernity oriented scholarship on technology and ethics prioritizes experimental sciences over technics, and technics over ethics. The body is not viewed as having any special axiological dimension and the machine becomes the moral agent. Perhaps this is the intention behind Bunge's proposal for the creation of the field of Technoethics: "echnologists should contribute to the overhauling of ethics, attempting to construct a technoethics as a science of right and efficient conduct"(1977, p. 107). This statement, which is consistent with the historical period in which it was given, should be reversed in the current period: ethics should contribute to the overhauling of technique, giving to it a positive anthropological value. In this way, Bunge's ideas were a precursor to other postmodern writings about science technology and ethics which followed. Technoethics highlights the central role of ethics in the overhauling of technics, giving to technology the possibility of respecting and promoting the whole person. It aligns with scholarly efforts to provide a more encapsulating multi-perspective inquiry into technology consistent with general ethical principles, historical knowledge, psycho-affective integrity, and values of justice and solidarity. In line with this, technoethics is compatible with the notion of humans as homo technicus.

REFERENCES

Bibeau, G. (2011). What is human in humans? Responses from biology, anthropology, and philosophy. *The Journal of Medicine and Philosophy*, *36*, 354–363. doi:10.1093/jmp/jhr025 PMID:21859676

Bunge, M. (1977). Toward a technoethics. *The Monist*, *60*, 96–107. doi:10.5840/monist197760134

Campbell, A. V. (2009). *The body in bioethics*. New York, NY: Routledge-Cavendish.

Cavalieri, P., & Singer, P. (1993). *The Great Ape Project: Equality Beyond Humanity*. London, UK: Fourth Estate.

Choza, J. (1988). *Manual de antropologia filosofica*. Madrid, Spain: Rialp.

De Andres Argente, T. (2002). *Homo cybersapiens: La inteligencia artificial y la humana*. Pamplona, Spain: EUNSA.

Deplaces, U. (2011). Technological Enhancements of the Human Body: A Conceptual Framework. *Acta Philosiphica*, *20*, 53–72.

Elliot, C. (2011). Enhancement technologies and the modern self. *The Journal of Medicine and Philosophy*, *36*, 364–374. doi:10.1093/jmp/jhr031 PMID:21903906

Fabro, C. (1955). *L'anima*. Roma, Italy: Studium.

Foster, J. (2001). A Brief Defense of the Cartesian View. In *Soul, Body and Survival* (pp. 15–29). Ithaca, NY: Cornell University Press.

Galatino, N. (2005). Koerper e Leib: tra determinismo biologico e determinismo culturale. In *La sfida del post-umano: Verso nuovi modelli di esistenza?* (pp. 49–65). Roma, Italy: Studium.

Galimberti, U. (2000). *Psichè e tecnè*. Milano, Italy: Feltrinelli.

Galvan, J. M. (2004). On technoethics. *IEEE Robotics and Automation Society Magazine*, *10*, 58–63.

Galvan, J. M. (2008). Creation and casuality: The case in Christian theological anthropology. In M. Negrotti (Ed.), Yearbook of the artificial: Nature, culture & technology: Natural chance, artificial chance (vol. 5, pp. 129.141). Bern, Switzerland: Peter Lang Verlag.

Galvan, J. M. (2011). La tecnoetica e le speranze umane. In *Scienza, Tecnologia e Valori Morali: Quale futuro?* (pp. 175–185). Roma, Italy: Armando Editore.

Gingras, Y. (2005). Éloge de l'homo technologicus. Saint-Laurent, Canada: Les Ed.s Fides.

Guardini, R. (1994). *Ethik: Vorlesungen an der Universität München, 1950-1962*. Padenborn, Germany: Matthias-Gruenewald Verlag.

Irrgang, B. (2005). *Posthumanes Menschsein?* Weisbaden, Germany: Franz Steiner Verlag.

Lain Entralgo, P. (1991). *El cuerpo humano: Una teoria actual*. Madrid, Spain: Espasa.

Lain Entralgo, P. (1995). *Alma, cuerpo, persona*. Barcelona, Spain: Circulo de Lectores.

Luppicini, R. (2008). Introducing technoethics. In *Handbook of research on technoethics* (pp. 1–18). Hershey, PA: Idea Group. doi:10.4018/978-1-60566-022-6.ch001

Luppicini, R. (2010). *Technoethics and the evolving knowledge society*. Hershey, PA: Idea Group. doi:10.4018/978-1-60566-952-6

Luppicini, R. (2013). *Handbook of research on technoself: Identity in a technological society* (Vol. 1-2). Hershey, PA: IGI Global.

Meilaender, G. C. (1995). *Body, soul, and bioethics*. Notre Dame, IN: University of Notre Dame Press.

Polo, L. (1993). *Presente y futuro del hombre*. Madrid, Spain: Rialp.

Polo, L. (1996). *Etica: Hacia una version moderna de los temas clasicos*. Madrid, Spain: Union Editorial.

Puech, M. (2008). *Homo sapiens technologicus: Philosophie de la technologie contemporaine, philosophie de la sagesse contemporaine*. Paris: Éditions Le Pommier.

Schaler, J. A. (Ed.). (2009). *Peter Singer Under Fire: The Moral Iconoclast Faces His Critics*. Chicago, IL: Open Court.

Schramme, T. (2008). Should we prevent non-therapeutic mutilation and extreme body modification? *Bioethics*, *22*(1), 8–15. PMID:18154584

Signore, M. (2006). *Lo sguardo della responsabilità: Politica, economia e tecnica per un antropocentrismo relazionale*. Roma, Italy: Studium.

Spaemann, R. (1991). *Moralische Grundbegriffe*. München, Germany: C.H. Beck Verlag.

Torralba Roselló, F. (2005). *Qeé es la dignidad humana?* Barcelona, Spain: Herder.

Warwick, K. (2003). Cyborg morals, cyborg values, cyborg ethics. *Ethics and Information Technology*, *5*(3), 131–137. doi:10.1023/B:ETIN.0000006870.65865.cf

Wojtyła, K. (1982). *Persona e Atto: Città del Vaticano*. Vatican State, Italy: Libreria Editrice Vaticana.

ADDITIONAL READING

Allenby, B., & Sarewitz, D. (2011). *The technohuman condition*. Cambridge, MA: The MIT Press.

Bar-Cohen, Y., & Hanson, D. (2009). The coming robot revolution: Expectations and fears about emerging intelligent, humanlike machines. New York, NY: Springer Science + Business Media.

Benford, G., & Malartre, E. (2007). *Beyond human: Living with robots and cyborgs*. New York, NY: Tom Doherty Associates.

Brooks, R. A. (2003). *Flesh and machines: How robots will change us*. New York, NY: Vintage Books.

Carr, N. (2011). *The shallows: What the internet is doing to our brains*. New York, NY: W. W. Norton & Company.

Castells, M. (2004). *The power of identity*. Oxford, UK: Blackwell.

Coleman, B. (2011). *Hello avatar: Rise of the networked generation*. Cambridge, MA: MIT Press.

Gergen, K. (2009). *Relational being: Beyond self and community*. New York, NY: Oxford University Press.

Kurzweil, R. (1999). *The age of spiritual machines: When computers exceed human intelligence*. New York, NY: Viking.

Kurzweil, R. (2005). *The singularity is near: When humans transcend biology*. New York, NY: Viking.

Levy, D. (2007). *Love and sex with robots: The evolution of human-robot relationships*. New York, NY: HarperCollins.

Lin, P., & Allhoff, F. (2008). Against unrestricted human enhancement. *Journal of Evolution & Technology*, *18*(1), 1–7.

Perkowitz, S. (2004). *Digital people: From bionic humans to androids*. Washington, DC: Joseph Henry Press.

Rahimi, S. (2000). Identities without a reference: Towards a theory of posthuman identity. *Journal of Media and Culture*, *3*(3).

KEY TERMS AND DEFINITIONS

Cyborg: Cyborgs (cybernetic organism) are beings with both biological and artificial parts.

Homo Technicus: Homo technicus is a term used to describe the historical condition of human beings as intertwined within an advancing technological society.

Posthuman: The concept of posthuman describes a speculative future being capable of embodying different identities and understanding the world from multiple perspectives.

Technoethics: Technoethics is defined as an interdisciplinary research field concerned with all ethical aspects of technology within a society shaped by technology. It deals with human processes and practices connected to technology which are becoming embedded within social, political, and moral spheres of life. It also examines social policies and interventions occurring in response to issues generated by technology development and use. This includes critical debates on the responsible use of technology for advancing human interests in societ (Luppicini, 2008).

Technoself: Technoself describes the evolving configurations of human-technological relationships that continually shape the human condition and what it means to be a human being.

Chapter 2
The Relevance of Value Theory for the Ethical Discussion of Human Enhancement

Tobias Hainz
Hannover Medical School, Germany

ABSTRACT

The aim of this chapter is to provide a practical introduction to the central issues of value theory in order to demonstrate their relevance for the ethical discussion of human enhancement technologies. Among the value-theoretical issues discussed are value lexicality, the monism-pluralism dichotomy, and incommensurability. A particular enhancement technology analyzed from a value-theoretical perspective is radical life extension, the direct and intentional extension of the maximum human life span. Several examples are given to show how value-theoretical concepts are implicitly reflected in arguments for and against human enhancement. At the end of the chapter, it should be clear that value theory can and should make stronger contributions to the ethical discussion of human enhancement and that, in this discussion, an increased awareness of value-theoretical issues is desirable.

INTRODUCTION

Value theory or, as it could also be called, axiology is a hybrid part of philosophy that has a considerable formal foundation but is also relevant for philosophy's practical subdisciplines, such as political philosophy, normative ethics, or, even more practical, applied ethics. The formal foundation of value theory becomes evident as soon as one recognizes the existence of issues like locating different values on a value scale, comparing different value scales to each other, and theoretically distinguishing between different kinds of value. When dealing with the formal foundation of value theory, logical and semantical problems are much more in the foreground than the practical applicability of value theory. However, as soon as one takes the importance of these theoretical or formal issues for granted and focuses on the more material aspects of value theory, its relevance for practical philosophy cannot be ignored. Applied ethics in particular constantly deals with issues

DOI: 10.4018/978-1-4666-6122-6.ch002

like autonomy, naturalness, freedom, or justice, all of which can be regarded as values. Value theory, therefore, is located between theoretical and practical philosophy inasmuch as it is theoretically well-informed but cannot be separated from the practical areas of philosophy, such as applied ethics.

One particular topic in applied ethics is human enhancement, that is, the improvement of human capacities, broadly construed, through non-therapeutical interventions. The ethical discussion is very diverse but can be summarized as a discussion of the moral status of human enhancement: Is it ethically permissible or not? Are there any reasons why the application of enhancement technologies by free and rational people to themselves should be prohibited? Could parents even be morally obliged to use preimplantation genetic diagnosis or genetic engineering and, thereby, enhance their offspring? Arguments for and against human enhancement refer to autonomy, well-being, distributive justice, individual rights of future people, the naturalness of the human body, and further aspects that have to be taken into account. It should, therefore, be evident that value-theoretical issues are implicitly present in the ethical discussion of human enhancement. Yet, these issues are rarely made explicit, which is why an increased awareness of the interconnections between genuine value theory and the ethical discussion of human enhancement appears to be a *desideratum*.

This article attempts to point out at least some of these interconnections by introducing the most central issues within value theory and relating them to the ethical discussion of human enhancement in order to show *how* applied ethics can make use of value theory and *why* it should indeed favor a more explicit inclusion of value-theoretical considerations when analyzing a topic like human enhancement. Since a thorough discussion of the interconnections between value theory and the ethical discussion of human enhancement is beyond the scope of this article, it was inevitable to choose a particular value-theoretical issue as a prime example while mentioning other issues in a more superficial way. The example that will be discussed in detail will be the lexical ordering of values or, in short, value lexicality. Other aspects of value theory and their relevance for the ethical discussion of human enhancement that will be discussed more superficially will be the dichotomy between value monism and value pluralism as well as incommensurability.

VALUE THEORY AND HUMAN ENHANCEMENT: AN OVERVIEW

Since this article will not deal with controversies within value theory but with its possible application to the ethical analysis of human enhancement, it seems advisable to provide a mostly general perspective on those value-theoretical issues that are of importance for the aim of this article without actually arguing for or against a specific position on a controversial issue. This background section will, therefore, neglect controversial debates within value theory that would deserve a separate analysis. Instead, it will provide a general overview of value theory in order to enhance the understanding of the relevant issues and their location within the value-theoretical landscape. This overview of central topics in value theory will necessarily be selective and non-exhaustive. Its purpose is to give the reader an impression of which issues value theory is concerned with and to introduce the topics that will later be revisited when the applicability of value theory to the ethical discussion of human enhancement will be demonstrated. Furthermore, a similar overview of human enhancement will be included in this section. The purpose of this overview will be to clarify how to conceive of human enhancement.

The first part of this section will feature an introduction to the most basic aspects of value theory that will later be connected to the ethical analysis of human enhancement. In the second part, value lexicality will be discussed in greater detail since it shall serve as the prime example of the practical relevance of value theory. The third and final part will feature a brief introduction to the concept of human enhancement.

Central Aspects of Value Theory

The subject matter of value theory are, of course, *values*, but regardless of this semantical triviality, there are problems of considerable difficulty to determine the nature of this subject matter. First of all, one should at least have a basic understanding of what values are in themselves. One possible way to describe what values are is to describe them as things that are good. It should, though, be avoided to read this 'good' as referring to moral goodness because this interpretation would lead to the implausible view that a good – or virtuous – person is a value herself. Furthermore, this kind of goodness should not be confused with attributive goodness (see Geach, 1956), like in the statement 'This is a good bulb', which means something like 'This bulb is good for illuminating the room because it works well'. Instead, it should be regarded as goodness without any further qualification, so that the statement 'Freedom is good' does not create confusion by suggesting that freedom is good in any specific sense. As soon as one utters this statement, one only expresses the opinion that freedom is a value.

While this brief elaboration on the nature of values might still appear to be rather uninformative, it is certainly preferable to an agnosticist stance towards values: "I myself do not know what 'values' are; like William Bennett, when I hear the word 'values', I reach for my Sears catalogue." (Millgram, 2005, 273) In some cases, it might be in order to remain silent about the the nature of a phenomenon like values, but as long as one sticks to the suggested reading of 'good' as a predicate without any further qualification, one should get a clearer impression of what values are than if one leaves the concept of values unanalyzed.

One further question regarding the nature of values is whether they are *objective* or *subjective*. Objectivism about values is the view that values exist independently from the state of people's minds, that is, they exist as values regardless of whether people actually want to acquire them or not. If, for example, one is an objectivist about values, freedom – as a 'hot candidate' for being included on a list of values – is valuable even if some people agree that freedom should be limited. These people would simply be misguided to a certain extent. Subjectivism about values is the opposite view, namely, the view that the existence of values as values depends on the state of people's minds. Only things that people actually want to acquire or at least regard as desirable under some circumstances qualify as values. If, for example, nobody cared about justice, justice would not be a value. This is only one specific issue at the very center of value theory, but its importance should be obvious: if objectivism is true, people can actually make mistakes when judging things as valuable, and other people can make correct judgments. Therefore, those people who fall prey to incorrect judgments could eventually be made responsible for the actions that result from these judgments. On the other hand, if subjectivism is true, it is much harder to accuse people of making incorrect judgments about which things qualify as values or not. It might still be possible to accuse them of making implausible judgments, but this accusation is much weaker and might also have less dramatic practical consequences. Given the interpretation of values as things that are good, it is easy to reformulate the positions of objectivism and subjectivism: objectivism claims that there are things that are good independently from the content of people's minds whereas subjectivism claims that the goodness of things depends on the content of people's minds.

One should also be aware that there is no consensus in the debate about value *monism* and value *pluralism*. According to value monism, there is only one genuine value because all other values are ultimately reducible to this basic value. This position is, for example, reflected in classical hedonism (see Bentham, 2007; Mill, 1871) that regards hedonic pleasure as the only basic value. Value pluralists, on the other hand, deny the existence of only one basic value, for example, by proposing that human welfare is a value that consists of several components each of which is irreducible to the other ones. This type of value pluralism has been described as the "hybrid theory" of welfare (Heathwood, 2010, p. 652). Value pluralism often appears to be the intuitively favorable position, but as we will see below, it creates further difficulties. If value monism were true, though, the guidance function of morality would mostly appear to be an empirical issue rather than a matter of principle: in order to determine the morally right action, we would have to ask which of the alternative actions would bring about the most intense instantiation of the only existing value. If, however, value pluralism were true, we would not only have to discuss these empirical issues but also which of the available values deserves to be instantiated at the expense of the other available values.

Value pluralism is strongly connected to the problem of *incommensurability*. If there are more things that qualify as values than just one basic value, it could turn out to be the case that some of these values are incommensurable, that is, they "cannot be precisely measured by a single 'scale' of units of value" (Chang, 1997, p. 1). In other words, if incommensurable values exist, these incommensurable values cannot be weighed against each other, one cannot reasonably claim that the instantiation of one of these values to a certain degree is more desirable than the instantiation of another of these values to another degree. One could not even say that indifference between the instantiation of one or another of these values is the most reasonable perspective. Incommensurable values simply cannot be ranked.

It should be obvious that the possibility of value incommensurability implies at least one practical problem: suppose you have to choose between the instantiation of value *A* and the instantiation of value *B*. If *A* and *B* are incommensurable, you cannot rely on the nature of these values alone in order to make a reasonable judgment regarding your choice. You have to rely on further, maybe very complicated, considerations if you wish to arrive at a prudentially and morally justified decision. A situation of this kind can, for example, be created if one has to choose between freedom as one value and justice as another value – if one assumes, for the sake of this analysis, that freedom and justice are incommensurable. One would have to analyze the consequences of instantiating freedom and justice, respectively, in order to arrive at a justified decision, and this analysis can turn out to be quite complicated. Since incommensurability can only exist at all if value pluralism is true, one could wish for it to be false, so that value monism is true. Yet, it would be an obvious fallacy to assume that value monism is true because value pluralism creates the possibility of incommensurability, which is why dealing with both the theoretical and practical consequences of incommensurability is one of the most important issues in value theory.

Finally, it is common and important to distinguish between *intrinsic* and *extrinsic* values. Intrinsic values are values that are good in themselves or for their own sake; their goodness cannot be explained with reference to their positive contribution to the instantiation of another value. Although one might argue about the exact definition of 'intrinsic values' (see, for example, Bailey, 1979), formulations like 'in themselves' or 'for their own sake' seem to be indispensable when one attempts to explain the nature of intrinsic values. It should, though, be noted that the position that intrinsic values exist at all has been attacked in the history of philosophy (see, for example, Beardsley, 1965). On the other hand, extrinsic values are not good in themselves inasmuch as they are only good because they contribute to the instantiation of other, intrinsic values. While this brief explanation of the dichotomy is far from being exhaustive, it should be acceptable as a first approximation. There is, however, considerable disagreement on which things should be regarded as intrinsic values. Classical hedonists like Jeremy Bentham (2007) or John Stuart Mill (1871) – who also qualify as value monists – would probably claim that hedonic pleasure alone is an intrinsic value whereas all other values are merely extrinsic. Yet, one can also find entire lists of things the inventors of these lists claim to be intrinsic values, an example being the list of William Frankena (1963, p. 72).

Furthermore, it should be noted that although intrinsic values are usually regarded as much more important when determining the moral rightness or badness of an action, extrinsic values are not entirely irrelevant. For, depending on what one considers as intrinsic values, some extrinsic values gain or lose relevance inasmuch as they are more or less closely connected to an eventual instantiation of these intrinsic values. To give an example, if one is a value monist who believes that hedonic pleasure alone is intrinsically valuable, two potential extrinsic values like health and virtuousness might differ with regard to their actual relevance as long as one can establish a stronger relation between one of these values and hedonic pleasure. (Intuitively, one might believe that the connection between health and hedonic pleasure is stronger than the connection between virtuousness and hedonic pleasure, but this is at least partly an empirical matter.) The extrinsic value that exhibits a stronger connection to hedonic pleasure can count as more relevant for practical considerations.

Value Lexicality

Being the prime example for the application of value theory to the ethical discussion of human enhancement in this article, *value lexicality*, deserves to be discussed in greater detail than the value-theoretical issues that were discussed in the previous section. An even longer elaboration on value lexicality that is the precursor to this article can be found in Hainz (2012a).

Imagine you are presented with the following choice: if you kill a random stranger you have never met before and who nobody cares for, you will be given one million dollars. If you refuse, nothing will happen. As I believe, most people would refuse to kill the stranger, not only because they are afraid of being punished but also because they believe that killing him is deeply immoral. Although they might justify their belief in different ways, for example, with reference to the rule that it is always wrong to kill an innocent person, to the stranger's dignity as a human being, or to the stranger's desire to live being frustrated in the case of his death – the belief those people hold is identical. One particular attempt to defend this belief is not to refer to any deontological rule or to the consequences of killing the stranger but to the value of the two dominant phenomena involved in this situation: money and human life. A pretheoretical line of defense could sound like this: "I refuse to kill the stranger because his life is valuable. Yes, money is valuable, as well, but there is no amount of money that can outweigh this stranger's life. Even if you offered me one billion dollars, I would still refuse to kill him because I am a moral person and taking human life for money is immoral. Human life does not have a price but is priceless."

A person who gives such an explanation for her refusal to kill the stranger can be conceived of as someone who believes that both money and life qualify as values but that there is no amount of money that is more valuable than human life. This belief can be stated in the more general form detached from the initial example that certain values cannot be outweighed by others, regardless of the amount of these other values. There exists an ordering, according to people who hold this belief, such that certain values are located on a higher level than other values which cannot reach this level, not even in principle. This ordering shall be called a *lexical ordering*, and values that are more valuable than others and cannot be outweighed by them in principle shall be called *lexically more valuable* with regard to the values they are compared to or having *lexical priority* to them.

Three philosophers who can be credited with theorizing about value lexicality are John Rawls, John Stuart Mill, and James Griffin. While Rawls is usually regarded as the philosopher who introduced the expression of a 'lexical ordering', Mill famously treats the topic of value lexicality when he postulates the existence of different kinds of pleasure. Finally, Griffin's account of value lexicality can be regarded as the most lucid one.

In *A Theory of Justice*, Rawls discusses serial or lexical orderings when he considers the so-called 'priority problem', that is, the problem how to balance different principles of justice. He suggests "that we may be able to find principles which can be put in what I shall call a serial or lexical order. [...] This is an order which requires us to satisfy the first principle in the ordering before we can move on to the second, the second before we consider the third, and so on. A principle does not come into play until those previous to it are either fully met or do not apply. A serial ordering avoids, then, having to balance principles at all; those earlier in the ordering have an absolute weight, so to speak, with respect to later ones, and hold without exception." (Rawls, 1971, pp. 42-43) This passage can be thought of as a *locus classicus* of the concept of lexicality. According to Rawls, there might be principles that are lexically prior to other principles, and if we encounter a situation in which two principles are relevant, with one of them being lexically prior to the other one, we have to satisfy the lexically higher principle first, regardless of what the lower principle tells us. An example of two principles that are put in a lexical ordering are the two principles of justice Rawls formulates later. The first, lexically prior principle requires "each person [...] to have an equal right to the most extensive basic liberty compatible with a similar liberty for others" (Rawls, 1971, p. 60), whereas the second principle tells us that "social and economic inequalities are to be arranged so that they are both (a) reasonably expected to be to everyone's advantage, and (b) attached to positions and offices open to all" (Rawls, 1971, p. 60). If these principles are ordered lexically, then no frustration of the first principle can be justified with a satisfaction of the second principle, even if it is satisfied with extreme intensity. The first principle always has to be considered and met before one turns to the second principle.

Mill also suggests that certain things can be put in a lexical ordering when he claims that some kinds of pleasure have lexical priority to others: "Of two pleasures, if there be one to which all or almost all who have experience of both give a decided preference, irrespective of any feeling of moral obligation to prefer it, that is the more desirable pleasure. If one of the two is, by those who are competently acquainted with both, placed so far above the other that they prefer it, even though knowing it to be attended with a greater amount of discontent, and would not resign it for any quantity of the other pleasure which their nature is capable of, we justified in ascribing to the preferred enjoyment a superiority in quality so far outweighing quantity as to render it, in comparison, of small account." (Mill, 1871, p. 12) Mill famously claims that if we rely on the judgment of an expert with regard to the quality of two kinds of pleasure, this expert's judgment can provide us with sufficient information to put these kinds of pleasure in a lexical ordering. The criterion by which these kinds of pleasure are ordered has an objective as well as a subjective component. The subjective component is the expert's judgment, based on his personal experience, but there is also an objective component inasmuch as everyone should be able to have access to the judgment and to comprehend it. An obvious difference between Mill and Rawls is that Mill speaks of kinds of pleasure being lexically ordered whereas Rawls deals with principles. This difference, however, should not be problematic since not only principles but also values and possibly other kinds of things can serve as elements in a lexical ordering. There is no logical contradiction in speaking of a 'lexical value ordering' and, referring to another ordering, a 'lexical ordering of principles'.

Griffin considers lexicality as one specific kind of incommensurability, given a broad interpretation of what 'incommensurability' can refer to. His technical term for lexicality is 'trumping', but he also uses the notion of lexical priority in his explication of 'trumping,' which is why we are justified to stick to the notion of 'value lexicality'. Griffin explains value lexicality as follows: "It takes the form: *any* amount of *A*, no matter how small, is more valuable than *any* amount of *B*, no matter how large. In short *A* trumps *B*; *A* is lexically prior to *B*." (Griffin, 1986, p. 83) Being very precise and fairly unambiguous, Griffin's account of value lexicality shall serve as the basis for its subsequent application.

It has to be noted that regarding lexicality as a specific subtype of incommensurability is a peculiarity of Griffin's loose reading of what 'incommensurability' means. Griffin (1997) himself discusses this problem in detail and suggests that 'incommensurability' can and should be understood as 'incomparability', the claim "that there are values that cannot be got on *any* scale, that they cannot even be compared as to 'greater', 'less', or 'equal'" (Griffin, 1997, p. 35). Although it is merely a problem of terminological choice, it has to be addressed for the sake of conceptual clarity. One may, of course, conceive of incommensurability as incomparability, but if one makes this choice, one must not regard lexicality as a subtype of incommensurability because values in a lexical ordering *are* comparable.

A final question is how to put values in a lexical ordering. We need a reliable criterion because a lexical ordering must not be arbitrary if it shall guide our actions and be applicable to practical problems. I believe that the following criterion can serve this purpose: for two values *A* and *B*, *A* is lexically more valuable than *B* if the existence of *A* is a necessary prerequisite for the existence of *B*. It is a purely descriptive and factual question whether one thing is a necessary prerequisite for another thing, so that no normative considerations are involved in the creation of a lexical ordering if one sticks to this criterion. Applying it to the initial example of killing a stranger for an amount of money shows that it works at least in this particular case: only living persons can own money because ownership is not a property corpses can exhibit. It is a necessary prerequisite to be alive to own any amount of money. If one accepts this criterion, then one can and should also accept that life is lexically more valuable than money. It even works on a larger scale if one assumes that without life 'money' would be no money but only pieces of paper and metal. Living people create money by declaring that those pieces of paper and metal *are* money, and if there were no living people, no money would exist. There might be counterexamples to this criterion, but I sincerely believe that it deserves consideration. However, if it is found to be insufficient, there might be another criterion that is more reasonable. The account of value lexicality presented here works with this criterion but is not necessarily dependent on it. As long as there is *any* criterion that serves the purpose of non-arbitrarily putting values in a lexical ordering, the concept of value lexicality is still significant and interesting.

What Is Human Enhancement?

Savulescu, Sandberg & Kahane (2011) differentiate between several approaches to human enhancement, two of them being particularly interesting: according to the "not-medicine approach", enhancement is "[a]ny change in the biology or psychology of a person which increases species-typical normal functioning above some statistically defined level" (Savulescu, Sandberg & Kahane, 2011, p. 5). This approach relies on Christopher Boorse's account of health and disease (Boorse, 1975; 1977) that should be taken into consideration if one attempts to deal with the not-medicine approach in detail. However, it will be neglected here since the not-medicine approach is comprehensible without a deeper discussion of Boorse's account of health and disease. The not-medicine approach proposes that statistics can tell us what normal functioning for a typical member of any species is, and if there is an intervention that raises the functioning of a member of this species over the statistically normal level, then this intervention is not medical treatment but enhancement. It would be medical treatment if it restored the functioning of a member from a lower level to the normal level.

The other approach offered is the welfarist definition that regards enhancement as "[a]ny change in the biology or psychology of a person which increases the chances of leading a good life in the relevant set of circumstances" (Savulescu, Sandberg & Kahane, 2011, p. 7). This approach is much broader than the not-medicine approach since it includes medical treatment as well as non-medical intervention. It is also not strictly descriptive like the not-medicine approach because it involves the concept of 'leading a good life'.

Both approaches have their merits and deserve to be discussed more thoroughly, but a choice between them has to be made. To my mind, the not-medicine approach has several advantages over the welfarist approach – not in general but with respect to the subject of this article. Since we are interested in the relevance of value theory for the ethical discussion of human enhancement, it would be a kind of prejudice to rely on an approach to human enhancement that already regards it as beneficial. From this point, it would only be a small step to the assumption that human enhancement positively contributes to the instantiation of extrinsic or even intrinsic values. Therefore, we should assume that it is an open question whether it is beneficial or not which is why we need a descriptive account of human enhancement, and the not-medicine approach is such an account. Consequently, I will stick to the not-medicine account and regard human enhancement as any intervention into the biology of an organism that raises its functioning over the statistically normal level of a typical member of the human species.

Given this approach to human enhancement, one can think about many existing or merely possible specific enhancement technologies, such as physical enhancement or doping, cognitive enhancement that may target mental capacities as well as emotions, or even moral enhancement that seeks to improve our ability to make moral judgments.

VALUE THEORY AND HUMAN ENHANCEMENT

Since we now have a basic understanding of what value theory is and which questions it deals with, this section will focus on its relevance for the ethical discussion of human enhancement.

Value Lexicality and Radical Life Extension

One topic that has emerged in contemporary bioethics is known as *radical life extension* (see for example Horrobin, 2006, pp. 279-281), that is, the direct and intentional extension of the maximum human life span through biotechnological interventions. Although we know that the average human life span can be altered through healthy nutrition, widespread availability of medical drugs, or improved standards of hygiene, it is still unknown whether the maximum human life span can be altered at all. There are people who claim that a considerable extension of the maximum human life span is at least biologically possible (Gems, 2003, p. 31), some of them being extremely optimistic (De Grey & Rae, 2007), whereas their opponents believe that such claims are not supported by solid biological reasons

(Hayflick, 2004). However, despite the insecurities about the biological possibility of radical life extension the question whether a world in which such life extension technologies exist would be better than a world without them is important and a justified field of study for bioethicists. If radical life extension is possible, then we should take action as soon as possible, either in order to fuel the development of such technologies – in the case that they are regarded as desirable – or in order to prevent them – in the case that they are regarded as undesirable.

Many arguments have been advanced in the discussion about radical life extension. I will only sketch one particular argument that is of special importance for the topic of this article: the argument from overpopulation. Several bioethicists claim that radical life extension is undesirable or at least problematic because the widespread availability of life extension technologies would lead to severe overpopulation, scarcity of resources, and a resulting decrease in individual welfare (Glannon, 2002; Singer, 1991) whereas other philosophers are skeptical about the force of this argument (Miller, 2002; Mackey, 2003).

One of the most interesting objections to the argument from overpopulation is put forth in Cutas (2008): "If the right to life does not have a deadline, then its requirements make the argument from overpopulation irrelevant: even if extending life span does prove to create overpopulation related issues, these have to be address [sic!] without violating the right. Neither the right to come into existence nor the right to reproduce can rightly trump the right to life of present rights holders." (p. e7) Cutas assumes that if there is a temporally unrestricted right to life at all, then even the worst possible consequences in terms of overpopulation do not justify the prohibition of radical life extension because the right to life trumps such concerns. It is striking that Cutas uses the term 'trumps' that is Griffin's notion for value lexicality, as well. Rephrasing Cutas' objection in the spirit of value lexicality should, thus, not prove to be too difficult.

According to Cutas, there are two values involved, life and general welfare. Prohibiting radical life extension would preserve a high level of welfare because the negative effects of overpopulation would be avoided, but such a prohibition would also violate the right to life. On the other hand, the right to life would be respected if there were no prohibition on radical life extension, but then we would encounter a severe decrease in welfare. If there were a non-lexical ordering the right to life and welfare could be put in, then a significant amount of welfare could outweigh and justify the violation of the right to life. However, Cutas claims that the right to life and welfare are ordered lexically, that is, the right to life is lexically more valuable than welfare. This is the reason why even a significant increase in welfare does not justify the violation of the right to life and why a prohibition of radical life extension argued for by proponents of the argument from overpopulation is not justified, as well.

I believe that this is a proper reconstruction of this debate that solves one possible problem: Cutas could run into the objection that her claim that the right to life trumps a decrease in welfare lacks justification because she does not provide a reason why rights cannot be outweighed by other considerations. But if we accept that life and welfare are both valuable things, then we can argue that life is lexically more valuable than welfare since it is a necessary prerequisite. This reconstruction of the debate on the argument from overpopulation against radical life extension, therefore, shows that the theoretical concept of value lexicality cannot only be applied in order to structure, interpret, or reorganize existing arguments, but also in order to enrich a debate by including a genuinely new element. In this case, Cutas' rights-based objection to the argument from overpopulation is reformulated by using the value-theoretical toolbox, but additionally, a possible refutation of her objection is anticipated.

The Problem of Plurality

It is an unusual feature of the ethical discussion of human enhancement that not just a few closely connected, but many more or less unrelated potential values play an important role. To provide a non-exhaustive list of potential values that are mentioned in arguments for and against human enhancement: personal autonomy (Hildt, 2009), justice (Davis, 2004; Farrelly, 2010), naturalness (Fukuyama, 2002), giftedness (Sandel, 2009), the right to life (Cutas, 2008), well-being (Savulescu, Sandberg & Kahane, 2011), morality (Douglas, 2008) and health (Hainz, 2012b).

This feature of the discussion is one reason for its complexity since one cannot simply evaluate the existing arguments from one specific perspective, for example, a rights-based perspective without having to presuppose that certain values take priority over others. Adopting a specific perspective seems to imply that one gives priority to a certain value this perspective is based on, but the decision to give priority to a certain value should be justified. Moreover, as soon as one decides to adopt a specific perspective regarding human enhancement, one implicitly enters the realm of value theory – at least if one attempts to launch an argument for or against human enhancement from the perspective one has adopted. One example was already given in the previous section where Cutas' rights-based argument for radical life extension was reconstructed as an argument about the lexical priority of the right to life over general welfare. By arguing that the right to life trumps general welfare, Cutas does not only subscribe to the position that the right to life has lexical priority over general welfare but also to value pluralism and to the claim that the right to life as well as general welfare are indeed values.

In many cases, the ethical discussion of human enhancement appears to be a debate on how to weigh arguments against each other. Yet, from a value-theoretical perspective, this debate often tends to neglect the problem of how to compare values instead of how to compare arguments. One example is an article by Pijnenburg & Leget (2007) against radical life extension: although the authors acknowledge that "[l]ife is an intrinsic good" (Pijnenburg & Leget, 2007, p. 587), they argue that public funding for research on radical life extension may be problematic for reasons of distributive justice and that a radically extended life might be less meaningful than an ordinary life. Their arguments, therefore, are based on the assumption that both justice and meaning within one's life are values, and they at least implicitly assume that these arguments can be raised as objections to arguments that are based on the assumption that life is a value. However, these assumptions raise considerable value-theoretical issues, such as, how we should compare life as an intrinsic value to justice and meaning, both of which could be intrinsic but also extrinsic value. I do not deny that comparing life to justice and meaning is impossible or unreasonable, but it should be noted by raising arguments of this kind, one implicitly commits oneself to the view that these values can be compared to each other.

Problems of this kind only arise if one assumes that value pluralism is true, but as it seems, this assumption is almost ubiquitous in the ethical discussion of human enhancement. One could, therefore, not truly solve the problem by simply postulating that value monism is true because this assumption would just not acknowledge the reality of contemporary ethical reasoning. Yet, a general awareness of the problems value pluralism creates could be sufficient for investigating whether all of these potential values that feature in the ethical discussion of human enhancement indeed qualify as values or whether there is a kind of 'value inflation'. One could, for example, adopt a value-theoretically informed perspective in order to argue that justice or meaning are merely extrinsic values that do not deserve the same kind of consideration as life as an intrinsic value. However, one could also try to make the case for 'demoting' life by arguing that life *simpliciter* does not qualify as an intrinsic value since it is a purely descriptive phenomenon with no evaluative component. One could argue that only if one speaks of 'life' with further qualifications – for example, 'healthy life' or 'personal life' – that contain a normative component, life can qualify as an intrinsic value.

This is just one suggestion how one could deal with the problem of plurality in the context of human enhancement from a value-theoretical perspective. Yet, further examples can be easily imagined and show that value-theoretical contributions to the ethical discussion of human enhancement can help to adopt novel perspectives to existing arguments and that some arguments are based on implicit value-theoretical assumptions that deserve to be re-evaluated.

Incommensurable Values

The problem of plurality creates a further possible problem that was already alluded to in the previous section: since the ethical discussion of human enhancement features the inclusion of many values that are rather unrelated to each other, the following scenario is not too improbable: one argument favors a specific enhancement technology because it would contribute positively to the instantiation of an intrinsic value. Another argument, however, challenges the first argument by claiming that the enhancement technology in question would contribute negatively to the instantiation of another value. If the values mentioned in these arguments are incommensurable for some reason, a stalemate is created.

This general problem can be illustrated: in a recent article, Schaefer, Kahane & Savulescu (2013) argue for certain forms of cognitive enhancement that would be a positive contribution to a person's autonomy. Value-theoretically speaking, autonomy features as a value in this argument that would be instantiated to a higher degree by the application of the enhancements in question. However, one can imagine the usual counterarguments against human enhancement in general that appeal to the preservation of human nature (Fukuyama, 2002). The question is whether autonomy and naturalness – if one accepts that both of them are values, maybe even intrinsic values – can be weighed against each other: is there a degree of autonomy that would justify the application of a certain enhancement technology if it were sufficient for reaching this degree, even though its application would also imply a loss of naturalness? If the answer to this question is 'yes', then autonomy and naturalness are commensurable. However, the answer could also be 'no', so that autonomy and naturalness would be incommensurable or naturalness would have lexical priority over autonomy. How should we decide which response to these questions is correct?

Although providing a solution to this problem is well beyond the scope of this article, it should be clear that it cannot be solved within the ethical discussion of human enhancement but is an issue of value theory. The problem of incommensurability should not be underestimated because arguments for and against human enhancement could always be the foundation of public policies that are not just relevant for philosophical discourse but might have an immediate impact on the lives of individuals. However, if some arguments cannot actually be weighed against each other because the values they refer to are incommensurable, policy decisions that are seemingly based on these arguments possess a certain degree of arbitrariness and, therefore, lack justification.

The problem of incommensurable values can be regarded as a further example that underlines the relevance of value theory for the ethical discussion of human enhancement. First, value theory should enable us to identify potentially incommensurable values, and second, it should also give practical guidance for dealing with incommensurability in practical contexts. However, participants in the ethical discussion of human enhancement should also be aware of the threat of incommensurability and its possible effects on the argumentative structure of the debate and, furthermore, on future policies.

FUTURE RESEARCH DIRECTIONS

This broad overview of the relation of value theory to the ethical discussion of human enhancement suggests several directions towards future research could be directed. It can serve as the basis of recommendations for value-theoretical projects as well as projects in the realm of applied ethics.

It should have become clear that although value theory deals with truly foundational questions of moral philosophy, its concepts and various assumptions are much more relevant for practical issues, human enhancement in particular, than one might believe after a first impression. However, we still lack information on how to actually use value-theoretical concepts in order to solve practical problems or at least provide us with instruments that are of assistance in solving them. Since this article concentrated on raising awareness with regard to the relevance of value theory for the ethical discussion of human enhancement, one of its limitations is that it does not provide a toolbox of value-theoretical instruments that are ready for application to practical issues. However, the article at least shows why such a toolbox would be beneficial. Because of the generality of these suggestions, I will describe a kind of heuristic that could serve as an example for the development of further heuristics or instruments. This heuristic relies on the concept of value lexicality and on my tentative proposal that for two values *A* and *B*, *A* takes lexical priority over *B* if and only if *A* is a necessary prerequisite of *B*. It consists of several steps that should lead to an informed conclusion.

1. When assessing arguments about human enhancement, find out whether they feature any things that can count as values. – This is the obvious starting point for any ethical discussion that involves the concept of value lexicality. If there are no values mentioned in the arguments in question, then one cannot apply the concept at all and has to rely on something else.
2. If the arguments in question feature the values *A, B, C, ... Z*, then investigate whether there is a connection between those values such that some of them are necessary prerequisites for the others. – If there is no such connection, then the concept of value lexicality cannot be applied. However, one does not have to stick to this criterion because it might still be debatable. Someone who favors another criterion for value lexicality can still proceed with the analysis if his particular criterion is met by *A, B, C, ... Z*.
3. If the connection described in 2. can be established, then a lexical ordering of *A, B, C, ... Z* is possible and determined by the nature of this connection. – This is the important preliminary conclusion that leads to a judgment on the arguments in question. It is also striking that no further argument supporting the lexical ordering is needed because it follows directly from the purely descriptive fact that some values are necessary prerequisites for others and the acceptance of this criterion.
4. Of those arguments in question, the argument that recommends acting upon the lexically highest value is the *prima facie* best argument. – Finally, putting *A, B, C, ... Z* in a lexical ordering shall lead to an assessment of the arguments featuring them. Although it is obvious that no debate can be solely decided by applying the concept of value lexicality, it can be claimed that if other things are not taken into consideration, the argument that features the lexically highest value deserves special credit. All of the arguments in question recommend acting upon certain values, but the purely descriptive criterion of some values being necessary

prerequisites for others provides a guideline for putting the arguments themselves in an ordering (though not a lexical one).

The suggestions in the previous paragraphs as well as the heuristic already allude to recommendations for future ethical discussions of human enhancement. Since it could be shown that value theory can make valuable contributions to these discussions, ethicists should analyze issues related to human enhancement with increased awareness of their value-theoretical foundations. They could, for example, not only exchange arguments for and against particular enhancement technologies or human enhancement in general but also analyze, for example, whether these arguments feature intrinsic or extrinsic values, have to deal with incommensurability, or can be reformulated as lexicality arguments. This strategy of adopting a value-theoretically informed perspective on human enhancement could, then, enable experts in value theory to develop these heuristics and instruments that are actually beneficial for the ethical discussion.

CONCLUSION

This article provided a brief introduction to value theory with a focus on its applicability to the ethical discussion of human enhancement. It was shown that there are various value-theoretical issues – such as value lexicality, value pluralism, and incommensurability – that are implicitly reflected in this discussion. However, these issues are hardly analyzed explicitly, which is why an increased awareness with regard to the relevance of value theory for the ethics of human enhancement is desirable. It is possible and reasonable to formulate several proposals that are, on the one hand, directed towards value theorists but also, on the other hand, towards ethicists who deal with human enhancement. Given the considerable practical importance of value theory, value theorists should attempt to develop heuristics or instruments that are theoretically informed but can also be applied to practical issues without being too difficult to handle. Ethicists who participate in the debate about human enhancement are well-advised to craft their arguments with an increased awareness of possible value-theoretical issues that are included in the premises of their arguments. Ideally, these proposals lead to a more robust connection between value theory and applied ethics as well as, in the camp of value theory, an increased commitment to practical applicability and, in the camp of applied ethics, an increased willingness to analyze the value-theoretical foundations of the discussion of human enhancement.

REFERENCES

Bailey, J. A. (1979). On intrinsic value. *Philosophia*, *9*(1), 1–8. doi:10.1007/BF02379981

Beardsley, M. C. (1965). Intrinsic value. *Philosophy and Phenomenological Research*, *26*(1), 1–17. doi:10.2307/2105465

Bentham, J. (2007). *An introduction to the principles of morals and legislation*. Mineola, NY: Dover Publications.

Boorse, C. (1975). On the distinction between disease and illness. *Philosophy & Public Affairs*, *5*(1), 49–68.

Boorse, C. (1977). Health as a theoretical concept. *Philosophy of Science*, *44*(4), 542–573. doi:10.1086/288768

Cutas, D. E. (2008). Life extension, overpopulation and the right to life: Against lethal ethics. *Journal of Medical Ethics*, *34*, e7. doi:10.1136/jme.2007.023622 PMID:18757626

Davis, J. K. (2004). Collective suttee: Is it unjust to develop life extension if it will not be possible to provide it to everyone? *Annals of the New York Academy of Sciences*, *1019*, 535–541. doi:10.1196/annals.1297.099 PMID:15247081

De Grey, A., & Rae, M. (2007). *Ending aging: The rejuvenation breakthroughs that could reverse human aging in our lifetime*. New York, NY: St. Martin's Press.

Douglas, T. (2008). Moral enhancement. *Journal of Applied Philosophy*, *25*, 228–245. doi:10.1111/j.1468-5930.2008.00412.x PMID:19132138

Farrelly, C. (2010). Equality and the duty to retard human ageing. *Bioethics*, *24*(8), 384–394. doi:10.1111/j.1467-8519.2008.00712.x PMID:19222442

Frankena, W. K. (1963). *Ethics*. Englewood Cliffs, NJ: Prentice-Hall.

Fukuyama, F. (2002). *Our posthuman future: Consequences of the biotechnology revolution*. New York, NY: Picador.

Geach, P. T. (1956). Good and evil. *Analysis*, *17*(2), 33–42. doi:10.1093/analys/17.2.33

Gems, D. (2003). Is more life always better? The new biology of aging and the meaning of life. *The Hastings Center Report*, *33*(4), 31–39. doi:10.2307/3528378 PMID:12971059

Glannon, W. (2002). Identity, prudential concern, and extended lives. *Bioethics*, *16*(3), 266–283. doi:10.1111/1467-8519.00285 PMID:12211249

Griffin, J. (1986). *Well-being. Its meaning, measurement and moral importance*. New York, NY: Clarendon Press.

Griffin, J. (1997). Incommensurability: What's the problem? In R. Chang (Ed.), *Incommensurability, incomparability, and practical reason* (pp. 35–51). Cambridge, MA: Harvard University Press.

Hainz, T. (2012a). Value lexicality and human enhancement. *International Journal of Technoethics*, *3*(4), 54–65. doi:10.4018/jte.2012100105

Hainz, T. (2012b). Aging as a disease. In M. G. Weiss & H. Greif (Ed.), *Ethics – society – politics: Papers of the 35th International Wittgenstein Symposium* (pp. 109-111). Kirchberg: Austrian Ludwig Wittgenstein Society.

Harris, J. (2007). *Enhancing evolution: The ethical case for making better people*. Princeton, NJ: Princeton University Press.

Hayflick, L. (2004). Anti-aging is an oxymoron. *Journal of Gerontology*, *59A*(6), 573–578.

Heathwood, C. (2010). Welfare. In J. Skorupski (Ed.), *The Routledge companion to ethics* (pp. 645–655). Abingdon, UK: Routledge.

Hildt, E. (2009). Living longer: Age retardation and autonomy. *Medicine, Health Care, and Philosophy*, *12*, 179–185. doi:10.1007/s11019-008-9162-y PMID:18668344

Horrobin, S. (2006). Immortality, human nature, the value of life and the value of life extension. *Bioethics*, *20*(6), 279–292. doi:10.1111/j.1467-8519.2006.00506.x

Mackey, T. (2003). An ethical assessment of anti-aging medicine. *Journal of Anti-aging Medicine*, *6*(3), 187–204. doi:10.1089/109454503322733045 PMID:14987433

Mill, J. S. (1871). *Utilitarianism*. London: Longmans, Green, Reader, and Dyer.

Miller, R. A. (2002). Extending life: Scientific prospects and political obstacles. *The Milbank Quarterly*, *80*(1), 155–174. doi:10.1111/1468-0009.00006 PMID:11933792

Millgram, E. (2005). Incommensurability and practical reasoning. In E. Millgram (Ed.), *Ethics done right: Practical reasoning as a foundation for moral theory* (pp. 273–294). Cambridge, UK: Cambridge University Press. doi:10.1017/CBO9780511610615.011

Nielsen, L. W. (2011). The concept of nature and the enhancement technologies debate. In J. Savulescu, R. ter Meulen, & G. Kahane (Eds.), *Enhancing human capacities* (pp. 19–33). Malden, MA: Wiley-Blackwell. doi:10.1002/9781444393552.ch2

Pijnenburg, M. A. M., & Leget, C. (2007). Who wants to live forever? Three arguments against extending the human lifespan. *Journal of Medical Ethics*, *33*(10), 585–587. doi:10.1136/jme.2006.017822 PMID:17906056

President's Council on Bioethics. (2003). *Beyond therapy: Biotechnology and the pursuit of happiness*. Washington, DC: Author.

Rawls, J. (1971). *A theory of justice*. Cambridge, MA: Harvard University Press.

Sandel, M. J. (2009). The case against perfection: What's wrong with designer children, bionic athletes, and genetic engineering. In J. Savulescu, & N. Bostrom (Eds.), *Human enhancement* (pp. 71–89). New York, NY: Oxford University Press.

Savulescu, J., Sandberg, A., & Kahane, G. (2011). Well-being and enhancement. In J. Savulescu, R. ter Meulen, & G. Kahane (Eds.), *Enhancing human capacities* (pp. 3–18). Malden, MA: Wiley-Blackwell. doi:10.1002/9781444393552

Schaefer, G. O., Kahane, G., & Savulescu, J. (2013). Autonomy and enhancement. *Neuroethics*. doi: doi:10.1007/s12152-013-9189-5

Singer, P. (1991). Research into aging: should it be guided by the interests of present individuals, future individuals, or the species? In F. C. Ludwig (Ed.), *Life span extension: Consequences and open questions* (pp. 132–145). New York, NY: Springer.

ADDITIONAL READING

Allhoff, F., Lin, P., Moor, K., & Weckert, J. (2010). Ethics of human enhancement: 25 questions & answers. *Studies in Ethics, Law, and Technology, 4* (1), article 4.

Anderson, E. (1993). *Value in ethics and economics*. Cambridge, MA/London: Harvard University Press.

Bailey, R. (2013). For enhancing people. In M. More, & N. Vita-More (Eds.), *The transhumanist reader. Classical and contemporary essays on the science, technology, and philosophy of the human future* (pp. 327–344). Malden, MA/Oxford: Wiley-Blackwell. doi:10.1002/9781118555927.ch32

Bostrom, N. (2005). Transhumanist values. In F. Adams (Ed.), *Ethical issues for the twenty-first century* (pp. 3–14). Philosophy Documentation Center.

Bostrom, N. (2008). Dignity and enhancement. In The President's Council on Bioethics (Ed.), Human dignity and bioethics (pp. 173-206). Washington, DC.

Bostrom, N., & Roache, R. (2007). Ethical issues in human enhancement. In J. Ryberg, T. Petersen, & C. Wolf (Eds.), *New waves in applied ethics* (pp. 120–152). Basingstoke: Palgrave Macmillan.

Bradley, B. (2006). Two concepts of intrinsic value. *Ethical Theory and Moral Practice*, *9*(2), 111–130. doi:10.1007/s10677-006-9009-7

Chang, R. (Ed.). (1997). *Incommensurability, incomparability, and practical reason*. Cambridge, MA/London: Harvard University Press.

Ferrari, A., Coenen, C., & Grunwald, A. (2012). Visions and ethics in current discourse on human enhancement. *NanoEthics*, *6*(3), 215–299. doi:10.1007/s11569-012-0155-1

Griffin, J. (1996). *Value judgement. Improving our ethical beliefs*. Oxford: Clarendon Press.

Hanks, C. (Ed.). (2010). *Technology and values. Essential readings*. Malden, MA/Oxford: Wiley-Blackwell.

Hansson, S. O. (2001). *The structure of values and norms*. Cambridge: Cambridge University Press. doi:10.1017/CBO9780511498466

Harman, G. (2000). *Explaining value and other essays in moral philosophy*. Oxford: Clarendon Press.

Harris, J. (2010). Moral enhancement and freedom. *Bioethics*, *25*(2), 102–111. doi:10.1111/j.1467-8519.2010.01854.x PMID:21133978

Hartman, R. S. (1967). *The structure of value: foundations of scientific axiology*. Carbondale, IL: Southern Illinois University Press.

Huemer, M. (2010). Lexical priority and the problem of risk. *Pacific Philosophical Quarterly*, *91*, 332–351. doi:10.1111/j.1468-0114.2010.01370.x

Kagan, S. (1998). Rethinking intrinsic value. *The Journal of Ethics*, *2*, 277–297. doi:10.1023/A:1009782403793

Kourany, J. A. (2013). Human enhancement: making the debate more productive. *Erkenntnis*. Retrieved November 15, 2013. doi:10.1007/s10670-013-9539-z

Lemos, N. M. (1994). *Intrinsic value. Concept and warrant*. Cambridge: Cambridge University Press. doi:10.1017/CBO9780511663802

Moore, G. E. (2004). *Principia ethica*. Mineola, NY: Dover Publications.

Rabinowicz, W. (2008). Value relations. *Theoria*, *74*, 18–49. doi:10.1111/j.1755-2567.2008.00008.x

Rescher, N. (1969). *Introduction to value theory*. Englewood Cliffs, NJ: Prentice-Hall.

Rescher, N. (2004). *Value matters. Studies in axiology*. Heusenstamm: Ontos.

Sandel, M. J. (2007). *The case against perfection. Ethics in the age of genetic engineering*. Cambridge, MA/London: Belknap Press.

Savulescu, J., & Bostrom, N. (Eds.). (2009). *Human enhancement*. Oxford/New York, NY: Oxford University Press.

Savulescu, J., ter Meulen, R., & Kahane, G. (Eds.), *Enhancing human capacities*. Malden, MA/Oxford: Wiley-Blackwell. doi:10.1002/9781444393552

KEY TERMS AND DEFINITIONS

Extrinsic Value: A thing that is good inasmuch as it positively contributes to the instantiation of a thing that is good in itself.

Human Enhancement: The non-therapeutical improvement of human capacities through external interventions.

Incommensurability: The phenomenon that two or more values cannot be weighed against each other.

Intrinsic Value: A thing that is good in itself.

Radical Life Extension: The direct and intentional extension of the maximum human life span.

Value Lexicality: The phenomenon that one – lexically prior – value cannot be outweighed by another value, regardless of the latter value's amount or intensity.

Value Monism: The position that only one basic value exists.

Value Pluralism: The position that more than one basic value exist.

Value Theory: A branch of philosophy that deals with theoretical as well as practical issues regarding values.

Chapter 3
Personal Liability and Human Free Will in the Background of Emerging Neuroethical Issues:
Some Remarks Arising from Recent Case Law up to 2013

Angela Di Carlo
Scuola Superiore Sant'Anna of Pisa, Italy

Elettra Stradella
University of Pisa, Italy

ABSTRACT

In this chapter, the authors analyse the issues connected to emerging neurotechnologies, in particular their effects on (legal) concepts like capacity, liability, testimony, and evidence, and also on fundamental constitutional rights and freedoms like the right to autonomy and the right not to be treated without consent (in the general framework of the principle of human dignity). Starting from preliminary remarks on the key concepts of neuroethics/technoethics, neurolaw/technolaw, the authors investigate how personal liability is changing in the framework of new scientific developments. The chapter underlines that neurolaw challenges some of the traditional legal institutions in the field of law (e.g., criminal law). From the point of view of ethics, the chapter concludes that neuroethics is not challenged by the data coming from the use of emerging neurotechnologies, but human self-perception is strongly affected by it.

DOI: 10.4018/978-1-4666-6122-6.ch003

PRELIMINARY REMARKS: NEUROETHICS AND TECHNOETHICS AND NEUROLAW AND TECHNOLAW

The aim of this chapter is to analyse some of the issues connected to emerging neurotechnologies, considering in particular their effects on (legal) concepts like capacity, liability, testimony, and evidence.

Some preliminary clarifications should be made regarding the relationship between neuroethics and technoethics. Starting from concrete issues we can underline the "intersection" between the various "N-Ethics": the growing attention to the ethical questions regarding both neuroscientific applications in general and all the emerging technologies in the field of biomedical research in particular, such as nanotechnologies, bionics, and neural interfaces, as well as innovative biomedical applications, such as biomechatronic prostheses, hybrid bionic systems, and bio-mechatronic components for sensory and motor augmentation.

All these "forms of technique," in various ways connected with biomedical applications, are generating (and will increasingly generate) new ethical, legal and safety implications, such as human enhancement and human tele-operation but they also affect already known issues, such as surveillance, privacy, and dual-use.

Starting from these implications it is possible to find common ground uniting neuroscience, robotics, and different emerging technologies, and, at the same time, to define a uniform but pluralistic scenario of technoethics. Although there exist many differences between different technologies in terms of legal relevance, the main standard to be used could be the "enhancing standard", concerning how these technologies can affect human functions (looking at every human "capability," in the language of Sen and Nussbaum, from cognitive capabilities to the power to exert one's own rights and freedoms). This standard allows us to put together technological developments like neurorobotics and bionics, neural interfaces, robotics for neurorehabilitation, assistive robotics, prosthetics, gerontechnologies, and generally neuroscience goals. Indeed, they not only affect ethical issues like the controversial philosophical concepts of identity, autonomy, and self-assessment, but actually bear on fundamental rights, simultaneously enhancing some of them and infringing on the true implementation of others. An important example is the right to privacy. Many robots are already endowed with the ability to perceive, store and use sensitive data related to human beings, such as biometric information, but also details concerning, for example, users' consumer preferences, habits, or emotional states; at the same time, as we will see, the *vulnus* to privacy is raised as a potential peril deriving from the use of neurotechnologies (Farah, 2002; Wolpe *et al.*, 2005).

In this chapter we will not deal with the problem of regulating emerging technologies and designing a common framework of ethical and legal principles. Instead our point of view will be grounded on the political and social application of sciences and technologies providing society with new instruments to prevent crimes and to administer justice.

The "ethical intersection" is clear with particular reference to the relationship between neuroethics and technoethics, and, consequently, between neurolaw and technolaw. The inclusive vocation of technoethics could be quite plainly drawn from the application of the syllogism: "Humans are naturally technical (*homo technicus*); humans are naturally ethical; technology is naturally ethical." The thesis is that humankind is technical by nature. Technology is not an addition to humanity but is just one of the ways, maybe one of the more significant ways, in which humans are distinguished from animals because human beings are forced to interact by working with the material cosmos in order to produce technology (Galvan, 2003).

This is connected with the second statement: *humans are naturally ethical,* from which the third would derive.

In order to argue this thesis (which is, of course, contestable: the fact that humans are naturally technical and ethical does not automatically make the technique – and technology – ethical, because these could be two different spheres of action) we can underline that Aristotle distinguishes ethics from technique for a merely linguistic reason. The two verbs, *prassein* (to act) and *poiein* (to do) could be substantially the same, because in any event humans when they act produce something that leads them to self-fulfillment. Indeed, technology is ethics (insofar as it is self-fulfillment) and then it is "good," because the "evil" can lie only in the utilization of the object.

While for Aristotle and the Greeks in general authentic behaviour is exclusively represented by the disinterested knowledge of what already exists, Christianity may individuate the achievement of God (the true goal of ethical action) in the making, building and operation of technology. Francis Bacon in 1600 prophesied the technological era, identifying knowledge with making: it is the action of making that definitively becomes the constitutive element of human action – that is, ethical action – because it is strictly connected with self-fulfillment. Kant, moving from the idea that humans are naturally ethical and ethics are formally identical in everybody (the celebrated categorical imperative), finds the value of human action just in the intention on which it is based: there is not ethics in doing, but ethics remains in modalities of doing itself.

These reflections help us to understand that technoethics could well summarize and bring together into a unique notion all the complexity deriving from scientific developments. At the same time they imply that neuroethics, as part of technoethics, is the ethics concerning the premises of the "technical action" made by humans. These premises concern the functioning of the brain, the genetic pool, and the mechanisms that control human behaviour.

A specular reflection can be seen to the relationship between technolaw and neurolaw, in the field of the social and legal implications of the above-mentioned syllogism. Technolaw and neurolaw shall be seen in the light of the change in the Hegelian point of view with respect to the Kantian approach. Hegel accuses Kant of abstractness and traces ethics back to a synthesis of legality and morality concretized in the state, in which the human being realizes her potential through (ethical) action. In this framework ethics stays in the progress and (spiritual) making represents the coming true of the human being.

When was neuroethics born? Its arrival could be attributed to "Neuroethics: Mapping the Field," a conference that took place in San Francisco, on 13–14 June 2002, on the initiative of Stanford University and the University of California. For the first time this word was used and the concept defined as:

The study of ethical, legal and social issues that arise when scientific discoveries on the brain enter into medical practice, in legal interpretations and in health and social policies. These discoveries are coming in the field of genetic, brain imaging and in diagnosis and prediction of diseases. The task for Neuroethics is to investigate how medics, judges and lawyers, politics and insurers, and public opinion at large, deal with these topics.

The effective growth of neuroethical issues and of neuroethics as an issue, could be connected to some (hard) judicial cases since 1981 that have put in evidence the difficulties encountered by judges in deciding whether or not to admit some neurotechnologies in evidence (brain images in particular) (Khosbin & Khosbin, 2007); some argue that the discoveries made in the last twenty years will constitute an important challenge for the legal system (Garland & Glimcher, 2006).

Already in *United States v. Hinckley* (525 F.Supp. 1324 (1981)) the defendant (who had made an attempt to assassinate US President Ronald Reagan) put forward a defence based on his insanity caused by the atrophy of a zone of the brain visible in Computerized Tomography (CT). The court admitted this evidence and expert witnesses called by Hinckley confirmed that from data that appeared from scanning his brain it could be reasonably thought that the atrophy caused schizophrenia. The jury thus found Hinckley not guilty on the grounds of insanity.

The genetic basis for crime, one of the most relevant issues in neurolaw, was directly faced for the first time less than twenty years ago, when the National Institute of Health in USA withdrew its funding from a conferencc on genetics and crime. The idea shocked some experts who considered that this kind of study aimed to underline ethnic and racial differences, potentially leading to the danger of eugenics. The president of the Association of Black Psychologists at the time declared that such research was in itself "a blatant form of stereotyping and racism" (Cohen, 2011). Criminologists have for a long time largely rejected or ignored genetics and biological causes for crime, concentrating on social causes. But now they are warily returning to this subject, exploring how genes might strengthen the risk of committing a crime and displaying violent behavior.

From the strictly legal perspective, the international legal debate has focused especially on possible uses of neurotechnologies for forensic purposes. The topic has found the strongest attention in the United States, where a significant and debated case law has considered the role of scientific evidence (and, in particular, technological evidence) in court proceedings (Goodenough &Tucker, 2010; Jones & Shen, 2011). As we will see below, this attention on the legal implications of neuroscience and its applications has spilled over into the European context, entering both (criminal) case law and many multidisciplinary research projects. We can mention above all the European Association for Neuroscience and Law (EANL), led in Italy by the University of Pavia and involving neuroscientists, legal scholars, and ethicists from various European countries.

Neuroscience, neurotechnologies and law are nowadays interrelated from two different perspectives: the application perspective and the normative or prescriptive perspective (the latter in its philosophical sense). The former regards all the legal, technical and procedural issues concerning the use of neurotechnologies as scientific evidence in a trial. The normative perspective asks us to investigate the cultural "aptitude" of neuroscience and neurotechnologies to rebuild some legal features and calls into question some of the foundations of the modern concept of legal liability.

HOW PERSONAL LIABILITY IS CHANGING IN THE FRAMEWORK OF NEW SCIENTIFIC DEVELOPMENTS

The prescriptive perspective raises a crucial question: does neuroscience really offer to law (criminal law above all) a new window of opportunity to reform itself? And, if so, would it be desirable and constitutionally oriented to create a reformed criminal justice system that views crime essentially as a brain disease (Erickson, 2012)? Some scholars argue that neuroscience does not offer anything new that cannot be incorporated into existing substantive criminal law doctrine. In the case of an agent who lacks control over his or her behaviour because of a psychological disorder, criminal law in most jurisdictions already affords a plea of non-responsibility, coherently with the principles of deterrence and the retributive character of our legal systems in the field of criminal justice.

It seems to be an issue of the possibility of finding a difference between a behaviour determined by a psychological disorder and a behaviour determined by a neurological injury.

[I]t seems quite reasonable to examine the brains of offenders in search of abnormalities that can be linked to their legally relevant conduct. If an offender has some frontal lobe disorder, and

studies show that frontal lobe disorder is somehow linked with aggression and violence, then it seems perfectly sensible to commit them in order to prevent future tragedies. But this trajectory invariably leads us down the road of examining and classifying brain differences and engaging the power of law before any conduct has occurred. This enterprise is not really about care and treatment but preventive detention. It is an old maxim that the law does not punish mere thoughts – acts are required. But the promise of neuroscience to tell us that certain citizens are predisposed to act in certain ways is nearly irresistible [...]. Legislatures could easily craft statutes which declared such persons as dangerous, but mentally abnormal offenders in need of confinement and treatment. More broadly, there is little reason to believe that neuroscience would not touch on most legal issues involving the capacity of legal agents. (Erickson, 2012, p. 18)

This very strong critique on the implications of the relation between neurotechnologies and law fails to consider some elements. First of all, unfortunately it is not true that law never punishes mere thoughts. What about "hate crimes" and all the criminal features that (above all) in many European countries punish expression that contrasts with the core values of the legal system and aimed at injuring other people's rights? Second, the forensic use of neurotechnologies is mainly addressed to mitigation of the liability of the subject, not to increase his/her punishment or prevent his/ her conduct (these eventual implications of neurotechnologies are simply futuristic, for the moment). Third, while there is probably no difference between a psychological disorder and a neurological injury in order to describe the relation between conduct and will and to decide the nature and degree of the relative liability, it is true that some difference shall be recognized between a "disease" and a permanent and *ab origine* condition derived from the composition and the characterization of the genetic pool.

Some have underlined that neuroscience could dissipate ancient illusions such as free will and that the retributive aims proper to criminal justice risk to be replaced with deterrence, prevention and treatment (Greene-Cohen, 2004). Indeed deterrence (a) is traditionally part of the aims of criminal law, and (b) actually raises some problems with regard to "diseased subjects" that would not be able "to be deterred."

Another point of view, recently expressed in some interesting essays, integrates neuroscientific goals and the potentialities of neurotechnologies within already existing legal categories. This idea, which seems quite popular and is certainly well discussed, starts from the statement that although it is clear that the brain enables the mind, there is not yet any real confidence about how this happens, and about the connection between brain and action (Morse, 2007). Morse, who rejects the scenario of neuroethics and neurolaw as an issue, argues that

[T]he genuine problem of free will is metaphysical and often spawns confusion. Roughly, it refers to whether human beings possess the ability or power to act uncaused by anything other than themselves, which is referred to as libertarian freedom of the will [...]. Solving the free will problem would have profound implications for responsibility doctrines and practices, but, at present, the problem plays no proper role in forensic practice or theory because this ability or its lack is not a criterion of any civil or criminal law doctrine. Forensic psychiatry and psychology address problems genuinely related to responsibility, including consciousness, the formation of mental states such as intention and knowledge, the capacity for rationality, and compulsion, but they never address the presence or absence of free will. People sometimes use "free will" loosely to refer to genuine responsibility doctrines, but this simply distracts attention. (Morse, 2011, p. 26)

It is important to underline this distinction between free will and responsibility, *rectius* between free will and capacity. As it has been written, since a general capacity to be rational is the primary condition for liability, the lack of capacity is the primary excusing condition.

We will say something in the next paragraph about the US case of *Roper v. Simmons* and its implications: anyway it is quite obvious that infants, and persons with mental diseases or having particular conditions that determine a lack of self-consciousness (and so, of capacity), are subjected to a different treatment from the legal point of view. The connection between lack of capacity and responsibility looks different depending on which type of responsibility we are considering; moral, social, political, or legal. And even in the framework of legal responsibility, criminal liability should not be assimilated to other forms of liability.

Moreover, compulsion or coercion is also an excusing condition, but it is something different from lack of rational capacity, and this difference can be useful in order to investigate the possibility to make the above-mentioned distinction between psychological disorder, brain disease and genetic predisposition.

In particular we shall consider three different conditions: a literal compulsion (when the person's movement is a mechanism, the product of mechanistic causes and cannot be ruled by the agent's desires, beliefs, and intentions); and a metaphorical compulsion, in a double sense (when the agent acts intentionally, but in response to some hard choice imposed on the agent, externally or internally) (Morse, 2007).

In the case of material compulsion the excuse derives from something like a lack of action (the action is not ascribable to the person in her desires and intentions because it is not "made" – in the stronger sense of the word – by the person herself). This is what happens in some brain diseases. In the case of metaphorical compulsion, when the compulsion has an internal origin, it is impossible to individuate a responsibility too, and this is the case of many psychological disorders.

But what about genetic predispositions? Do they determine lack of rationality? Shall they determine a sort of "brain defence" a mitigating circumstance as it has happened in some cases with the so-called "cultural defences," when people lacking the common system of values of the state in which they live are excused for criminal conduct that is "culturally oriented" and caused by their own beliefs (Stradella, 2009)?

We may consider the role of the genetic pool from one side, and of brain diseases from another side, but in any case these phenomena seem something different from the concept of free will, which concerns the intrinsic nature of the human being and his or her rationality.

Another question about how neurotechnologies are able to affect law regards the "communication" of neuroscientific discoveries. Very complex instruments and results are often described in a simplistic manner. Above all brain imaging techniques offer the opportunity "to access other people's pain, by making it visible, and to some extent, measurable" (Bottalico, 2011). Recently a patent on pain detection, entitled "Objective Determination of Chronic Pain in Patients," was conferred to Dr Robert England, an orthopaedic surgeon in California, whose method involves the use of Functional Magnetic Resonance Imaging (Camporesi-Bottalico, 2011).

The evolution of personal liability may also be affected by another very sensitive field of neurotechnology: brain-machine interfaces (BMIs). BMIs offer the possibility of a new generation of technologies that allow users to control devices directly via the nervous system. They can be used to develop new communication pathways to restore and augment sensory and motor function in disabled individuals. At the core of BMI system design is a controller that must accommodate the seamless interaction between systems of neurons with electronics and robotics (Wessberg, 2000;

Lebedev, 2006; Micera *et al.*, 2010). From the neurolaw point of view there are many implications of this kind of neurotechnology: the expression of informed consent by people who suffer from locked-in syndrome, the uncertainty of future implications for the patient's brain, the redefinition of the biological and mental boundaries of any individual (Bottalico, 2011).

Even the issue of personal liability is involved in BMI developments, because they constitute a sort of intermediation between the human brain and intentionality and concrete action or the manifestation of self-determination.

An important difference shall be delineated between technological developments like BMIs or other forms of devices designed to enhance human capabilities and neurotechnologies directed to increase scientific knowledge in the field of human functioning and behaviours. Both seem to concern ELSI (the acronym used to express the complex field of ethical, legal and social issues connected to technological developments), but while technological enhancement and other forms of intermediation between humans and the world directly affect human action, neurotechnologies in the field of genetics allow us to know and forecast (to what extent is the true question) human action.

The incidence on determination of liability in legal practice and forensic field is not easy to assess. In the report *Genetics and Human Behaviour: the Ethical Context* (2002), the group of experts constituted by the Nuffield Council of Bioethics in the UK conclude that starting from the available knowledge it is not possible at this time to state the existence of genetic conditions that can justify an exclusion of criminal liability, but it is possible to take into consideration some of these conditions in order to determine the punishment and eventually mitigate (Santosuosso, 2009).

Moreover, from the point of view of comparative studies, there is a serious issue regarding the regulation of the use of neurotechnologies and neuroscientific developments. Criminal law and procedures have a strong country-based connotation; on the contrary, science has a transnational dimension and its intrinsic nature determines that legal solutions, in the field of scientific discoveries and applications, are less state-bound than ever before. This is a quite complicated issue that seems to require an intersection between different legal systems, which may be feasible due to the integration of traditional sources of law with new types of norms: ethical norms and soft law above all, even if in the field of criminal law there are many criticisms (Bernardi, 2009).

NEUROTECHNOLOGIES AND BEHAVIOURAL GENETICS IN RECENT CASE-LAW: NEW PERSPECTIVES FOR NEUROLAW

To understand the potential role of neuroscience in case law we can start from two different cases that are significant, we think, although for different (or opposite) reasons. Only the second is a judiciary case.

First is the story of Phineas Gage, a classic of neuroscience: indeed it seems to reveal that behaviour is not exclusively the product of personality and will (in its psychological meaning), but mechanisms that control the functioning of the brain play a very important role. Phineas Gage (1823-1860) was a US railway worker. He and his crew were excavating rocks to make way for the railroad. Gage was preparing for an explosion, using a tamping iron to compact an explosive charge in a borehole. As he was doing so, the iron produced a spark that ignited the powder, and the resulting blast propelled the tamping iron straight through his head. Remarkably, Gage survived, being conscious and walking within minutes. Back at Gage's nearby lodgings, Harlow, the physician who attended at the scene, removed small bone fragments from the wounds, replaced larger fragments that had been displaced by the passage of the tamping iron, and closed the large wound at the top of Gage's head with adhesive straps. Several days later, one of the wounds became infected and Gage fell into a semi-comatose state, but soon recovered and started to lead an apparently normal life, only "apparently" because his relatives and friends began to notice worrying changes in his behaviour (neurophilosophy.wordpress.com). It is interesting to quote the description of the manifestations of Gage's injury made by the physician some years later, found in the *Bulletin of the Massachusetts Medical Society*:

His contractors, who regarded him as the most efficient and capable foreman in their employ previous to his injury, considered the change in his mind so marked that they could not give him his place again. He is fitful, irreverent, indulging at times in the grossest profanity (which was not previously his custom), manifesting but little deference for his fellows, impatient of restraint of advice when it conflicts with his desires, at times pertinaciously obstinent, yet capricious and vacillating, devising many plans of future operation, which are no sooner arranged than they are abandoned in turn for others appearing more feasible. In this regard, his mind was radically changed, so decidedly that his friends and acquaintances said he was 'no longer Gage.'

This case has been used (and, in some opinion, abused) ever since it first appeared. Even if it is quite difficult to separate the facts of this story from the fictions, certainly Gage's case reveals that complex functions such as decision-making and social cognition are largely dependent upon some specific areas of the brain, and specifically frontal lobes.

What would have happened if Gage had killed someone or had manifested aggressive and violent behaviours? How should the judge have taken into account the terrible facts we have just narrated?

The second starting point is the US Supreme Court decision in the case of *Roper v. Simmons*, (543 U.S. 551 (2005)). In this case, the Supreme Court held the unconstitutionality of capital punishment for crimes committed while under the age of 18, overruling previous sentences on offenders above or at the age of 16 (*Stanford v. Kentucky*, 492 U.S. 361 (1989)). Using the "evolving standards of decency" test under the Eighth Amendment of the United States Constitution, the Court held that it was cruel and unusual punishment to execute a person who was under the age of 18 at the time of the offence. Justice Kennedy cited many sociological and scientific research studies showing that youngest have a lack of maturity and sense of responsibility compared with adults, and are also more vulnerable to negative influences, having less control over their environment. Five years later in *Graham v. Florida* (560 U.S. 48 (2010)) the US Supreme Court, with Justice Kennedy again writing for the majority, went one step further ruling that it is against the Constitution, under the Eight and the Fourteen amendments, for juvenile to be sentenced to life in prison without parole for a crime other than homicide. As in *Roper v. Simmons*, the Court ruled explicitly citing developments in psychology and brain science, raising the question to what extent advances in neuroscience should influence the sentencing of adolescents (Buchen, 2012).

The cases mentioned above seem to be a relevant starting point for some concluding remarks. Indeed they seemingly represent two situations that can be assimilated in the relation between human will, freedom and responsibility, but they actually lead to different approaches and show the ambivalent nature of neuroethics and neurolaw. Gage's case illustrates that the brain and its biological mechanisms can strongly (co-) determine human behaviour, and some conduct can be ascribed to its functioning more than to the moral assessment made by the individual with regard to a specific situation. The "good" Gage and the "bad" Gage were simultaneously the same individual and two different persons: they were the same individual because both before and after the accident Gage acted according to his own ethical nature, aiming at what he considered such as the best for him. But they were two different persons as well, since an empirical evaluation of the person, made by his own behaviours and character, would have led one to distinguish the "first" Gage from the "second".

Crossing this experience with *Roper v. Simmons* – and with the latter *Graham v. Florida* – we notice that the transformation of character and perception of reality can derive also from a "natural" and "ordinary" evolution of the individual during their life. In *Roper v. Simmons* a different kind and degree of responsibility is ascribed to young people in comparison with adults. Nevertheless it is obvious that youth cannot be considered as a pathology or a psychological disorder, nor that the brain of infants has a specific functioning (at least as concerns its bodily mechanisms).

Maybe we should infer that responsibility does concern the iteration between self-determination and action: when something interferes with the linear process that leads from thought towards action, then we can discuss the quantification and the qualification of responsibility, and maybe we shall discuss a neuroethics and neurolaw question. In both the case of Gage and in *Roper v. Simmons* that iteration was not linear, and the determination of self was deviated by something: free-will and determinism did not matter, but simply a reasonable assessment of the *suitas* of the action depending on the kind of distortion that has intervened.

It is interesting to look at the recent case-law in order to investigate how these issues have been treated by Courts, and which answers have been found in neurotechnology and its discoveries. Although Courts faces an increasing number of defendant and lawyers introducing neuroscientific evidence in trials, case-law still show a variety in the way these issues influence the sentencing of defendants.

The case of Christopher Tiegreen illustrates, for instance, that "trying to explain the brain and human behavior can clash with how the legal system determines culpability, competency and the manner in which such cases should be handled (Davis, 2012).

Tiegreen was involved in a car accident and, after a month, he emerged a different person from the coma. He suffered from an injury to the frontal lobe of his brain and became violent toward his family. One day, he slipped away from the apartment where he was supposed to be under 24-hour supervision and attacked a young women and her baby. During the trial, where Tiegreen was charged with aggravated assault, criminal attempt to commit rape and sexual battery, and third-degree child cruelty, his attorney tried to demonstrate the relation between Tiegreen's violent behavior and his brain injuries, asking to have him declared mentally incompetent.

Despite the introduction of neuroscience evidence in the trial, in May 2012 the Court of appeal of Georgia upheld the jury's verdict which found Tiegreen mentally competent to stay in trial and be sentenced for the alleged crimes. In Tiegreen case the Court clearly failed to consider the above mentioned case of Phineas Gage, denying a direct relevance to the brain injuries suffered by the defendant.

A criminal case decided in Italy by the Court of Como in 2011 received a great deal of publicity (Neuroethics and Law Blog, May 2011).

In that case (the s.c. Albertani case), a young Italian woman was charged with the murder of her sister and the kidnapping and attempted murder of her mother. The Court of Como found the accused guilty but only partially responsible, mitigating the sentence from 30 to 20 years imprisonment, of which at least 3 years are to be spent in a mental hospital. The defending attorney presented a defence of insanity based on psychiatric, neuroscientific and genetic tests, in order to demonstrate scientifically that the accused was partially insane at the time of the crime. The experts appointed by the Court and by the parties presented very conflicting conclusions and consequently the Court asked for further experts to be consulted.

The new experts declared the presence of a dissociative identity disorder, mainly revealed by memory deficits detected both through behavioural analysis and memory tests (the Autobiographical Implicit Association Test [IAT] and Time Antagonistic Response Alethiometer [TARA], used to establish whether an autobiographical memory trace is encoded in the respondent's brain). They tried to demonstrate the accused's incapacity to distinguish between right and wrong and control her impulsivity, and did it using neurotechnologies such as EEG and VBM. Voxel Based Morphometry (VBM) showed a lack of integrity and functionality of the anterior cingulated cortex, linked to obsessive-compulsive disorder and aggressiveness, where the person lacks the capacity to substitute the automatic behaviour with a different one. A strong tendency toward aggressive behaviour was illustrated on the basis of the presence of a genotype related to the MAOA-uVNTR polymorphism. Some literature describes this genetic pattern as an "unfavourable" condition for individuals raised in antisocial contexts (Caspi *et al.,* 2002; Pietrini-Bambini, 2009): genetic pool thus would be a sort of *multiplier* of external conditions in the definition of the individual's tendencies to be aggressive, violent and inclined to commit crimes.

This testimony was considered the most convincing of the three. The judge explained why neuroscience and genetics should be entitled to play a role in criminal procedure and emphasized modern psychiatry's difficulty in recognizing mental insanity with the necessary precision and reliability, in defining precise diagnosis of mental pathologies, and in evaluating personal abilities to distinguish between right and wrong. She stated that this case does not introduce a new deterministic method of deriving specific criminal attitudes from brain morphology: moreover this has been underlined in many occasions by Pietrini, the neuroscientist who is most involved with these issues.

In particular Pietrini was engaged in another case, where the Court of Appeal of Trieste reduced the sentence given to a convicted murderer by a year because he had genes linked to violent behavior: this was the first time neurotechnology had directly affected case law in Italy. The facts were that in 2007 an Algerian citizen who had lived in Italy for many years admitted having stabbed and killed a Colombian living in Italy who had insulted him because of the make-up the Algerian was wearing for religious reasons.

During the trial the defendant's lawyer asked the court to take into account that her client may have been mentally ill at the time of the murder. After considering three psychiatric reports, the judge agreed that the defendant's psychiatric illness was a mitigating factor to some extent and sentenced him to around three years less than he would otherwise have received.

On appeal, at the Court of Appeal of Trieste they were asked forensic scientists for a new independent psychiatric report to decide whether he should commute the sentence further. Pietrini (molecular neuroscientist at the University of Pisa) and Sartori (cognitive neuroscientist at the University of Padua), conducted a series of tests and found abnormalities in brain-imaging scans and in five genes that have been linked to violent behaviour – including the gene encoding the neurotransmitter-metabolizing enzyme monoamine oxidase A (MAOA). A study led by a geneticist at the Institute of Psychiatry, King's College, London, in 2002 had found low levels of MAOA expression to be associated with aggressiveness and criminal conduct of young boys raised in antisocial and violent environments (Feresin, 2009).

In their report, Pietrini and Sartori concluded that the defendant's genes would have made him more inclined to develop violent conduct if provoked. Pietrini stated that there is "increasing evidence that some genes together with a particular environmental insult may predispose people to certain behaviour": in this sense he underlined the combined role of genetics and environments, debunking the charges of being deterministic. But the question seems to become more urgent whether some link exists between genes and environment, and to what extent the ethical nature of humans is influenced by these two elements.

In this case, on the basis of the genetic tests, the judge removed a further year from the defendant's sentence, arguing that the defendant's genes "would make him particularly aggressive in stressful situations."

Another relevant case comes from the Netherlands. In 2007 a 63-year-old man stabbed a friend nine times, killing her. The defendant declared that he was annoyed by the victim's behavior. Moreover he declared that he saw that the victim was losing a lot of blood and when she regained consciousness and tried to get up he stabbed her again. At the time, he was intoxicated with alcohol and cocaine.

Some experts were asked to assess the defendant's capacity at the time of the event, and one of them, a behavioral neurologist, stated that his behavior during the incident was affected by damage to his frontal lobes, and this brain damage caused him to be unable to control his impulses and reflect on his actions in particularly difficult situations. It is interesting to underline that the expert stated that the suspect's brain damage had interfered with his *free will* (Klaming, 2011).

In this case the judge decided that although the defendant's behavior was affected by frontal lobe damage, he did not completely lack insight into the consequences of his actions. Indeed, the defendant was aware of the possibility that the victim would die as a result of the harm that he was inflicting on her. On the other hand, the judge, following the neurologist's report, considered the defendant severely diminished in responsibility due to frontal lobe damage, which would had interfered with his free will. He was sentenced for manslaughter to 18 months imprisonment, plus detention during Her Majesty's pleasure (District Court Amsterdam 2008, http://www.rechtspraak.nl, quoted by Klaming, 2011).

One of the most recent example of how neuroscience is challenging criminal trials –punishment in particular – is offered by the case of dr. Domenico Mattiello, an Italian pediatrician accused of pedophilia towards some of his patients. "His previous behavior was completely normal. He was a pediatrician for 30 years and he saw tens of thousands of children and never had any problem", stated Pietro Pietrini, the neuroscientist of the Albertani case mentioned above. But then something happened and Dr. Mattiello who had always behaved properly suddenly changed so drastically.

In the opinion of the neuroscientists Pietrini and Sartori, the explanation comes from a full psychiatric and biological analysis including a brain scan which shows a tumor growing at the base of Mattiello's brain. They both consider the case of the Italian physician similar to the one of an American schoolteacher who started to commit acts of pedophilia at the age of 40-year-old. In 2002 he was charged with this crime towards his step-daughter and during the trial medical examinations revealed that he had an egg-sized tumor in a part of the brain involving decision making. Once the tumor was removed, his criminal tendencies disappeared but it was necessary a second surgical intervention for the tumor to be completely removed and allowing him home.

Although there are not yet medical evidence proving a general correlation between a brain cancer pathology and tendencies to pedophilia attitudes, Pietrini and Sartori believes that also in the case of dr. Mattiello the existence of the tumor may have played a role in altering his behavior towards children. Thesis which had not convinced the public prosecutor, and the Court of Venice as well, that in January 2013 found Domenico Mattiello guilty and sent him to prison for five years.

Two are the remarkable points of this case-law. First, neuroscientific evidence are still facing skepticism in European courts. Secondly, the influence of neuroscience evidence strongly depends on the crime the attendant was charged with. In case of the so called "acquired pedophilia" the tendency to mitigate the punishment seems reasonably to make an offence, irrespective of solid neuroscience evidence, to the social perception of such a crime.

We will now offer some concluding remarks reflecting on the cases just described. We have raised some concerns about the use of neurotechnology in case law, already illustrated by scholars, in order to look for the best balance between the role of technological developments in discovering human nature and functioning and the protection of fundamental rights and freedoms, as well as the general principles of justice.

A first concern regards the fact that the assessment of intentionality or responsibility is at an early stage of its development (Klaming, 2011): this means that there is no certainty on the instruments we are considering in the paper, and actually it is not possible to draw a direct correspondence between data coming from neurotechnologies and the authentic will of the authors whose data belong to.

With regard to data, an important concern is on privacy issues (Farah, 2002; Wolpe *et al.*, 2005). Indeed data deriving from neurotechnologies and in particular from neuroimaging can reveal a large amount of information about an individual, most of them regarding "sensitive topics" and private aspects such as personality, mental illness, predispositions, character (Canil & Amin, 2002; Childress *et al.*, 1999).

The privacy issue becomes much more relevant when the use of neurotechnologies has legal purposes, and the *vulnus* to privacy risks being directed to a general monitoring by public power and institutions, in a preventive perspective actually not consistent with fundamental common constitutional principles of our legal and political systems. Another linked question regards who should have access to the data obtained by neurotechnologies.

This chapter raises questions more than offering solutions and answers to the complex issue of the use of neurotechnologies in law and the relation between neuroscience, ethics and legal liability. From the constitutional point of view, the only answer that it seems possible to frame is enucleated in the general principle of counterbalance between rights and interests: *privacy versus security*, for instance, and *retributive justice versus protection of weak people in society*.

Neurolaw could be an instrument, and in this sense it will be useful for judges and courts in order to have a larger store of knowledge in determining responsibilities. Perhaps neurolaw will be able to challenge some of the traditional legal institutions in the field of (above all criminal) law. From the point of view of ethics, neuroethics will not be so much challenged by the data coming from the use of emerging neurotechnologies, but maybe human self-perception will be strongly affected by it.

REFERENCES

Arnaudo, L. (2010). Diritto cognitivo: Prolegomeni a una ricerca. *Politica del Diritto, 1*.

Bernardi, A. (2009). Soft law e diritto penale: antinomie, convergenze, intersezioni. In *Soft law e hard law nelle società postmoderne*. Turin, Italy: Giappichelli.

Boella, L. (2008). *Neuroetica: La morale prima della morale*. Milan, Italy: Raffaello Cortina.

Bottalico, B. (2011). *European Center for Law, Science and New Technologies*. Retrieved from http://www.unipv.lawtech.eu

Bottalico, B. (2011). Neuroscience and law in a nutshell. *Diritti Comparati*. Retrieved from http://www.diritticomparati.it

Buchen, L. (2012). Arrested Development. *Nature*, *484*, 304–306. doi:10.1038/484304a PMID:22517146

Camporesi, S., & Bottalico, B. (2011). Can we finally 'see pain'? Brian imaging techniques and implications for the law. *Journal of Consciousness Studies*, *18*(9-10), 257–276.

Canil, T., & Amin, Z. (2002). Neuroimaging of emotion and personality: Scientific evidence and ethical considerations. *Brain and Cognition*, *50*(3), 414–431. doi:10.1016/S0278-2626(02)00517-1 PMID:12480487

Childress, A. R., Mozley, D., McElgin, W., Fitzgerald, J., Reivich, M., & O'Brien, C. P. (1999). Limbic activation during cue-induced cocaine craving. *The American Journal of Psychiatry*, *156*(1), 11–18. PMID:9892292

Cohen, P. (2011, June 19). Genetic basis for crime: A new look. *The New York Times*.

Davis, K. (2012). Brain Trials: Neuroscience is Taking a Stand in the Courtroom. *ABA Journal*.

De Cataldo Neuburger, L. (2010). Aspetti psicologici nella formazione della prova: Dall'ordalia alle neuroscienze. *Diritto Penale e Processo, 5*.

Erickson, S. K. (2012). The limits of neurolaw. *Houston Journal of Health Law & Policy*, 11.

Farah, M. (2002). Emerging ethical issues in neuroscience. *Nature Neuroscience*, *5*(11), 1123–1129. doi:10.1038/nn1102-1123 PMID:12404006

Feresin, E. (2009). Lighter sentence for murderer with bad genes. *Nature*. doi:10.1038/news.2009.1050

Forza, A. (2010). Le neuroscienze entrano nel processo penale. *Rivista Penale, 1*.

Galvan, J. M. (2003). On technoethics. *IEEE Robotics and Automation Society Magazine*, *10*(4), 58–63.

Garland, B., & Glimcher, P. W. (2006). Cognitive neuroscience and the law. *Current Opinion in Neurobiology*, *16*, 130–134. doi:10.1016/j.conb.2006.03.011 PMID:16563731

Goodenough, O. R., & Tucker, M. (2010). Law and Cognitive Neuroscience. *Annu. Rev. Law Soc. Sci.*, *6*, 61–92. doi:10.1146/annurev.lawsocsci.093008.131523

Greene, J. D., & Cohen, J. D. (2004). For the law, neuroscience changes nothing and everything. *Philosophical Transactions of the Royal Society of London. Series B, Biological Sciences*, *359*, 1775–1778. doi:10.1098/rstb.2004.1546 PMID:15590618

Khosbin, L. S., & Khosbin, S. (2007). Imaging the mind, minding the image: An historical introduction to brain imaging and the law. *American Journal of Law & Medicine*, *33*, 171–192. PMID:17910156

Klaming, L. (2011). The influence of neuroscientific evidence on legal decision-making: the effect of presentation mode. In *Technologies on the stand: Legal and ethical questions in neuroscience and robotics*. Njimegen, The Netherlands: WLP.

Lebedev, M. A., & Nicolelis, M. A. L. (2006). Brain-machines interfaces: Past, present and future. *Trends in Neurosciences, 29*(9), 536–546. doi:10.1016/j. tins.2006.07.004

Morse, S. J. (2007). The non-problem of free will in forensic psychiatry and psychology. *Behavioral Sciences & the Law*, *25*, 203–220. doi:10.1002/bsl.744 PMID:17393403

Morse, S. J. (2011). NeuroLaw exuberance: A plea for neuromodesty. In *Technologies on the stand: Legal and ethical questions in neuroscience and robotics*. Njimegen, The Netherlands: WLP.

Owen, J., & Shen, F. (2011). Law and Neuroscience in the United States. In *International Neurolaw – A Comparative Analysis* (pp. 349–380). Academic Press.

Pellegrini, S., & Pietrini, P. (2010). Siamo davvero liberi? Il comportamento tra geni e cervello. *Sistemi Intelligenti*, *22*, 281–293.

Pellegrini, S., & Pietrini, P. (2010). Verso un'etica. molecolare? *Giornale Italiano di Psicologia*, *37*, 841–846.

Pietrini, P., & Bambini, V. (2009). Homo ferox: The contribution of functional brain studies to understanding the neural bases of aggressive and criminal behavior. *International Journal of Law and Psychiatry*, *32*, 259–265. doi:10.1016/j.ijlp.2009.04.005 PMID:19477522

Rossini, P. M., Micera, S., Benvenuto, A., Carpaneto, J., Cavallo, G., Citi, L., & Dario, P. (2010). Double nerve intraneural interface implant on a human amputee for robotic hand control. *Clinical Neurophysiology*, *121*(5), 777–783. doi:10.1016/j.clinph.2010.01.001

Santosuosso, A. (2008, December 19). Il dilemma del diritto di fronte alle neuroscienze. In Atti del convegno Le neuroscienze e il diritto. Academic Press.

Santosuosso, A. (Ed.). (2009). *Le neuroscienze e il diritto*. Como-Pavia, Italy: Ibis.

Santosuosso, A. (Ed.). (2011). *Diritto, scienza, nuove tecnologie*. Padua, Italy: Cedam.

Stradella, E. (2009). Recenti tendenze del diritto penale simbolico. In *Il diritto penale nella giurisprudenza costituzionale*. Turin, Italy: Giappichelli.

Wessberg, J., Stambaugh, C. R., Kralik, J. D., Beck, P. D., Laubach, M., & Chapin, J. K. (2000). Real time prediction of hand trajectory by ensembles of cortical neurons in primate. *Nature*, *408*, 361–365. doi:10.1038/35042582 PMID:11099043

Wolpe, P. R., Foster, K. R., & Langleben, D. D. (2005). Emerging neurotechnologies for lie-detection: Promises and perils. *The American Journal of Bioethics*, *5*(2), 40–48. doi:10.1080/15265160590923367 PMID:16036657

KEY TERMS AND DEFINITIONS

Constitutional Rights and Freedoms: Freedoms and rights recognized and guaranteed by national constitutions.

Free Will: The ability to decide what to do independently and without any influence.

Law: the system of rules of a certain country, group, or area of activity.

Neuroethics: Ethics concerning the use of scientific findings about the human brain in medical practice, case-law, legal interpretations and public policies.

Neurolaw: The emerging field of studies that investigates the impact of findings in neuroscience on legal rules, standards and case-laws.

Technoethics: Technoethics is an interdisciplinary field of research concerning moral and ethical implications of technological advances.

Chapter 4
Artificial Ethics:
A Common Way for Human and Artificial Moral Agents and an Emergent Technoethical Field

Laura Pană
Polytechnic University of Bucharest, Romania

ABSTRACT

A new morality is generated in the present scientific and technical environment, and a new ethics is needed, an ethics which may be found in both individual and social morality, which can guide a moral evolution of different cultural fields and which has the chance to keep alive the moral culture itself. This chapter points out first the scientific, technical, and philosophical premises of artificial ethics. The specific, the status, and the role of artificial ethics is described by selecting ethical procedures, norms, and values that are suitable to be applied both by human and artificial moral agents. Moral intelligence as a kind of practical intelligence is studied and its role in human and artificial moral conduct is evaluated. Common features of human and artificial moral agents are presented. Specific features of artificial moral agents are analyzed. Artificial ethics is presented as part of the multi-set of artificial cognition, discovery, activity, organization, and evolution ways. A meta-ethical survey establishes the place of artificial ethics within the group of new and emergent ethical fields of the computer culture. Natural and artificial evolution are studied from an interdisciplinary and even from an intercultural perspective, and the co-evolution of human and artificial moral agents is sketched by means of technological and social prognosis.

DOI: 10.4018/978-1-4666-6122-6.ch004

PREMISES FOR AN ARTIFICIAL ETHICS

A new morality, freer in spirit but at the same time more strictly regulated, is generated in the present scientific and technical environment, by social and cultural evolutions which keep their own moral dimension in more implicit ways. The new ethics which may found both individual and social morality, which can guide a moral evolution of different cultural fields such as science and technology, and which can also maintain and continue the moral culture itself, will be an artificial ethics when compared to the traditional one.

When we think of artificiality we have to take into account two basic truths: 1) society is the first artificial "object", and all humanly established realities are consequently artificial; 2) perhaps the majority of contemporary results of human thinking and acting can be named artificial in comparison with all the past human performances.

For the specific field of moral culture, P. Danielson (1998) showed that "important parts of morality are artificial cognitive and social devices" (p. 292). From a larger cultural perspective, even J. Bentham, who described "the whole fabric of morals and legislation" can be cited. Another authoritative argument is furnished by Aristotle: he writes in his *Politics* (II. 5) about the automatic tools and installations of Daedal and even about the mentally controlled tripods created by Hephaestus, which served the "band" of gods. We cannot forget, in this context, Plato's *Republic*, where the oldest cyberneticist offered the first description of society as a human and artificial design and product, together with a project of a perfect but human organization of public affairs. As we know, he proposed and used a few efficacy means and criteria to build his "word-made" city.

If Aristotle emphasized the material and practical results of human or divine creativity, some very important and always necessary components, instruments and techniques of every activity – the intellectual ones - were earlier identified by the sophists who studied language as a „technique of techniques", by the Old stoic school that practiced the first elements of logical calculus and by Euclid's commentators who synthesized a lot of heuristic methods. Al-Khwarizmi later created working algorithms for any field of calculus (820 AD) and medieval mnemonics used linguistic algorithms in learning processes.

We are using here a general, but also not a soft meaning of the term "artificial". By "artificial" we mean any completely new social outcome of human activity, either theoretical or practical, such as a concept, a method or artifact, a social body, a system of relations or an institution, therefore, any kind of real mental construction or material achievement. At the same time, even if we are situated in the realm of values and spirituality, the term "new" is taken in its hardest sense and meanly means "unprecedented".

As an outstanding intellectual invention by which something entirely new was established, we can here mention a worthy successor of the above-named divine devices: the logical machine of R. Lull, made in order to produce all the possible (religious) knowledge by a mechanical method – his *Ars Combinatoria* (1275). This machine, made of paper, was effectively used for automatic but mechanical generation of religious truth.

We can include in the same class of „artifacts" or artificial products the first digital computer conceived by Pascal - the Pascaline (1642, 1645) - as well as his mechanical computer (1624).

The artificial universal language was coined by Kircher in *Polygraphia nova* (1663). In the same intellectual register we can add his *Ars magna sciendi sive combinatorica* (1669), alongside his cryptographic activities, his automata, as well as an image-projector or a magnetic clock and other magnificent inventions.

Leibniz was the more famous author of *Dissertatio de arte combinatoria* (1666) but also the inventor of a computing machine for the four mathematical operations, in different variants (the first in 1671). He also developed the binary computing system (1679/1701) and proposed the *characteristica generalis* (1682).

Perhaps as a follower of Descartes, who was not only a mathematician and a physicist, but also a biophysicist (he even had a dissection cabinet) and who proposed the mental experiment of the "brain in the tube", Leibniz also studied the mind, by applying a large-scale model – "the mill"; as we know, this idea and even the term were later used by Babbage.

In the moral field of culture, the novelty means something valuable – just like in all other creation fields - but also something that is imperatively necessary to be done.

As a specific feature of moral prescription itself, we can note its spiritual nature. A few current, important and useful aspects of spirituality are that it is:

1. Abstract and generally applicable;
2. Easily intelligible and easily formalized;
3. It has or may receive a precise operable character.

The spiritual nature of moral norms and rules is then a favoring condition that makes possible the practice of moral values both by humans and machines.

Some theoretical deficiencies of different great ethical systems and some practical difficulties of applying moral values in concrete conditions and by individual agents in social activities were frequently discussed by ethicists, in their common effort to establish a new foundational theory of moral choice, moral freedom and then of a deep moral conduct in an unprecedented cultural environment. Our paper on artificial intelligence and moral intelligence is one of these essays.

Possibilities and difficulties of a moral code implementation suitable for both human and artificial moral agents were already shown in some of our studies (Pană, 2005b; Pană, 2008b) and different levels of knowledge sharing, value understanding and practicing in human societies (Pană, 2004b) and by human and artificial agents were investigated (Pană, 2006b) at a Computing and Philosophy Conference, and later in a Cognitive Psychology-based study (Pană, 2008a), as well as in a collaborative article (Niculescu & Pană, 2010).

The specific and the main functions of Moral Intelligence as a condition of efficacy in human moral behavior and as a possible factor in a moral code implementation in intelligent human-artificial systems, activities and interactions that occur in moral context, were already signaled at an international conference held in USA (Pană, 2005b) and partially described in a paper on some current and possible features of moral behavior (Pană, 2005a).

Moral Intelligence will be the subject of a next paper, where we will analyze its internal structure and its place in a general system of intelligence types, as well as the properties which make it suitable and applicable by both human and artificial moral agents. The study will also discuss the theoretical value of this concept and the practical utility of this human quality that keeps human morality effective in spite of incoherence of moral theories and of all practical difficulties to perform either abstract or concrete moral prescriptions.

Our present work is a conceptual and interdisciplinary paper, which may also be considered as an inter-cultural product, because gathers and uses new knowledge and findings from a few distinct fields of culture such as science, philosophy and technology. The paper is elaborated at a meta-theoretical level, and for this reason it operates with some abstract and often original concepts, applies general principles such as the evolution principle in topical areas, and proposes new research themes.

As an example of such an inter- and meta-approach, the way in which we used the evolution principle for both a historical and philosophical interpretation of some intellectual inventions as crucial events for the field of information science and technology, at a conference on Cybernetics and systems, we mention our description of the main intellectual steps made in AI conceiving and implementing which are, in our vision: Intellimation, Computation, Automation, Cybernation, Datamation and Artificial Intelligence and, finally, Sociomation.

Artificial Ethics, its invention and implementation can be seen as an aspect of both AI development and Social Cybernetics completion. The latter may be improved and continued by methods and techniques offered by information science and technology and joined with achievements in social sciences, in which social cognition and intervention are studied and promoted. Sociomation may be the result of some very specific but united scientific and technical efforts.

The present paper is the newest development of the described research direction for the field of moral agent's theory, which was also preceded by studies in cognitive human and artificial agents viewed in their evolution. If the research on sociomation proposed a continuation of intellectual and technical achievements in the social area, the present one anticipates the possibility and argues for the necessity to develop and apply a new and more coherent, clear and concrete field of ethics, suitable for both human and artificial moral agents.

With this aim, the paper discusses and puts together some results of the venerable but mainly virtual classical ethics with an action-centered reflection on a specific value-system and with a multi-set of verisimilar and verifiable efficacy-oriented principles, norms and procedures. This attempt to constructively address some current ethical problems was anticipated by our critical research on Virtual Ethics and Virtue Ethics (Pană, 2005a) and will be completed by an extended version of this paper.

Artificial Ethics building will be, in fact, a product of the co-cognition and co-construction of human and artificial intelligent agents, in their "natural" or artificial evolution, which was also studied (Gregory, 2000) and here only mentioned. More probably a mixed, natural-artificial kind of moral agents will occur, on the background of different but concurrent processes of nowadays, such as the rising of artificial intelligence, artificial cognition and artificial discovery (see the last but one paragraph of this paper).

Here is perhaps the right place to mention that the present work is a result of studying and practicing a particular and important scientific perspective, which may be named *possi-diction,* that can be illustrated by and which includes *post-diction, retro-diction, prediction and pro-diction* as subordinated fields. As a thinking movement and as a social attitude this approach may also be named *possibilism* and integrates empiric research, theoretical synthesis, prospective thinking and ethical reflection. We have already used the same research kit in some previous writings on philosophy of nature and social philosophy, as well as in works on social and technological prognosis, some of them included among the additional readings.

The anticipated and later on sketched new morality, based on philosophical presuppositions, moral sciences and technical abilities will also be the result of a common evolution of the artificial cognitive and operative moral agents. These technical and cultural trends will be described, from a specific perspective, in our paper, which also proposes a new kind of ethics that will be an emergent field of technoethics.

WHAT ARTIFICIAL ETHICS IS AND WHAT WILL IT BE?

Artificial ethics is one of the newest scientific and technical activity fields. This domain of ethical research is born at the confluence of at least five other new fields of the contemporary ethical culture: information ethics, computing ethics, machine ethics and roboethics, as well as the global information ethics. Artificial ethics is con-generated with webethics which comes into being together with the recent concerns about the ethical issues raised by the virtual activities and communities on the web. The new research area proposed here, the Artificial Ethics, represents not only a new domain, but even a new level of the ethical activity, cognition and reflection.

The term artificial ethics was coined in 2004 and was intended to describe a field of artificial philosophy, discussed in a dedicated paragraph (Pană, 2004a, pp. 297-323), in our book on Philosophy of information and information technology, and constitutes the title of a three page section in the book, but the artificial character of morality was already outlined by Danielson (1998), who emphasized the possible degrees of creativity by which moral values are invented and moral culture is renewed.

Webethics or netethics are also names proposed by us as a result of some reflections concerning research made by Huberman and Stahl on collaborative knowledge groups as well as from the need to identify the specific values of the recently created virtual work groups and communities on the web that have a specific moral life as well as an appropriate netiquette.

The status and the role of Artificial Ethics may be established only by taking a glance at the present complexity of the ethical research field and at its newest domains. This description can only be a short and essentialist one, even if it also aspires to show a few of the now foreseeable trends of the field.

Artificial ethics is the term I find to be adequate for the new ethics which can be efficiently practiced by both human and artificial intelligent agents. Artificial ethics will rise by the common evolution of a few present fields of ethics such as computer ethics, information ethics, computing ethics and global information ethics. All these kinds of ethics are founded on the classical Philosophical Ethics, as well as on the newer Scientific Ethics.

Destined for computer and net workers with various professions, the ethics of computing is useful for all those who use a computer in the new artificial intellectual environment. Issues such as protection of software property, of user identity and privacy, and netiquette sharing and preservation are considered characteristic for this ethical field. Ethics of computing is not a professional ethics.

Accredited moral theories are studied and decision procedures for difficult moral problems are conceived in the field of computer ethics, created by professionals in philosophy, who use computers for theoretical and practical moral problems solving. P. Danielson revealed that using the computer in this field, we just extend the artificial feature of morality. The theoretical foundations of the field were established by Floridi (1999), Bynum (2000), Foridi and Sanders (2001) and the field was developed by Floridi and Sanders (2002).

Machine ethics concerns the computer itself if it runs intelligent programs because once turned into an intelligent machine, the information machine induces changes in the world, like humans do. All human activities have a moral dimension and a machine with similar possibilities needs moral functions. Machine ethics tends to become a research field of Artificial Intelligence, but it also needs a philosophical (ontological, axiological and pragmatic) foundation.

We may specify here that the term information machine was first used in the book entitled Philosophy of Technical Culture with a dedicated section (Pană, 2000a, pp. 430-438), a book which describes the internal structure of technical culture and its main components, its contemporary tendencies and new fields, such as information culture. A constituent part of technical culture, technical intelligence, may be recognized as a natural ancestor of artificial intelligence as of various emergent artificial intelligence types; among them, artificial moral intelligence may be the first, because it is imperatively necessary.

Global information ethics is a new level rather than a new domain of ethics, generated by the informatization and globalization of all significant human activities. The field was preceded and introduced by information ethics Foridi (1999;

2008). The first theoretician of global information ethics is T. W. Bynum (1998 and 2000); Foridi (2007) deepened its study. Ch. Ess (2006; 2013) addressed and analyzed important ethical aspects related to the present changes in knowledge, communication and culture. R. Cavalier (2005) pointed out the impact of internet on our moral lives and a network ethics was promoted by Foridi (2010).

Artificial ethics will not only be the result of a common moral evolution of human and artificial cognitive and moral agents, but also a part of artificial philosophy, recently generated through the birth of formal axiology, technical logic, information aesthetics or digital politics. Fr. Laruelle (1990) viewed artificial philosophy as the science of thinking, and even as the "unified theory of thinking" (p. 257), developed by mathematical and technical methods.

Another sources, means and parts of artificial philosophy were identified (Pană, 2005c) and a representation of the now developed fields of scientific, artistic and philosophical computing is included in a scheme of the present artificial virtual intellectual environment in papers such as (Pană, 2003a), and later a general map of philosophical, scientific and technological disciplines related to computer work, computer environment and computer culture was published in our already mentioned book on the Philosophy of information and information technology, more precisely in its section regarding information technology and artificial philosophy (Pană, 2004a, pp. 311-323).

A meta-ethical survey is undertaken with the task to establish the place of artificial ethics in the system of scientific and philosophical disciplines. It might also prove interesting to analyze meta-theoretical approaches which can be considered as founding cognitive and evaluative perspectives for the artificial ethics - consequentialism, prescriptivism, descriptivism, contextualism and possibilism (the latter accredited by cultural geography and developed by cultural ecology and not by philosophy) as well as the updated meliorism - with the aim to show the specific of Artificial Ethics.

The specific of artificial ethics is given by its:

1. Structure, as a result of a synthetic approach to understanding human and artificial agent's behavior;
2. Validity for all types of intelligent agents;
3. Status of moral invention.

By proposing and describing the new field of the Artificial Ethics, the paper gives an answer to the quintuple need to:

- remove the theoretical contradictions of past and present ethical doctrines;
- surpass the practical failures of human ethics;
- allow a moral code for various types of intelligent systems;
- ensure a conflict-free human-machine cooperation;
- follow the common evolution of human and artificial cognitive and moral agents.

The chapter also:

- distinguishes the values, principles, norms and rules functioning in this ethical domain;
- analyzes the main problems and promises of the new research field;
- anticipates and evaluates the future evolution of the field as well as of some other related research directions;
- brings up a topical, integrative and prospective vision about the studied fields of ethical and technical culture.

One of the most important purposes of this paper is to present the scope, the specificity and features, as well as the appropriate methodology and the sets of functions of a new field of the contemporary ethical culture, that of artificial ethics, in its necessary connection with the study of the present-day complexity of technoethics. A few aspects of the foreseeable evolution of this research and reflection field are also outlined.

We also wish to select and to discuss an adequate moral code to be shared by human and artificial ethical agents, to lay the theoretical foundation and to evaluate the practical impact of the technical implementation and of the social use of such a moral code.

Another aim of the paper is to outline the diverse methodological approaches which have to be integrated into a unique methodology to be able to treat the new ethical field, which also supposes the reunion of complex scientific, technical and even social activities.

As a product of philosophical reflection, Artificial Ethics is even a type of intellectual invention. It is actually a result of human action in all its forms, a complex system that is a cognitive, normative, instrumental and even technical system, the more so as we see:

- Technique as a set of procedures and methods that may orient any kind of activity and that
- All types of human activity are today mediated or realized by technology as well as that
- Information technology is implied in all kind of human activity.

Artificial Ethics is then both a conceptual and technical field of research and a theoretical and practical area of moral action.

Instruments, methods and techniques that are likely to be used in the field of artificial ethics are to be envisaged and developed by the common means of social cybernetics and the technical branch of cybernetics, now concerned with software agents, virtual work assistants, healthcare and social agents that are working both in virtual and computer environment and in real and human-computer interaction characterized environments.

Complex abilities and psychological aptitudes, as well as social skills needed by human and artificial moral agents, which are or will sharing an artificial ethics, will be studied by compiling certain results of cognitive psychology, computer modeled mental behavior, social agent's theory and practice, artificial life, softbots designing and using, intellectual techniques developing and inventing methods adopting or generating.

PROCEDURES, NORMS, PRINCIPLES AND VALUES IN ARTIFICIAL ETHICS: MORAL INTELLIGENCE IN HUMAN AND ARTIFICIAL MORAL CONDUCT

A new, artificial ethics is not only possible, but also necessary, because of the historical and current feebleness and failures of the human morality. Among other favorable conditions for an Artificial Ethics development we can also count the human and artificial evolution, a twofold evolution that nonetheless follows the same direction, as well as the more and more complex and effective man-machine, human-computer, human and robotic entities interaction and their present social meanings and consequences.

The interaction of intelligent (human and artificial) agents generates several new areas of research, including a new ethical research field. This new ethical domain can also be the result of a synthesis between other two newly emerging ethical disciplines: Computing Ethics and Machine Ethics. In this paper, the new ethical field will be called Artificial Ethics, but this both theoretical and practical research has at last two and equally important aims:

- To promote new ethical and meta-ethical solutions in the field of human ethics and
- To bring up an appropriate and operable value-set for both human and artificial intelligent moral agents.

To be suitable for both human and artificial moral agents is thus the most important feature of artificial ethics. The third but indirect result of our study can be an improvement of human morality itself, in the spirit of a renewed meliorism.

Even at its systematic level and in its scientific form, human morality is theoretically deficient and practically inapplicable. However, human moral conduct is often effective, because of moral intelligence development and employment. It is moral intelligence – and not necessarily moral consciousness – which ensures the coordination between the general moral values and the concrete conditions of moral action.

But in the till now practiced human conduct, to apply ethical values into individual behavior is a very complex process, and a permanent and often equal or increasingly remote connection between ethics and morality can be registered, because of the natural and right, but long and daring road from ideal values and general principles to particular norms and rules, and finally to the actual moral behavior.

How can moral values be put into practice by intelligent machines that use just an abstract form of intelligence? Can "natural evolution" of programs be adapted for the moral dimension of machine's action? Moral values are synthetic values which require a particular type of knowledge - practical knowledge; a special kind of knowledge is also necessary for their understanding – evaluative knowledge; their general and vague character calls for creative intelligence and their practice may require many psychical (individual) aptitudes and cultural (social) attitudes.

Understanding, reasoning, decision making, problem solving and heuristic conduct are needed in moral (human or artificial) conduct. The unitary expression of these functions/aptitudes is moral intelligence.

Task formulation, rules interpretation and consequences evaluation are other intellectual activities subsumed by moral intelligence. Therefore, moral intelligence cannot be simply dismissed as "moral cognition", not only because of its complexity but also because of its pro-active, operative nature.

Even if viewed as an operational branch of moral cognition, moral intelligence is a multitude of intellectual and practical activities. In turn, these activities subsume always used, but recently identified forms of thinking such as fuzzy thinking, fractal thinking, statistical and probabilistic, and even prospective and integrative thinking.

From such an operational and integrative perspective on moral cognition and action, we can evoke as an early and illustrative example Bentham's study on the main features and conditions of a legal and moral action; he starts his scientific work by exploring the whole of human mind: consciousness, motivation, state of mind, intentionality, emotion, passion as well as virtues and vices.

In a subsequent stage of research and in a less frequently used work, he conceives even *A Table of the Springs of Action* (1815). But his most relevant achievement for our purpose is his system of measuring the value or the utility of an action in order to realize happiness, by distinguishing a lot of characteristics of the generated pleasure or pain, such as intensity, duration, certainty or uncertainty, propinquity or remoteness, fecundity, purity and extent.

These variables or vectors of pleasure and pain, as simpler and clearer manifestations of satisfaction and suffering, make possible a quantitative evaluation and then a calculus that may be not only a philosophical, but also a scientific one. Bentham captures these properties of a complicated human experience even in an algorithmic or mnemonic structure – a little and cheerful piece of poetry, actually – but he also generates an entire list of instructions that allows the *felicific calculus* by which the moral value of each action can be determined.

But the results and consequences of an action are always easier to evaluate and this is the main way used in science, which is the opposite of the natural or „historical" course of any event or action. Science as explanation is, then, more post- or retro-diction and less pre-diction. To practically induce, guide or manage a moral conduct of a human being or a moral action of an artificial agent is so much more a matter of creative thinking.

Creative thinking in its technical form and also prospective thinking are thus other personal characteristics and more – sets of methods, techniques and skills - which are needed in artificial ethics. Forms of reasoning from non-sentential to situational and even affective ones are types of practical reasoning which are also useful and are studied by Sloman (1990). They may be involved in neural representation processes of the social space, described by Churchland (1998), a space in which we will see artificial moral agents reasoning and acting.

Moral intelligence is different from technical, political or economical intelligence, all of them being included in the group of practical intelligence forms. Moral behavior is a complex, practical, intellectual and spiritual behavior. Moral intelligence is operational at all these levels of moral conduct (action, cognition, spirituality) and moreover, it has concrete and abstract, imitative and creative, assertive and interpretative, persuasive and imperative, individual and group components. Therefore, moral intelligence is neither a simply practical intelligence form. Moral intelligence is not a form or a level, but a kind of intelligence: it is a synthetic kind of intelligence.

As a complex whole of aptitudes, activities and even intellectual techniques, moral intelligence regards the spiritual, cognitive and practical levels of moral conduct. If philosophically founded, deduced from scientific (moral) theories and technically implemented, artificial ethics can improve both human and artificial moral behavior.

Moral life is mainly a spiritual life; it is the kingdom of value-based choice, of moral reflection and moral freedom. Then, moral spirituality and activity has both philosophical and practical components and levels and it is, from this perspective too, a synthetic kind of intelligence.

As a decision form, moral decision has some characteristic features such as:

1. Universality (all complex decisions have a moral dimension);
2. Its foundation on specific values together with individual motives;
3. The combination between subjective evaluation of difficulty degrees and objective counting of success probability.

If we conceive motives computing as a simple motives inventory, identification, comparison and choice, as we are now doing, we are evaluating present or future acts through past decisions and we elude the construction of motives as constitutive substructure in the structure of an action. Motives are ... motivating and just secondarily justifying. Moreover, we always have to deal with a system of motives (not just to identify the right motive) and to use motivation as the internal and the most powerful incentive for action.

Decision itself is a process, a distinct and complex activity, not just a component of any activity. The specificity of moral action needs an analysis of some steps of moral decision as well as a study of a few types of moral decision, having situation and motivation as criteria. Do we have to "animate" cognitive and operative intelligent agents (knowbots and robots) or can task and value (finality) oriented, situational and concrete, theoretical and emotional intelligence be substituted by forward and feed-back (causality) governed, algorithmic and executive mechanisms? Singularity, un-repeatability and irreversibility of moral decisions are also to be pointed out.

Practicing artificial ethics by using moral intelligence supposes the use of a set of techniques. Some of these techniques are intellectual techniques such as value measurement, motives analysis, situation exploring and evaluation, decision processes foundation, resources evaluation and allocation, competences and responsibilities establishing, persuasive techniques, prospective techniques.

Moral reflection is conditioned by the construction of a scientific and philosophical superstructure of moral life and habitually represents a lacking level of morality in the case of human beings, too. Machine ethics is facilitated by the possibility to be philosophically founded, scientifically inferred (from ethical theories) and technically implemented.

A NEW VALUE-SYSTEM IS NEEDED OR CAN WE OPERATE WITH THE TRADITIONAL ONE?

To study and to build-up a new research field or even a new area of social practice we need a deep analysis of the present state and trends of the value-system that animates the cultural area which constitutes its background: this is the system of technical values, particularly the set of values associated with the knowledge-based information systems using virtual work groups and communities created on the web.

The specificity of technical values was not convincingly shown until now; it was, on the contrary, sometimes contested. Their pervasive tendency was signaled by H. Jonas, who also studied the main four cultural attitudes towards technique, and their destructive character was outlined by J. Ellul, who argued that technical values are removing other cultural values.

Their restrictive action towards professional responsibility was described by K. Bayertz, who has also launched the term „meta-responsibility". The specificity of technical values was contested, their resemblance with the scientific or economic ones being emphasized; the poverty of the technical axiological realm is always accused.

But the most important issue here is whether a new system of moral values is needed with the aim to build an artificial ethics or we can use the traditional one in the process of an artificial ethics building, an ethics that may be suitable both for human and artificial moral agents. Equally concerned in mind and action related philosophical issues, we formerly coined a value-centered vision which can reunite the various aspects of human endeavors (Pană, 1988). A value-centered approach – the more so as extended to the whole and to the evolution of a value-system – such as that developed in (Pană, 2004a) can also address human and artificial or even human-artificial interaction and evolution related topics.

Freedom must be allowed not just for human, but for artificial moral agents, because the degree of liberty directly determines the degree of responsibility. Moral responsibility implies moral conscience, but this is "selectively" developed even in the case of each human individual: only at the level of emotions and then at the level of habits induced from outside by the system of negative/positive sanctions or at the level of beliefs as well as that of reflections. Can instructions and rules, knowledge-based techniques and specialized semantic editors play the same roles in machines functioning?

Sincerity can be seen as a practical and moral variant of truth, which has its place in science, while the absolute can be studied as the central value of philosophy. Sincerity as a specific moral value has not the lie as a pair, just as truth can be opposed not only to the false or to different degrees of falsity, but also to a large cluster of incorrect social conducts, from dissimulation to hypocrisy, and from flattery to imposture.

Dignity may be thought as the central moral value, because:

1. It is a synthetic moral value; its realization and maintenance need the deep observation and the jointed practice of all other moral values.
2. Dignity ensures the common base of all types of moral and cultural behavior of a person as part of a community.

Happiness is, maybe, an even more important value. If dignity is the synthetic and central moral value, happiness is the finality of the whole moral conduct. Both Aristotle and Bentham used many sophisticated arguments in this matter, and Bentham, who invented the above analyzed felicific calculus, confessed in his Fragment on governance that every action strives, in the main, to generate happiness and this inward tendency of any action is taken by us as utility.

In order to achieve the maximum of happiness for all those implied in the accomplished action, the legislator or the moralist has α) to know human nature, β) to distinguish between different types of satisfaction and suffering and γ) to possess and develop the capacity to relate all the consequences of every evaluated action to a set of criteria. We can add, in our turn, that δ) the set of criteria itself has to be a system of criteria.

One of the posthumous works of Bentham, entitled *Deontology or a Science of Morals* (1834), is a written proof of his scientific and practical, but visionary and anticipatory ethics, in which the steps of his instructions for a scientific calculus of happiness show him as a forerunner of some systemic and operational methods used in various fields of contemporary science.

In the mentioned work, the author appreciates that the ethical error may be equated with an error in calculation and continues the analysis of some useful instruments in achieving happiness. Moreover, this work had been previously prepared by Jeremy Bentham (1824) within the framework of another book, one dedicated to the whole of human errors, the *Book of Fallacies.*

It is important to note here that almost two hundred years have passed and yet error is not always clearly distinguished from guilt and from sin. Thus the original syncretism of the three cultural fields (science, ethics and religion) persists, maybe because it was generated by the multifold essay to eliminate the evil that entered the world when humans dissociated themselves from nature and become associated as members of society, conceived as something totally different from nature and often seen as a super-nature.

This confusion will probably disappear only together with the disappearance of one or more of these three fundamental human attitudes in a still unforeseeable future. With them will also disappear, as very likely, some deeply human aptitudes, skills and talents which will not be more interesting and not even worthy to be experienced by some "evolved" humans or "advanced" technologies and machines.

Correctness may thus be taken as an important and even as the exclusive value of ethics – mainly for artificial ethics - but not only because it is suitable both for human and artificial morality (thanks to its increased clarity, intelligibility and applicability), but also because it may function with the same efficacy in all types of activity and at all levels of human or artificial conduct.

If we focus our attention on the system of values, correctness may also be considered as the adequate equivalent of the traditional "goodness" that evokes intentionality and subjectivity and even more - relativity to the context and even to some personal aptitudes and skills. For a larger discussion on the multitude of relations between different moral values see (Pană, 1988). But can the evoked conditions and difficulties of moral action be eliminated by the simple correctness and can this moral aspiration be as strong as goodness was in the history of morality?

Human morality is also a field of value invention. Accomplishment of new values implies many cognitive, affective and evaluative aptitudes. Even each moral value can be practiced only through some general or even special aptitudes, as well as imagination or intuition, and the central moral value can be realized by integrating the derived values and by the conjoint manifestation of all inner and cultivated necessary aptitudes.

Creativity is also a characteristic of moral conduct, which is a result of a mental and intellectual activity. It is defined sometimes even as a mental super-activity. It is, in addition, a singular process, conditioned by an original organization of several aptitudes and it is expressed by the unique quality and importance of its products.

The creative nature of the spiritual dimension of ethical behavior involves not just consciousness, but also unconsciousness. H. Ey (1998) explored the unconscious level of psychical life and described its structure (pp. 320-323). He has also shown that the unconscious is the infinite and permanent source of creativity (p. 355). Could a moral conduct-demonstrating machine also need the unconscious (mind)? Can this be populated to the same extent as the human unconscious? Can it be productively explored and exploited for the purpose of implementing creative conduct?

As moral agents, human and artificial intelligent agents show certain possible common features (ten of them analyzed below), some of them being possible to be ensured by moral intelligence. Artificial moral agents do not have to reproduce human moral qualities, but they have to receive or develop some functional, psychic and even spiritual abilities. A few "intelligence makers" appreciate that programs and even machines will be obtained by activities beyond engineering and by processes which are similar to primordial ones like garden tillage or bred baking (Hillis, 2001, p. 171) and foresee a "natural" evolution of artificial intelligence.

Moral intelligence and moral spirituality transcend biotical, psychical and even social behavior: they are cultural behaviors. But cultural behavior is still a practical behavior: cultural anthropology established that culture is the way in which values are experienced by people. Thus, artificial moral agents will need to be trained or even "educated" by processes similar to children's cultural formation.

Implementation of moral values, principles and norms raises difficult problems in the case of both human and artificial agents. But the realization of this task is helped by:

1. The synthetic nature of moral intelligence;
2. The multitude of the implied aptitudes;
3. The fact that moral values are generated by various activity fields, which also offer their specific means for their accomplishment;
4. The cooperation of entities;
5. The highly normative structure of moral conduct;
6. The possibility to develop moral techniques as variants of intellectual techniques;
7. The chance to apply a wide range of methods (philosophical, psychological, technical);
8. The opportunity to conceive and design all practical intelligence forms as clusters of some activities that are:
    a. value/norm guided;
    b. task/means evaluating;
    c. risk/benefit balancing;
    d. idealistic and not only realistic motivated and
    e. present/future oriented, and not just in terms of capacities/aptitudes that are only theoretically analyzed.

CHARACTERISTICS OF ARTIFICIAL MORAL AGENTS

Moral cognition, as a form of social cognition, may be a scientific knowledge and can be put into practice as a realization of a moral code, elaborated on the basis of a theory such as the Artificial Ethics presented here, by means of technical knowledge and design.

Moral cognition is structured at each level of knowledge: empirical, theoretical and metatheoretical. At the same time, all forms of human cognition, which are subject matters for the so called cognetics, also need, in our vision, a cognethics, a field of ethics which can concern the moral use of the results of all kinds of cognition.

Moral knowledge and technical knowledge are or are going to be put together to formulate and apply machine ethics, which is a common research field of philosophy and AI. Characteristics of artificial intelligent agents are studied in the latter domain. We can analyze from a philosophical point of view if these features presented below render these intelligence forms capable of moral action and "life".

Artificial moral agents can be studied as a sub-set of artificial intelligent agents. Classes of attributes of artificial agents such as a) sensing and acting, b) reasoning, c) learning and knowing, d) internal structure and e) diversity of their number, are distinguished by specialists in theory and methodology of intelligent agents, like Skolnicki and Arciszewski (2003).

Some of the features included in these classes are promising for moral activities because, according to the findings of the two authors mentioned above, these agents may act locally, may cooperate, are sophisticated, are trustful and can acquire knowledge.

The same authors add that individual agents do not model other agents and do not show internal states, but they also have a stable architecture and that they may work in a group. Swarm agents are locally active too, they share resources, have less autonomy but are more competitive and more mobile. They also react more directly and may discover roles in runtime, use fixed language and assume information to be true, but are less transparent and less reusable.

Human and artificial agents can be compared as intelligent agents. As shown by Skolnicki and Arciszewski (2003), while human intelligent agents have initiative, are both subjects and objects of action, are social agents and are reflective, artificial intelligent agents are only reactive, proactive and self-analytic. We can add that human agents are also omni-oriented and in(de)finitely perfectible, when artificial agents are only guidable and teachable.

Other features emphasized by computer scientists are not so favorable for ethical purposes. Thus, the intelligent agents created so far are not real-time, do not model other agents (but moral behavior is known as a learned or even as an imitated one); do not show internal states.

Newer agents like "swarm agents", however, show some useful traits because they share resources and may discover roles in runtime, but they have less autonomy; react more directly (while moral behavior is a reflective one); are less transparent; use fixed language; assume information to be true (when we know that "*Dubito, ergo cogito* ..." and more, that a fuzzy and not a bivalent logic is suitable in moral contexts); are less reusable (perhaps they cannot then fully use their capacity to learn).

As an outcome of our interdisciplinary research, that is grounded on various, philosophical, scientific and technical types of cognition and design, a set of specific properties of artificial moral agents was gradually identified by us (Pană, 2005b; Pană, 2006a; Pană, 2008a).

Artificial moral agents are or will be:

1. Individual entities (complex, specialized, autonomous or self-determined, even unpredictable ones);
2. Open and even free conduct performing systems (with specific, flexible and heuristic mechanisms and procedures of decision);
3. Cultural beings: the free conduct gives cultural value to the action of a human (natural) or artificial being;
4. Responsible and co-operative participants in group and community-centered activities. These qualities are based on the reciprocal dependence of liberty and responsibility, which requires an even stronger emphasis: the responsibility degree is directly dependent on the freedom degree or level. A single more nuance, but one often decisive may be added here: in a 0 degree liberty situation,

the responsibility is not always the same in humans, but it may be equally null in the case of an artificial (soft or material) agent.

5. Entities with "lifegraphy", not just with a "stategraphy";
6. Educable, not just teachable systems;
7. Endowed with diverse or even multiple intelligence forms, like moral intelligence;
8. Equipped not just with automatisms and intelligence, but with beliefs (cognitive and affective complexes);
9. Capable even of reflection (moral life is a form of spiritual, not just of conscious activity);
10. Components/members of some real (corporal or virtual) community.

These possible features of artificial moral agents make them more adequate to efficiently perform a determined moral code than human agents. Human morality has a complex structure which reproduces, rebuilds or anticipates the frame of social organization, with all its temporary or permanent weaknesses or troubles.

In addition, the present content of moral values is differentiated by cultural area, communities, social categories and professional groups, while artificial ethics can be derived from science-like or scientific ethics and other social sciences, technical sciences and sciences of the spirit such as mathematics.

Human morality as a whole was characterized by complex successive internal differentiations; among them we can distinguish some persistent dichotomies generated by elementary differences of biotic, psychic, social, historical and cultural kind.

Artificial ethics will be homogenous and global – a very universal ethics - with differentiations only for distinct fields of activity, types of tasks and degrees of complexity. Thus, artificial ethics will expectedly be the first effective ethics.

ARTIFICIAL ETHICS AS A KIND OF ARTIFICIAL COGNITION, ARTIFICIAL DISCOVERY, ARTIFICIAL INTERACTION AND INVENTION IN AN ARTIFICIAL ENVIRONMENT

Artificial cognition is present and useful both at the elementary and the high level of cognition. New cultural fields such as informational aesthetics, computing ethics and digital politics are now contributing to the birth of artificial philosophy as well (Pană, 2005c). Some new conceptual models have been developed and these models can be described by means of several new sets of traits and trends, as well as by means of many new forms of thinking (Pană, 2006d; Pană, 2006e).

A Romanian representative of cybernetics, St. Odobleja, created the first version of the generalized conception of cybernetics and demonstrated its multi- and inter-disciplinary character. In his work *Consonantal Psychology* (published in French, in 1938, and republished in America in 1941), he considered that cybernetics integrates theory, method and technique in a single body of knowledge and activity. He has defined cybernetics as a technique of artificial thinking, in a later conceived paper.

Odobleja (1982) anticipated that science would come to create ideas in labs. This will be possible, in his vision, because thinking is not inferred from the succession and association of ideas, but a self-generation process of ideas occurs, and this process can be explained by a consonantal process, similar to other physical and biotic resonance phenomena (p. 182). In his description of some common mechanisms and laws which operate in various domains of existence, Odobleja outlines that two principles are important: that of similitude or analogy and that of harmony or perfection.

Maybe the principle of similitude is also responsible for the impressive correspondence between his ideas and those formulated by J. Swift hundred years before, when he imagined a visit at the imaginary Academy of Logado, where both mathematical and philosophical or any other kind of intellectual achievements were possible to be obtained by humans without any specific knowledge and skills, but with the aid of a "thinking machine". As we have already shown, R. Lull aimed even formerly to deliver all the possible ideas in a certain cognition field.

If we use now the same principle, inspired by the above evoked thinkers and impressed by the present results of a computer-aided education and research, compared with the principles and productivity of creative work, maybe we will recognize the same type and degree of effectiveness of the projected information machine, which become more and more a computing machine, and we will build some more realistic, but maybe also more daring technical and social prognoses.

Artificial discovery is now described by computer scientists as consisting not only of stored knowledge discovered in knowledge bases and warehouses or as theorem demonstration and knowledge verification by virtual experiments in physics or chemistry, but as new knowledge generation sustained by knowledge-based information systems and techniques used by collaborative knowledge groups in some artificial and even virtual environments.

While cognition, in its various expressions, is more easily and likely to be seen as a systemic, methodical and in some respects even algorithmic process, carried on by team work in a well organized context in which artificial thinking and cognition may also be integrated, discovery is further on perceived as a more individual and even unique activity, needing personal qualities and requiring special skills, an activity in which the role of some special mental states and of some exceptional conditions and even of some hazardous events have their own place.

Discovery is not conceivable today without a technically dense research environment that includes, besides a specific toolkit, a potentially more important and huge set of intellectual techniques that may be practiced only by using advanced information systems. Under these conditions, even the meaning of the term "technique" is rethought and an important aspect of its often forgotten significations is revalued: technique is seen again as an intellectual method that may guide any kind of human activity. An efficient use of these intellectual and informational methods, techniques and skills explains why scientific discovery is now considered as the most efficient human activity and why its steps are taken as stages in a generally accepted model of efficacy.

The complex process of continuity between scientific discovery and technical invention may be described as a process where the place and role of specific aims, means and solutions are unequally distributed and balanced between intellectual invention and information techniques, in ways as those described in (Pană, 2006d), as well as between human and artificial intelligence and between natural and artificial cognitive agents.

Nowadays, Markov chains-based data and information identification, acquisition, structuring, sharing and storing methods and procedures are not the only ones applied; knowledge integration and even complex methods of facilitating knowledge interpretation, such as wavelets theory-based ones are also used by knowledge management systems with a large artificial segment (which may include automatic decision processes) and by business intelligence programs that can use all these facilities.

The whole research system is structured as a multitude of data, information and knowledge flows, nets and centers, which was depicted by us with the occasion of a Computing and Philosophy conference (Pană, 2007). This huge knowledge chain not only uses but is effectively based on, assisted by and controlled through a strong artificial infrastructure.

Artificial social activities, processes and structures are generated and widespread at a large scale nowadays and the entire society offers not only an "informational highway" (as that suggested by M. Dertouzos) or the necessary informational infrastructure, but also a whole technical and artificial material structure, as well as a technical and artificial, but virtual environment in which human activity is assisted and optimized by artificial agents, and in which powerful computing and managing activities are also performed in all types of social activities by intelligent knowledge-based artificial systems.

They influence human actions or even determine human actions, because they are:

1. Incorporated in industrial, organizational and cultural systems;
2. Hard and soft intelligent systems that are working, deciding, communicating, moving and acting automatically;
3. Used in all types of personal or social activities and because
4. Human activities are based, in addition, on computer furnished and displayed data, procedures and models that are standardized and that are *sine qua non* conditions of every professional or even of various common and daily activity such as learning, banking, transportation, communication or entertainment.

Consequently, we can say that the entire e-Society is, to some degree, an artificial society, and not only in a traditional sense, that shows that:

1. We are more and more surrounded and even overwhelmed by technical, fabricated, material and instrumental structures and processes, but also in the most concrete and current sense, which means that
2. The whole of our intellectual and spiritual activity is carried on by technical means and that
3. All these practical and theoretical activities are performed in a technical and artificial, but virtual environment as well.

Artificial society is, in this technical and intellectual context, a term justified enough to describe the multisided and multistage social space-time constituted of a growing number of micro- and macro artificial technical and intellectual societies. In a very topical sense, thanks to this entirely new context, we have and we can speak about societies of artificial, cognitive and practical agents and not only about a "society of mind" as anticipated by Minsky. However, even the more daring inquiries, those on artificial life, take as models and simulate, for the moment, manly natural communities and processes, as those described by authors like Bedau (1998).

The already built technopoles, composed of at least 25 intelligent buildings, each of them with an artificially maintained and completely controlled environment, may be seen as prototypes or just as seeds of future ... human settlements, sown in the fertile soil of the current technical environment.

In our opinion, not only the rhythm, but also the shapes of the future technical developments will be totally different than the current ones and therefore, a prognosis is difficult to be made, except if we will provide an increased confidence to some explorative methods of forecasting, to the detriment of those normative, and to those imaginative, versus the quantitative ones and when, if necessary, we will prefer the normative – synthetic – intuitive ones, and we will avoid the normative – morphologic – theoretical methods.

We have, in terms of prospective methods, a wide margin of choice, but in all cases it is necessary, in addition to the various methods, a methodology – preferable a systemic methodology -, that is one which can successfully accompany a prospective vision.

Artificial action and interaction are both products and sources of the whole social development. As we have seen, they are in fact clusters of different action types: cognition and discovery, moral values and their application by specific procedures, testing technical devices, inventing social structures or making political decisions. These activities are all individual or social, theoretical or practical; they are all also partially artificial.

Neumann's universal constructor can be studied as a first relevant example of a both human and artificial activity. J. Neumann conceived and designed it as a self-replicating machine in a cellular automata environment and aimed to demonstrate and to simulate the possibility of a universal construction and evolution. His proposal is actually a logical construction which was fully implemented after nearly fifty years by computer using, and is both an abstract and a concrete achievement, as well as both a human and an artificial achievement, because it is humanly conceived and designed, but may be and even needs to be artificially executed. Neumann's constructor performs not only an infinite but repetitive artificial construction, but also an open evolution, achieved by mutations and selections that reproduce natural evolution (that produces viable entities), and generate similar automata (that replicate themselves).

The second example is furnished by prospective studies that anticipate how a "terrestrial", yet artificial system and community will be built in a different space and time, such as in possible intra-galactic expeditions: only the necessary material components and maybe some energetic resources will be transported, but the planned artificial entities such as robots will be self-building entities; the competences associated with different activities and functions will be configured or transmitted as programs, which will be optimized even after possible terrestrial progress is achieved during the time necessary to get over cosmic distances and years. Under these conditions, a remote and de-phased, but a still jointed human-artificial evolution may occur.

A new cognition, action, interrelation and communication environment is already born and new work groups and communities are developed in the virtual global environment of the web, as we have already shown in a previous work. The specific needs and the appropriate means to satisfy these needs which evolve in this e-world are virtual as well and have all the chances to become universal human needs, which will be at the origin of the whole system of foreseeable common values.

But even the evolution of different human organization and community forms has different rhythms, duration and periodicity, and a common sense of evolution is hard to distinguish, both at the vision and practice level. The present globalization process is a general, but mainly a technological, economical and political phenomenon, and a less cultural and spiritual one, while it is precisely these two social fields that are essential for the human level of existence.

NATURAL AND ARTIFICIAL EVOLUTION: CO-EVOULTION OF HUMAN AND ARTIFICIAL INTELLIGENT AGENTS

At the social level of existence, natural and artificial forms of being are now co-existing and co-evolving, as we have already argued. These kinds of existence also are co-generative and co-functional. Some authors are even convinced that the successive generations of artificial cognitive systems, techniques and agents are also steps in human evolution. In our opinion, social evolution itself can now be natural, artificial and mixed in its every component dimension: biotic, mental, cognitive and even spiritual (Pană, 2008c).

We have to clarify that when we speak here about natural evolution we focus our attention on human evolution, which is nevertheless understood as a complex biotic, mental, social, historic and cultural evolution.

At the biotic, social and technical levels of evolution, finality as self-organization and self-development appears not as a ring-shaped connection between different aspects of a system or between evolution stages of a system, but it is an open evolution that can be modeled by specific methods. At the social and technological levels, these forms of asymmetric, even open evolution can be both anti-entropic and entropic, hazardous and directed (Pană, 2008a).

Social evolution has some specific forms, levels and means; it is a self-determined evolution if viewed as a whole and as a socio-cybernetic system, but it also may show uncertain and unforeseeable evolutions because:

1. Society is, mainly, a system of activities;
2. The most important conditions of human activity are done by other activities;
3. Human action is an instrument-mediated action;
4. These instruments belong to some different material, intellectual and spiritual spheres.

Through common activities in the present technical environment, human and artificial cognitive and moral agents are gaining some common characteristics and are co-participants in a continuous evolution process.

Human evolution can be analyzed at various levels of complexity and it shows today different kinds and degrees of possibilities from distinct perspectives. On the biotic level it is an open evolution as on the technical and spiritual levels, where human possibilities seem to be infinite, while from a psychological point of view, human evolution appears as closed.

This apparently pessimistic view on human behavior and its internal determination (motivation system) resides in a historical perspective on human nature, which has its own evolution, in its biotic, social and cultural dimensions, while its psychological (individual or social) dimension is now hypertrophied but at the same time also limited.

A few diagrams that show the importance and interaction of nine social factors were presented (Pană, 2008b) together with a representation of their effective or ineffective use, which generates the very specific, peak-to-peak movement of our history, in which the peaks are connected by steep, rapidly descending and ascending slopes.

Even if a few specialists in various culture fields and even in scientific disciplines demonstrated, from some partial or even general perspectives, that a sense can be associated only with a finite and closed system, another, perhaps a more consistent part of scholars, think that evolution and maybe development too are not only specific or particular, but general features of existence and we are living not only in a smart, but also in an "artful universe".

A Creative Space Theory was formulated by Andrzej P. Wierzbicki, a theory subsequently developed together with Y. Nakamori, who use the means of Computational Intelligence and who aim to anticipate and maybe to hasten the coming of the Knowledge Civilization Age, by models of creative processes implementing. The scope of creative action is centered on knowledge and technology creation by Wierzbicki and Nakamori (2005), who propose both micro- and macro-theories of creation and of creative environment, the latter - with more and more dimensions.

At this point we have to remind the reader another important contribution of St. Odobleja, who identified some general laws of creation, and who proposed even a "pedagogy of creation", on the basis of his „general cybernetics". He has pointed out some intensity, speed and frequency parameters of creation in the „little portative cerebral laboratory", but also some maybe more important spiritual requirements, such as to run for an ideal, to eliminate the fear of absurdity and to learn to meditate.

If we focus our attention only on some cognitive aspects of human-artificial co-evolution, we can still remark that nowadays various trends of cognitive evolution are occurring:

1. Human cognition is continued in its already sketched theoretical and methodological frameworks, but with outstanding new results;
2. New forms and levels of knowledge appear;

3. These new cognition ways are generated in the framework of new cognitive models;
4. They are integrated and developed by multi- inter- and trans-disciplinary research fields that reshape the structure and dynamics of our conceptual models. See, for more on these aspects (Pană, 2006c; Pană, 2006d).

At the same time, artificial cognitive means and techniques are conceived and used both at low and high intellectual and spiritual levels and these artificial instruments are involved in activities with an increasing rhythm and efficacy. The previous two processes are not independent and a third increasingly powerful tendency is now developing through their intertwined development, that of a human-artificial cognition, characterized by human initiative and aims, but accomplished by artificial methods, techniques and processes, that are used in an artificial intellectual environment, which is a technical and often virtual environment.

The accelerated rhythm and the growing social importance of the artificial agent's development is explained by the start point of their evolution, that of the upper level of human evolution, the intellectual one. This fact may constitute both the source and the basis of a belief, largely shared even if not explicitly evoked, mainly by computer scientists: the successive stages in the evolution of artificial agents are also steps in human evolution.

Humans evolve *with* and *by* their artificial creations - from our perspective - the growing set of artificial (cognitive, moral, robotic or "social") agents, and then we have, in fact, a common evolution or a co-evolution of human and artificial agents. Some psychologists and computer scientists have their own ideas and initiatives in this process (De Angeli, A. & Johnson, G. I., 2004). Are past and present models of knowledge development able to describe and to explain this complex common, natural and artificial, cognitive and practical evolution?

This is a hard problem, because we have here an evolution from a general artificial computing to the more applied scientific computing, from the explosively developed technical computing - that supports robotics, artificial intelligence, virtual reality and artificial life -, to the formerly practiced aesthetic computing and then from the philosophical computing to the more recently initiated ethical computing. All these are more or less slowly and timidly followed by the present digital education, communication, administration and politics with the entire cluster of other informational machine-based computing and communication activities that may eventually be mentioned as organizational computing.

Human brain, mind and knowledge were the real cognitive models until now and even if evolving knowledge models were built, these models were conceived separately for the human cognitive agent's development, respectively for the artificial cognitive agent's evolution. We now can see that at the social level of existence, natural and artificial forms of knowing and being coexist and co-evolve. By such complex emerging processes, new social systems will appear and maybe a new biotechnical species will develop.

To understand, to project and to optimize their common evolution we have to continue our study on the present complex cognition processes, especially those of social knowledge and prospective knowledge. Even individual and social self-cognition are important if we plan to built multi-agent systems, abuse-free human-artificial social agent's interaction or even an artificial self for some of these agents, which will cooperate with humans in a common, increasingly artificial environment.

Human-artificial agent's co-evolution is differently understood and anticipated by various kinds of specialists in technical or social branches of knowledge. This common evolution is therefore viewed in different ways:

1. As a technically directed evolution;
2. As evolution by simulation of the natural evolution of populations, that needs genera-

tion and management of complex structures and processes such as ecosystems, mutations, viruses and selection;
3. As self-structuring processes in ordered contexts: engineers will only create suitable conditions for a self-determined development of artificial activity and even of some artificial life forms;
4. As a learning activity, accomplished by cultural processes, like in the case of children's education.

Co-evolution of human and artificial cognitive agents also needs, under these last conditions, culture learning, values understanding, sharing and practicing.

What can be the model of artificial cognitive and active (moral, educational, health-care, communication and transportation services-carrying-on) agent's conceiving and building? Spirit, maybe? Which human and artificial needs must be induced and stimulated in humans and robots, in order to be able of a cultural conduct? As we see, not only the future and artificial, but even the present and human social agents have to be urgently formed and developed in the spirit of some durable and at the same time effective ways of thinking and doing.

As a more remote and ambitious common human and artificial project, the creation of new values - practical, intellectual and spiritual ones -, and then a common cultural activity can be anticipated. As we have already shown (Niculescu & Pană, 2010; Pană, 2006a), artificial learning, artificial discovery sequences, invention procedures and other scientific activities such as theorem demonstration, virtual experiments and even ethical decision procedures as well as forms of artificial philosophy were already implemented. Applied and used by humans and machines, who can meet in the middle of the road between the natural and the artificial, these types of activities, processes and techniques can facilitate a common, faster evolution.

REFERENCES

Bedau, M. A. (1998). Philosophical Content and Method of Artificial Life. In *The Digital Phoenix: How Computers are Changing Philosophy* (pp. 135–152). Oxford, UK: Blackwell Publishers.

Bynum, T. W. (1998). Global Information Ethics and the Information Revolution. In *The Digital Phoenix: How Computers are Changing Philosophy* (pp. 274–291). Oxford, UK: Blackwell Publishers.

Bynum, T. W. (2000). The Foundation of Computer Ethics. *Computers & Society*, *30*(2), 6–13. doi:10.1145/572230.572231

Cavalier, R. (Ed.). (2005). *The Impact of Internet on Our Moral Life*. Albany, NY: State University of New York Press.

Churchland, P. M. (1998). The Neural Representation of the Social World. In *The Digital Phoenix: How Computers are Changing Philosophy* (pp. 153–170). Oxford, UK: Blackwell Publishers.

Danielson, P. (1998). How Computers Extend Artificial Morality. In *The Digital Phoenix: How Computers are Changing Philosophy* (pp. 292–307). Oxford, UK: Blackwell Publishers.

De Angeli, A., & Johnson, G. I. (2004). Emotional Intelligence in Interactive Systems. In *Design and Emotion* (pp. 262–266). London: Taylor & Francis.

Ess, C. (2006). Universal Information Ethics? Ethical Pluralism and Social Justice. In E. Rooksby, & J. Weckert (Eds.), *Information technology and Social Justice* (pp. 69–92). Hershey, PA: Idea Publishing. doi:10.4018/978-1-59140-968-7.ch004

Ey, H. (1998). *Conscience*. Bucharest: Editura Ştiinţifică.

Floridi, L. (1999). Information Ethics: On the Theoretical Foundation of Computer Ethics. *Ethics and Information Technology*, *1*(1), 37–56. doi:10.1023/A:1010018611096

Gregory, R. (2000). *Future of Mind-Makers*. Bucharest: Editura Ştiinţifică.

Hillis, W. D. (2001). *The Pattern on the Stone: The Simple Ideas that Make Computers Work*. Bucharest: Editura Humanitas.

Laruelle, F. (1990). *Théorie des identitées, fractalité generalisée et philosophie artificielle*. Paris: P. U. F.

Niculescu, C., & Pană, L. (2010). Architecture of a Multi-Framework Set for Collaborative Knowledge Generation. In *Proceedings of the 11th European Conference on Knowledge Management* (pp.73-75). Academic Publishing Limited.

Odobleja, S. (1982). *Consonantal psychology*. Bucharest: Editura Ştiinţifică si Enciclopedică.

Pană, L. (1988). Cognition and action centered values in human conduct structuring. *Revue Roumaine des Sciences Sociales. Série de Philosophie et de Logique*, *32*(1-2), 11–18.

Pană, L. (2000a). Information machine, information action, information thinking and information man. In L. Pană, *Philosophy of Technical Culture* (pp. 430–452). Bucharest: Editura Tehnică.

Pană, L. (2000b). Moral Culture. In L. Pană, *Philosophy of Technical Culture* (pp. 104–112). Bucharest: Editura Tehnică.

Pană, L. (2003a). The Intelligent Environment as an Answer to Complexity. In *Proceedings of the IUAES Congress: XV ICAES 2K3 Humankind / Nature Interaction: Past, Present and Future*, (Vol. 2, p. 1198). International Union of Anthropological and Ethnological Sciences.

Pană, L. (2004a). Artificial Philosophy. In L. Pană,*Philosophy of Information and Information Technology*. Bucharest: Politehnica Press.

Pană, L. (2004b). Modeling some Evolutions of Value Systems from the Perspective of Technical Culture. In L. Pană (Ed.), *Evolutions in Value Systems under the Influence of Technical Culture*. Bucharest: Politehnica Press.

Pană, L. (2005a, March). *From Virtue Ethics to Virtual Ethics*. Paper presented at the Interdisciplinary Research Group of the Romanian Committee for History and Philosophy of Science and Technology of the Romanian Academy. Bucharest, Romania.

Pană, L. (2005b). Moral Intelligence for Artificial and Human Agents, Machine Ethics. In *Papers from the AAAI Fall Symposium Series*. AAAI Press.

Pană, L. (2005c). Philosophy of Artificial and Artificial Philosophy. *Academica*, *15*(34), 32–34.

Pană, L. (2006a). Knowledge Management and Intellectual Techniques – Intellectual Invention and Its Forms. In R. Trapl (Ed.), *Cybernetics and Systems: Proceedings of the Eighteenth European Meeting on Cybernetics and Systems Research*, (vol. 2, pp. 422-427). Austrian Society for Cybernetic Studies.

Pană, L. (2006b, June). *Values Inventing and Motives Computing: Elements of Artificial Ethics for Cognitive and Operative Moral Agents*. Paper presented at the 4th European Computing and Philosophy Conference. Trondheim, Norway.

Pană, L. (2007, June). *From Information Flows and Nets to Knowledge Groups and Works*. Paper presented at the 5th European Computing and Philosophy Conference. Enschede, The Netherlands.

Pană, L. (2008a). An Integrative Model of Brain, Mind, Cognition and Conscience. *Noema*, *7*(1), 120–137.

Pană, L. (2008b). Moral Intelligence: Elements of Artificial Ethics for Cognitive and Moral Agents. *NOESIS*, *33*(1), 39–51.

Pană, L. (2008c). The Preferential Sense as a Source of Natural and Artificial Evolution. In *Proceedings of the 14th International Congress of Cybernetics and Systems of WOSC* (pp. 984-993). Wroclaw, Poland: Wroclaw University of Technology and the World Organization of Systems and Cybernetics.

Skolnicki, Z., & Arciszewski, T. (2003). Intelligent Agents in Design. In *Proceedings of 2003 ASME International Design Engineering Technical Conferences & The Computers and Information in Engineering Conference*. Mason University.

Sloman, A. (1990). Motives, Mechanisms, Emotions. In M. Boden (Ed.), *The Philosophy of Artificial Intelligence* (pp. 231–247). New York: Oxford University Press.

Wierbicki, A. P., & Nakamori, Y. (2005). *Creative Space: Models of Creative Processes for the Knowledge Civilization Age*. Springer. doi:10.1007/b137889

ADDITIONAL READING

Pană, L. (2000). *Philosophy of Technical Culture*. Bucharest: Editura Tehnică. (In Romanian)

Pană, L. (2003b). The Technically Possible and the Specific Intellectual Space. In A. Bazac, G. G. Constandache, & B. A. Balgiu (Eds.), *Cognitive strategies and European Integration* (pp. 83–93). Bucharest: Politehnica Press.

Pană, L. (2004). *Philosophy of Information and Information Technology*. Bucharest: Politehnica Press. (In Romanian)

Pană, L. (2006c). Artificial Intelligence and Moral Intelligence, TripleC: Cognition, Communication, Co-operation, Open Access Journal for a Global Sustainable Information Society, Publisher: University of Salzburg, Austria, 4 (2), 254-264.

Pană, L. (2006d). Co-evolution of Human and Artificial Cognitive Agents. *Proceedings from Computers and Philosophy. An International Conference,* i-C&P 2006, 3-5 May 2006 – Laval, France (pp. 366-378). Intitute Technologique de Laval, Le Mans Université, France.

Pană, L. (2006e). Intellectics and Inventics. *Kybernetes. International Journal of Systems and Cybernetics*, *35*(7/8), 1147–1164. doi:10.1108/03684920610675148

Pană, L. (2008d). Sociomatic Systems, Studies and Reflections: A Challenge for Social and Intellectual Invention. In Robert Trapl (Ed.), *Cybernetics and Systems*, volume 2, Proceedings of the Nineteenth European Meeting on Cybernetics and Systems Research, University of Vienna, Austria, 25-28 March, 2008 (pp. 325–330), Austrian Society for Cybernetic Studies and Austrian Research Institute for Artificial Intelligence.

Pană, L. (2009a). Communication and inter-personalization in virtual working groups constituted on the Web. In B. A. Balgiu & G. C. Constandache (Eds.), *Communication – suggestion and influence: Interdisciplinary and Trans-disciplinary Aspects* (pp. 52–64). Bucharest: Sigma Publishing House. (In Romanian)

Pană, L. (2009b). *Possibility, Infinity and Predictability*. Bucharest: Politehnica Press. (In Romanian)

Pană, L. (2010). Crucial Intellectual Events in Information Science and Information Technology. *NOESIS*, *34*(1), 171–196.

KEY TERMS AND DEFINITIONS

Artificial: Any completely new social outcome of human activity, either theoretical or practical, such as a concept, a method or artifact, a social body, a system of relations or an institution, therefore, any kind of real mental construction or material achievement. Artificial Ethics itself is a kind of artificial cognition, artificial discovery, artificial interaction and invention in an artificial environment. When we are situated in the realm of values and spirituality, the term "new" is taken in its hardest sense and meanly means "unprecedented".

Artificial Ethics: Is one of the newest scientific and technical activity fields. This domain of ethical research is born at the confluence of at least five other new fields of the contemporary ethical culture: information ethics, computing ethics, machine ethics and roboethics, as well as of global information ethics. Artificial ethics is con-generated with webethics which comes into being together with the recent concerns about the ethical issues raised by the virtual activities and communities on the web. Artificial Ethics, its invention and implementation can also be seen as a domain of mutual completion and stimulation of two scientific and technical research areas, AI and Social Cybernetics. Artificial Ethics is then both a conceptual and technical field of research and a theoretical and practical area of moral action. To be suitable for both human and artificial moral agents is the most important feature of artificial ethics Artificial Ethics represents not only a new domain, but even a new level of the ethical activity, cognition and reflection.

Artificial Evolution: Is a process which takes place in a couple conjoined ways, such as a) a technically directed evolution; b) evolution by simulation of the natural evolution of populations, that needs generation and management of complex structures and processes such as ecosystems, mutations, viruses and selection; c) self-structuring processes in ordered contexts: engineers will only create suitable conditions for a self-determined development of artificial activity and even of some artificial life forms; d) a learning activity, accomplished by cultural processes, like in the case of children's education.

Artificial Moral Agents: Artificial moral agents can be studied as a sub-set of artificial intelligent agents. They may have classes of attributes of artificial agents such as a) sensing and acting; b) reasoning; c) learning and knowing; d) internal structure; e) variability of their number. The following set of specific properties of artificial moral agents was gradually identified by (Pană, 2005b; Pană, 2006a; Pană, 2008a). Artificial moral agents are or will be: 1) complex, specialized, autonomous or self-determined, even unpredictable entities; 2) open and even free conduct performing systems; 3) cultural beings: the free conduct gives cultural value to the action of a human (natural) or artificial being; 4) responsible and co-operative participants in group and community-centered activities; 5) entities with "lifegraphy", not just with a "stategraphy"; 6) educable, not just teachable systems; 7) endowed with diverse or even multiple intelligence forms, like moral intelligence; 8) equipped not just with automatisms and intelligence, but with beliefs; 9) capable even of reflection; 10) components/members of some real (corporal or virtual) community.

Artificial Philosophy: Is a recently conceived and generated field of philosophical culture, at the confluence of formal axiology, technical logic, information aesthetics or digital politics. Fr. Laruelle (1990) viewed artificial philosophy as the science of thinking, and even as the "unified theory of thinking", developed by mathematical and technical methods. Four other possible philosophical sources and future components or even fields of Artificial Philosophy were analyzed in (Pană, 2005c).

Co-Evolution (of Human and Artificial Moral Agents): Natural and artificial forms of being are now not only co-existing but also and co-evolving. More, these kinds of existence are co-generative and co-functional. Co-evolution of human and artificial agents is the process and the product of the co-cognition, co-generation, co-operation and co-construction of human and artificial intelligent agents, in their "natural" or artificial evolution. Through common activities in the present technical environment, human and artificial cognitive and moral agents are gaining some common characteristics and are co-participants in a continuous evolution process.

Emergent Ethical Fields: The interaction of intelligent (human and artificial) agents generates several new areas of research, including a new ethical research field. This new ethical domain can also be the result of a synthesis between other two newly emerging ethical disciplines: Computing Ethics and Machine Ethics.

Global Information Ethics: Is a new level rather than a new domain of ethics, generated by the informatization and globalization of all significant human activities. The field was preceded and introduced by *information ethics* Foridi (1999; 2008). The first theoretician of *global information ethics* is T. W. Bynum (1998 and 2000); Foridi (2007) deepened its study. Ch. Ess (2006; 2013) addressed and analyzed important aspects of ethics related to the present changes in knowledge, communication and culture. R. Cavalier (2005) pointed out the impact of internet on our moral lives and a *network ethics* was promoted by Foridi (2010).

Machine Ethics: Concerns the computer itself if it runs intelligent programs because once turned into an intelligent machine, the information machine induces changes in the world, like humans do. All human activities have a moral dimension and a machine with similar possibilities needs moral functions. Machine ethics tends to become a research field of Artificial Intelligence, but it also needs a philosophical (ontological, axiological and pragmatic) foundation.

Moral Intelligence: May be viewed as an operational branch of moral cognition. More analytically studied, moral intelligence is a multitude of intellectual activities. In turn, these activities subsume always used, but recently identified forms of thinking such as fuzzy thinking, fractal thinking, statistical and probabilistic, and even prospective and integrative thinking. Understanding, reasoning, decision making, problem solving and heuristic conduct are also needed in moral (human or artificial) conduct. The unitary expression of these functions/aptitudes is moral intelligence. Task formulation, rules interpretation and consequences evaluation are other intellectual activities subsumed by moral intelligence. This kind of human and artificial intelligence cannot be simply dismissed as "moral cognition", not only because of its complexity but also because of its pro-active, operative nature. Moral intelligence consists, in fact, of a multitude of intellectual and practical activities. Moral intelligence is different from technical, political or economical intelligence, all of them being included in the group of practical intelligence forms. Moral behavior is a complex, practical, intellectual and spiritual behavior. Moral intelligence is operational at all these levels of moral conduct (action, cognition, spirituality) and moreover, it has concrete and abstract, imitative and creative, assertive and interpretative, persuasive and imperative, individual and group components. Therefore, moral intelligence is neither a simply practical intelligence form. Moral intelligence is not a form or a level, but a kind of intelligence: it is a synthetic kind of intelligence.

Natural and Artificial Evolution: At the social level of existence, natural and artificial forms of being are now co-existing and co-evolving (Pană, 2006d). These kinds of existence also are co-generative and co-functional. Some authors are even convinced that the successive generations of artificial cognitive systems, techniques and agents are also steps in human evolution. In our opinion, social evolution itself can now be natural, artificial and mixed in its every component dimension: biotic, mental, cognitive and even spiritual (Pană, 2008c). When we speak here about natural evolution we focus our attention on human evolution, which is nevertheless understood as a complex biotic, mental, social, historic and cultural evolution. Artificial evolution was above defined. Among other favorable conditions for an

Artificial Ethics development we can also count the human and artificial evolution, a twofold evolution that nonetheless follows the same direction, as well as the more and more complex and effective man-machine, human-computer, human and robotic entities interaction and their present social meanings and consequences.

Technological and Social Prognosis: These two great branches of prognosis are combined in this chapter in two main reasons: first, because Artificial Ethics itself is a common achievement of technical and social sciences and second, because just like any other topical development in Artificial Intelligence, its development has a multisided and multistage influence over all activity fields. Both technical prognosis and social prognosis are complex scientific fields and we are using here a few specific methods and more results already obtained through previous research. This latter aspect confirms the interdisciplinary feature of the paper and illustrates the meta-theoretical level to which is drawn.

Chapter 5
Of Robots and Simulacra:
The Dark Side of Social Robots

Pericle Salvini
Istituto di BioRobotica, Italy

ABSTRACT

In this chapter, the author proposes a theoretical framework for evaluating the ethical acceptability of robotic technologies, with a focus on social robots. The author proposes to consider robots as forms of mediations of human actions and their ethical acceptance as depending on the impact on the notion of human presence. Presence is characterised by a network of reciprocal relations among human beings and the environment, which can either be promoted or inhibited by technological mediation. A medium that inhibits presence deserves ethical evaluation since it prevents the possibility of a mutual exchange, thus generating forms of power. Moreover, the impact of social robots on human beings should be carefully studied and evaluated for the consequences brought about by simulated forms of human presence, which have both physical and psychological dimensions and are still unknown, especially with respect to weak categories, such as children, elderly, and disabled people.

INTRODUCTION

In the following, I will attempt to draw a line between what is acceptable and what is not from an ethical point of view with regard to the technological enhancements of human beings through robotic technologies, with a focus on social robots. With such an objective, I may appear to be a technophobe, a Luddite or a conservationist. Quite the opposite. I agree with the definition that humans are technical by nature, even if it may sound to be a contradiction in terms, but I also agree with the truism that "not all progress is good or necessary". We cannot deny that technology and science are core aspects of the human nature. Nevertheless, it is also unquestionable that there are other forces, driven by scientific interest and economics, which push scientific and technological developments towards choices that are not always integral to the survival of human species.

DOI: 10.4018/978-1-4666-6122-6.ch005

This chapter responds to the needs and objectives of the ethics of technology, which are called technoethics or roboethics and these are:

1. To identify the dangers and benefits that come out from the research and application of advanced robotic technologies and systems;
2. To develop tools and knowledge which allows us to direct the development of robotic technology in a sustainable way for the human being (present and future generations) and the natural environment (Veruggio and Operto, 2010).

The benefits provided by robotic technologies are manifold and visible: from factories automation to robotic surgery, search and rescue operation, security, space and underwater exploration, assistance to elderly and disabled people, just to name a few of the most popular and current applications. However, there are also concerns surrounding the use of robots, especially with respect to their level of autonomy (e.g. autonomous, semi-autonomous or teleoperated), the task to be performed (e.g. warfare, care, surgery, logistics, etc.), and the typology of users involved (laypeople, children, elderly and disabled people, etc.).

Why Should We Care About Technologies?

There are many ways to demonstrate that care should be taken about technological and scientific progress, which in some way may also be applied to robotic technologies. I have chosen three: two from philosophy and one from history. French philosopher Paul Virilio introduces the concept of the "accident of the future" (Virilio, 1997). According to Virilio: 'Every time a technology is invented, take shipping for instance, an accident is invented together with it, in this case the shipwreck, which is exactly contemporaneous with the invention of the ship. The invention of the railway meant, perforce, the invention of the railway disaster. The invention of the airplane brought the air crash in its wake' (Virilio, 2000: 32). Virilio is especially interested in the real-time communication technologies, which, according to him, present new alarming characteristics: one of which is that the accident is no longer limited to a specific here-and-now, but is delocalized, taking place everywhere. As a consequence, the accident of the future will be integral, meaning it will be a general accident that involves all mankind. Examples of "almost integral accidents", according to Virilio, are radioactivity leakage and a virus in an electronic network, which make the "globalization" effect of the accident clear. Virilio's argument is a warning to us against the increasingly dependence on technology, and, at the same time, a recommendation to invest more effort in technological risk assessment in order to diminish (not to eliminate!) potential drawbacks: 'to examine the hidden face of new technologies, before that face reveals itself in spite of us' (*Ibid.*: 40).

We can also try and answer the question of why should we care, by looking at the way opened up by Hans Jonas. Jonas introduces the principle of responsibility, which holds human beings responsible for the preservation of life with respect to the current generation of human beings and those that follow along with the natural environment. Indeed, according to Jonas, 'with certain developments of our powers the nature of human action has changed, and since ethics is concerned with action, it should follow that the changed nature of human action calls for a change in ethics as well: [...] in the more radical sense that the qualitatively novel nature of certain of our actions has opened up a whole new dimension of ethical relevance for which there is no precedent in the standards and canons of traditional ethics' (Jonas, 1985: 1). Autonomous or teleoperated, connected or disconnected from the human body, robots clearly fall in the category of technologies which transform the nature of human action by allowing one to act from a distance, to enhance its actions, and even to act without physical or cognitive efforts.

If Virilio is telling us that every new discovery brings with it a new danger, similarly Jonas is saying that the new possibilities offered by scientific and technological developments put forward the problem of the consequences that "we do not know yet": 'The gap between the ability to foretell and the power to act creates a novel moral problem. With the latter so superior to the former, recognition of ignorance becomes the obverse of the duty to know and thus part of the ethics' (*Ibid.*: 8).

However, both Virilio's and Jonas's remarks come into their own when we look at history, which is my third and final way of demonstrating a need for an ethics of technology and science. In other words, as is pointed out by Patrick Lin and colleagues, the history of scientific and technological progress is overflowing with examples of products, the result of technological and scientific research, which have gone wrong: DDT (DichloroDiphenylTrichloroethane) and asbestos are two cases in point (Lin et al. 2011). Again we find ourselves in the realm of the unknown as the effects, or accidents, are caused by what we are unable (or unwilling) to predict. Of course, this goes back to safety, risk assessment and human fallibility, but also, as pointed out by Virilio, to a sort of dark side of technology, which will never be possible to illuminate.

What about robots? Though, as mentioned before, robotics is a form of human enhancement which can modify the way we act in the world, it is not possible, at least in the current historical moment, to consider the risks brought about by robotics as potentially catastrophic as those caused by nuclear power, environmental pollution, climate change, or genetically modified food, just to name a few. As a matter of fact, it is necessary to distinguish between fictional from factual fears when addressing the dangers of robotics. If we exclude science fiction futures in which robots will take over the world and enslave human beings, what remains to be afraid are a few issues which depend on how we – human beings – will make use of robotics technologies. For instance: autonomy in lethal decision making, a 21th Century robotic divide, the dual use of robotics technology, the disappearance of moral and legal responsibility, the reduction of labour opportunities, the protection of privacy, etc.

In what follow, I will present the theoretical framework used to evaluate the ethical sustainability of robotics, which I will discuss with respect to social robots. The framework is based on the concepts of mediation and presence. The third section will be dedicated to a few terminological clarifications and to the characterisation of social robots as simulacra of presence. Finally, in the forth and final section, I will point out the major ethical issues emerging from the application of the framework to social robots.

THE THEORETICAL FRAMEWORK: PRESENCE AND ROBOTIC MEDIATION

The theoretical framework I propose is based on two key concepts: human presence as given by reciprocal relations, and technology as a form of mediation of human action and cognition.

My conceptualisation of presence takes inspiration from art, in particular theatre. Elsewhere (Salvini, 2006), I adapted Roger Copeland's definition of theatrical presence ('presence in the theatre has [...] to do with [...] the way in which the architectural and technological components of the performance space either promote or inhibit a sense of "reciprocity" between actors and spectators' (Copeland, 1990) and proposed the following definition: 'the way in which natural and/or artificial (i.e. cultural, technological...) factors/conditions either promote or inhibit a sense of reciprocity between human beings or between an environment and a person' (*Ibid.*).

According to such definition, it is possible to distinguish between two forms of presence: one *immediate*, where the body is the source and carrier of presence, and *mediated*, where the body is still the source of presence, but technology has replicated and replaced it in its function of carrier.

I see presence, whether immediate or mediated, as characterised by networks of reciprocity, that is, by dialogic exchanges or two-way channels. In a presence condition, either with respect to other human beings or to the natural environment, a human being is always subject and, at the same time, object of actions and perceptions: e.g. I can see, but I am also part of the visible; I can touch but I am also part of the tangible; and so forth. Maurice Merleau-Ponty's notion of intertwining or the chiasm can be used to illustrate my argument: '[m]y body as a visible thing is contained within the full spectacle. But my seeing body subtends this visible body, and all the visible with it. There is reciprocal insertion and intertwining of one in the other' (Merleau Ponty, [1964] 1968).

The second key concept of my theoretical framework is mediation; in other words, I argue that all technologies are a form of mediation of a human action. There are at least two reasons why a robot, for instance, can be compared to a medium. First, a medium, by definition, is something that stays in between: 'a state that is intermediate between extremes; a middle position' (WordNet). Whether it is a robotic hand or leg directly linked with the nervous system (i.e. an hybrid bionic system or cyborg), a surgical or military tool controlled at a distance by a human operator (a tele-robotic system), or an autonomous robot designed for personal care or industrial applications, the robot always stays in between a human's action or function and its fulfilment (i.e. between a desire or need and its realisation or satisfaction). An autonomous robot, with respect to bionic and tele-robotic systems, is a form of mediation independent from real-time, human control (it is a *simulacrum* of presence, as I shall explain in the next section).

Secondly, in his well-known analyses of media, Marshall McLhuan pointed out that 'the content of a medium is always another medium' (McLhuan, 1985 [1979]). In other words, a medium is always an extension of another medium: e.g. the car is an extension of our legs. Similarly, a robotic prosthesis, a remotely controlled vehicle or the arm of an industrial robot can be considered, though in different ways, as extensions of our limbs, perceptions and actions.

Therefore, by replacing, restoring, replicating and even augmenting a function, robotic technologies are mediating and extending human agency. It goes without saying that technological mediation is a form of human enhancement.

I propose to consider technological mediations as acceptable or non-acceptable, depending on its impact on the notion of human presence. As a matter of fact, I argue that the condition of presence implies always an ethical dimension – one of responsibility – arising from the awareness of being in a reciprocal relation with another human being or the environment. Tom Lombardo maintains, '[r]eciprocity has not only served as a primary mechanism for the creation of biological and social complexity, it provides a universal principle upon which human values and ethics are defined. Reciprocity is the foundation of the concepts of justice, equity, and perhaps even human care and kindness' (Lombardo, 1987). However, as I shall point out, in many cases, the mediation of presence brought about by new technologies is based on the loss of reciprocal relations

In what follows, I will focus my attention on social robots and I will explain why they can be considered as simulacra of presence.

SOCIAL ROBOTS AS *SIMULACRUM* OF PRESENCE

A social robot is defined as a robot 'able to communicate and interact with us, understand, and even relate to us, in a personal way. It is a robot that is socially intelligent in a human-like way' (Breazeal 2002). In other words, a social robot is an autonomous machine designed to interact with human beings in a human-like way. Before turning to the issue of robots and simulacra, I need to clarify first the terminology that I am using. Well aware of the various and contradictory uses of the term *autonomy*, especially if one looks at the humanities (Schmidt and Kraemer, 2006; Decker, 2007), I simply define an autonomous robot as 'a machine that collects information from the surrounding environment and utilises them to plan specific behaviours which allows it to carry out actions in the operative environment' (Lin et al, 2011). Therefore, bionic systems, tele-robotic systems and pre-programmed robots or automata are excluded from this definition. As a matter of fact, a pre-programmed robot has limited or no sensing capabilities, due to the fact that it operates in structured and known environments. Likewise, tele-robotics systems and bionics always imply the control, though partial or complete, by a human being and therefore cannot be considered as autonomous in the sense outlined above.

Autonomous robots can be considered as the utmost form of enhancement and extension of a human being' capabilities, since they allow one to act in the world (i.e. to carry out a task), without actual physical or cognitive intervention. The etymology of the word robot is significant in pointing out its teleology. The origin of the word robot, credited to Karel Capeck's play *R.U.R.* (1923), comes from the Slavonic language *robota*, which means "slave" (Online Etymology Dictionary). However, as I shall point out, with an autonomous robot we are no more dealing just with an extension of a human function, as in the case of tele-robotics systems or bionics, but with a simulacrum of "presence": a medium that takes decisions independently from a human being, according to the interactions between its own programme and its sensory capabilities.

As far as the adjective *social* is concerned, I refer to robots specifically designed for being engaged in social interactions with human beings.

Social robotics originated in research in Artificial Intelligence, Human-Computer Interaction and Human-Robot Interaction as a way to facilitate communication between users and robotic machines, in tasks where the users were no more expert operators, but laypeople and the robots' operative environments were no more factories, but unstructured and dynamic environments. The basic idea was, and for some researchers still is, to facilitate the interaction among robots, human beings and the operative environments by providing robots with the morphology, capabilities, and behaviours of human beings and by exploiting the human tendency to anthropomorphisation and socialisation with media (Reeves and Nass, 1998).

However, it seems to me that there exists also a different line of research, one in which the focus has gradually shifted from designing robots with social properties functional to the accomplishment of a task, to designing robots in which the social capabilities are the main function. No more a vehicle for facilitating interactions, sociability has become an end in itself, a sort of virtuosity of the machine.

In interpersonal relations, social interaction means the capacity to be friendly and pleasant to people. Applied to a robot, the adjective "social" implies several kinds of abilities, among which, cognitive and physical skills as well as social and emotional capabilities, such as behaving according to social norms or expressing and understanding emotions and feelings (for a survey on social robotics see Fong, Nourbakhsh, and Dautenhahn, 2003). However, interacting with human beings is a very complex and dynamic activity which can be hardly foreseen and translated into lines of code.

As I said at the beginning, an autonomous robot is a special case of mediation since it can replicate and extend human functions without the direct intervention of a human operator, that is, in an autonomous way. Therefore, I argue that an autonomous robot can be defined as a simulacrum of presence.

A simulacrum is 'a representation of a person (especially in the form of sculpture)' (WordNet). According to Baudrillard there are three orders of simulacra, corresponding to three historical periods: the first one, associated with the pre-modern period is characterized by the image as a clear counterfeit of the real; the second one is characterised by images as imitations and copies of the real by means of mechanical reproduction and this period is associated with modernity; and finally the third one, associated with post-modernity, is characterised by the simulacra, which proceeds the real and make any distinction no more possible: 'it is no longer a question of imitation, nor duplication, nor even parody. It is a question of substituting the signs of the real for the real" (Baudrillard, 1994: 2).

In the history of machine development, it is possible to trace a path leading towards the simulacrum by looking back at Norbert Wiener's distinction between the industrial and cybernetic revolutions (Wiener, 1988 [1954]). According to him, during the industrial revolution machines were considered 'purely as an alternative to human muscle' (*Ibid.*: 136). With the *cybernetics* revolution machines started to be considered as an "alternative to human intelligence". According to Wiener the cybernetic revolution was characterized by the vacuum tube and the feed-back (*Ibid.*: 153). The latter, central to Wiener's theory of communication and control, is 'the property of being able to adjust future conduct by past performance' (*Ibid.*: 33).

Although Wiener never used the word *simulacrum*, in my opinion, a similar idea is contained in his well-known parallel between cybernetic machines and living organisms: 'It is my thesis that the physical functioning of the living individual and the operation of some of the newer communication machines are precisely parallel in their analogous attempts to control entropy through feed-back' (*Ibid.*: 26-27).

On the same line of Wiener, according to Baudrillard, 'one can clearly mark the difference between the mechanical robot machines, characteristics of the second order, and the cybernetic machines, computers, etc., that, in their governing principle, depend on the third order' (Baudrillard, 1994: 126). The governing principle is clearly the attempt to control entropy thorough feed-back.

The current *simulacrum revolution* is characterized by machines that are capable of reproducing human muscle (power), intelligence ("judgment capabilities"), and, in the case of social robots also the "warm" qualities of human beings, namely: faculty for sensation, faculty for feelings, life, humanity, intelligence (Cerqui and Arras, 2003).

Of course, robotics technology is still far from delivering us robots that can substitute the real, as those provided by since fiction. However, there are already significant steps in that direction, such as the development of human-like capabilities thanks to advancement in hardware and software components, as well as in the design of application scenarios in which the robot is deliberately meant to be a *sign of the real*. Telepresence robots, in which a sort of robotic clone replaces and extends the presence of its human operator (Sakamoto and Hishiguro, 2009) and robotic pets, like the seal robot *Paro*, which is successfully used in clinical studies to treat autism and dementia senile (Shibata et al. 2010), are two cases in point.

THE DARK SIDE OF SOCIAL ROBOTS

There is currently much debate surrounding social robots use with elderly, children or disabled in care, companionship, entertainment and education applications; even in the case of clinical applications with patients affected by autism or senile dementia, there is no agreement among scholars. One of the main arguments used by detractors is about authenticity: 'For an individual to benefit significantly from ownership of a robot pet they must systematically delude themselves regarding the real nature of their relation with the animal. It requires sentimentality of a morally deplorable sort. Indulging in such sentimentality violates a (weak) duty that we have to ourselves to apprehend the world accurately' (Sparrow, 2002). On the other hand, among the scholars in favour of the use of social robots in care applications Jeason Borenstein and Yvette Pearson maintain that 'as long as there is no intention to deliberately deceive or neglect dementia patients through the use of a robot, the fact that some patients may form erroneous beliefs about a robot caregiver – a process over which other agents may have little control – does not necessarily amount to being disrespectful to a care recipients' (Borenstein and Pearson, 2010).

In the theoretical framework proposed in this paper I have argued that a social robot is a kind of *mediation*, that is, an extension and enhancement of human actions and, at the same time, a *simulacrum of presence*, since by means of self-regulation (i.e. autonomy) it can replicate reciprocal relations. Drawing on this framework, it seems to me possible to point out at least two major ethical issues – besides authenticity – that make problematic the deployment of social robots: the risk of establishing unbalanced power relations among individuals and the risk of potential negative consequences on the psychological health of human subjects, especially with respect to weak categories of users, such as children, elderly and disabled people.

Likewise any kind of mediation that disrupts the reciprocity of presence, autonomous robots may create situations of power, by preventing the possibility of any real response. Therefore, its ethical acceptance or non acceptance should be evaluated depending on each specific application. For instance, handling hazardous material via a telerobotic system, lifting heavy loads thanks to a robotic exoskeleton or killing by means of a drone are all cases in which the reciprocity of presence is deliberately replaced by unilateral relations. In other words, by inhibiting the channels of reciprocity, technological mediation removes the subjects from his/her entanglement with other human beings and/or the environment and favours a subjective perception and unilateral action. This is very well explained by French critic Jean Buadrillard, who in his gloomy analysis of media, argues that: 'The totality of the existing architecture of the media founds itself on this pattern definition: they are what always prevents response, making all processes of exchange impossible [...]. This is the real abstraction of the media. And the system of social control and power is rooted in it' (Baudrillard, 1981: 170). Baudrillard's notion of responsibility should not be confused with Jonas's obligation towards nature and future generations. On the contrary, according to Baudrillard, the lack of responsibility corresponds to an impossibility to reciprocate. Indeed, power, according to Baudrillard, 'belongs to the one who can give and cannot be repaid. To give, and to do it in such a way that one is unable to repay, is to disrupt the exchange to your profit and to institute a monopoly' (*Ibid.*: 170).

What is happening with social robots is that robots are used to replace human beings in their social interaction with other human beings. Feelings, emotions and cognitive process are detached from a person and embodied and encoded into robots, which are then used for interactive and collaborative tasks with human beings.

Such a trend towards the design of human-like capabilities in robots includes also reciprocity. Reciprocity is among the six psychological benchmarks, together with autonomy, imitation, intrinsic moral value, moral accountability, and privacy proposed to measure the 'success in building increasingly human-like robots' (Kahn et al. 2007). According to the authors, reciprocity is one of the fundamental aspects of human life, and it is central to moral life and moral development. The authors ask: 'can people engage substantively in reciprocal relationships with humanoids?' The authors point out that for substantial reciprocity they mean interactions of the same complexity of those occurring between human beings. In other words, something different from 'hybrid unidirectional form, where the human is able ultimately to control or at least ignore the humanoid with social and moral impuntiy' (*Ibid.*).

One may question whether robots capable of disobeying the orders given by their owners would have any useful application and it is even questionable whether they will be completely safe.

However, even if robots were capable of responding to human beings by engaging in "substantial" reciprocal relations, the ontological difference between a person and a robot will corrupt the exchange, which will be only apparently reciprocal. The robot will be always responding according to its programme and therefore the interaction will be again unidirectional, even if of a better quality. The capabilities of robots to replicate reciprocal relations, even at the social level, corresponds to what Baudrillard defines as 'forms of response *simulation*, themselves integrated in the transmission process' (Baudrillard, 1981:170). As explained by the French philosopher, 'reversibility has nothing to do with reciprocity' (*Ibid.*: 181), since 'it integrates the contingency of any such response in advance'. Behind the appearance of genuine presence and reciprocity, there is a programme, lines of code written by another human being, according to mathematical models of human behaviour which will never be able to predict the complexity and unpredictability of human beings.

The effects brought about by the *simulation of presence* in tasks which involve the "warm qualities" of human beings have not been fully investigated yet. In other words, we know very little about the social, psychological and anthropological consequences brought about by long term interactions between human beings and robots endowed with physical, cognitive, and even social capabilities.

An interesting case study is *Paro*, the robotic seal (Parorobots, 2011), which elicits strong feelings of attachment in its owners, even outside therapeutic applications (Shibata, Kawaguchi, Wada, 2010). The interaction with this kind of robots is different from the interaction with the fictional characters of the arts or the psychological engagement with liminal objects like the Teddy Bear, which is still characterised by the play with the difference between the factual and the fictional. According to Baudrillard: 'Something has disappeared: the sovereign difference between them that was the abstraction charm. For it is the difference which forms the poetry of the map and the charm of the territory, the magic of the concept and the charm of the real' (Baudrillard, 1983: 3). On the contrary, with artificial entities like *Paro*, such difference is becoming more subtle, and the "willing suspension of disbelief" is superseded by the increasing reality of the simulacrum along with an involuntary compulsion to do something for them. As pointed out by psychologist Sherry Turkle, social robots can be described as "nurturing machines", that is, 'a machine that presents itself as dependent' (Turkle, 2007). According to her, 'nurturing creates significant social attachments'. Indeed, Turkle believes that the problem with robots it is not about their capabilities but about our vulnerabilities. She finds that '*the psychology of engagement that comes from interacting with social robots*', especially if the human is a weak person, such as a child or an elderly person, '*creates an effective illusion of mutual relating*'. While a child is aware of the projection onto an inanimate toy and can engage or not engage in it at will, *a robot that demands attention by playing off of our natural responses may cause a subconscious engagement that is less voluntary.*

CONCLUSION

In conclusion, the design and development of robots capable of establishing reciprocal relations with human beings (i.e. simulacra of presence), as in the case of social robots, should be carefully evaluated, since they may generate unbalanced power relations (e.g. new forms of social control) and have unknown effects on human users, (e.g. feelings of attachment, which, besides being non-authentic, may also be not voluntary).

As pointed out by Michael Decker, a human should be replaced by a robot only after a careful evaluation of several kinds of replaceability: first of all technical, socio-economic, legal and finally also ethical (Decker, 2007). According to Decker, ethical replaceability should be evaluated in terms of means-ends. On the one hand: 'whether the ends involved in the usc of autonomous robots are ethically justifiable' and on the other 'whether robots should or should not be implemented as means to achieve these ends' (*Ibid.*: 317). I would add to Decker's arguments one based on "responsibility", according to Baudrillard's meaning. Indeed, delegating an autonomous machine to fulfil a certain task, instead of ourselves, it could be a way to withdraw one from the moral and legal responsibility of presence, and in so doing to establish an unbalanced relation of power. The risks related to a loss of responsibility, in legal and moral terms, have already been addressed by a few scholars (Matthias 2004, Marino and Tamburrini 2006, Robins 1996).

In the specific case of social robots, the replication of reciprocal relations is a simulation based on predictions and it may never substitute the complexity and unpredictability of human relationships, at least at the time being. As pointed out by Turkle, complexity and unpredictability are fundamental aspects to human experience. Therefore, the effects caused by prolonged interaction with simulacra of presence may bring about unknown consequences to our understanding of being human: 'Relationship with computational creatures may be deeply compelling, perhaps educational, but they do not put us in touch with the complexity, contradiction, and limitations of the human life cycle. They do not teach us what we need to know about empathy, ambivalence, and life lived in shades...' (Turkle, 2007).

REFERENCES

Baudrillard, J. (1981). Requiem for the Media. In *For a Critique of the Political Economy of the Sign*. St. Louis, MO: Telos Press.

Baudrillard, J. (1983). *Simulations, Semiotext(e), Inc*. Columbia University.

Baudrillard, J. (1994). *Simulacra and Simulation: The Body in Theory: Histories of Cultural Materialism*. University of Michigan Press.

Borenstein, J., & Pearson, Y. (2010). Robot caregivers: Harbingers of expanded freedom for all? *Ethics and Information Technology*, *12*, 277–288. doi:10.1007/s10676-010-9236-4

Breazeal, C. (2002). *Desining socialble robots*. The MIT Press.

Cerqui, D., & Arras, K. O. (2003). Human Beings and Robots: Towards a Symbiosis? A 2000 People Survey. In *Proceedings of International Conference on Socio Political Informatics and Cybernetics* (PISTA'03). Orlando, FL: PISTA.

Copeland, R. (1990). The presence of mediation. *The Drama Review*, *34*(3).

Dario, P., Laschi, C., & Guglielmelli, E. (1999). Design and experiments on a personal robotic assistant. *Advanced Robotics*, *13*(2).

Decker, M. (2007). Can Humans Be Replaced by Autonomous Robots? Ethical Reflections in the Framework of an Interdisciplinary Technology Assessment. In *Proceedings of ICRA'07 Workshop on RoboEthics*. Rome, Italy: ICRA.

Fong, T., Nourbakhsh, I., & Dautenhahn, K. (2003). A survey of socially interactive robots. *Robotics and Autonomous Systems*, 42.

Jonas, H. (1985). *The imperative of responsibility: In search of an ethics for the technological age*. The University of Chicago Press.

Kahn, P. H. Jr, Ishiguro, Friedman, Kanda, Freier, Severson, & Miller. (2007). What is a Human? Toward Psychological Benchmarks in the Field of Human-Robot Interaction. *Interaction Studies: Social Behaviour and Communication in Biological and Artificial Systems*, *8*(3), 363–390. doi:10.1075/is.8.3.04kah

Lombardo, T. J. (1987). *The reciprocity of Perceiver and Environment: The evolution of James J. Gibson's ecological psychology*. Hillsdale, NJ: L. Erlbaum Associates.

Marino, D., & Tamburrini, G. (2006). Learning robots and human responsibility. *Int. Rev. Inform. Ethics, 6*.

Matthias, A. (2004). The responsibility gap: Ascribing responsibility for the actions of learning automata. *Ethics and Information Technology*, 6.

McLhuan, M. (1994). *Understanding Media: The Extensions of Man*. The MIT Press.

Merleau-Ponty, M. (1968). The Intertwining - The chiasm. In *The Visible and the Invisible*. Evanston, IL: Northwestern University Press.

Online Etymology Dictionary. (n.d.). Retrieved from http://www.etymonline.com

Parorobots. (n.d.). Retrieved from http://www.parorobots.com/

Patrick, Keith, & George. (2011). Robot ethics: Mapping the issues for a mechanized world. *Artificial Intelligence*, *175*, 5–6.

Reeves, B., & Nass, C. (1998). *The media equation: How people treat computers, television, and new media like real people and places*. CSLI Publications.

Robins, K. (1996). *Into the Image: Culture and politics in the field of vision*. London: Routledge. doi:10.4324/9780203440223

Sakamoto, D., & Hiroshi. (2009). Geminoid: Remote-Controlled Android System for Studying Human Presence. *Kansei Engineering International, 8*(1).

Salvini, P. (2006). Presence: A Network of Reciprocal Relations. In *Proceedings of PRESENCE 2006: The 9th Annual International Workshop on Presence*. PRESENCE.

Schmidt, C. T. A., & Kraemer, F. (2006). Robots, Dennett and the Autonomous: A Terminological Investigation. *Minds and Machines*, *16*(1). doi:10.1007/s11023-006-9014-6

Shibata, T., Kawaguchi, Y., & Wada, K. (2010). *Investigation on people living with Paro at home*. Paper presented at 2010 IEEE-RO-MAN. Viareggio, Italy.

Sparrow, R. (2002). The March of the Robot Dogs. *Ethics and Information Technology*, *4*(4), 305–318. doi:10.1023/A:1021386708994

Turkle, S. (2007). Authenticity in the age of digital companions. *Interaction Studies: Social Behaviour and Communication in Biological and Artificial Systems*, *8*(3). doi:10.1075/is.8.3.11tur

Veruggio, G., & Operto, F. (2008). Roboethics: Social and Ethical Implications of Robotics. In *Springer Handbook of Robotics*. Springer-Verlag. doi:10.1007/978-3-540-30301-5_65

Virilio, P. (1997). *Open Sky*. London: Verso.

Virilio, P. (2000). *From Modernism to Hypermodernism and Beyond*. Sage Publications.

Wiener, N. (1954). *The Human use of Human Beings – Cybernetics and society*. Da Capo Press.

WordNet. (n.d.). Retrieved from http://wordnetweb.princeton.edu/perl/webwn

KEY TERMS AND DEFINITIONS

Autonomy: According to the robotics engineer's standpoint, autonomy is the possibility to carry out a task without human help, for a prolonged period of time and in a dynamic and unstructured environment; according to the philosopher's point of view, autonomy is mainly related to free-will.

Ethical, Legal and Social Implications of Robotics (ELSI of Robotics): The study of the ethical, legal and social implications brought about by research and deployment of robotics technologies and systems.

Human Presence: The network of reciprocal relations characterising being in the world.

Power Relation: A relation characterised by an unbalanced distribution of powers or by the impossibility to respond back.

Reciprocity/Reciprocal Relations: A two-way relation or an exchange as opposed to a one-way relation or a distribution.

Robotic Mediation: The ways in which robotics technologies mediate, that is, stay in between human beings' actions and perceptions and the world.

Simulacrum: A copy of the real which becomes a substitution of the real.

Simulation: The action of substituting the real with its sign.

Social Robot: Robotics artifacts, often with a pet or human-like appearance, designed and developed to elicit and establish social relations with human beings.

Chapter 6
Technoethics:
Nature and Cases

Peter Heller
Manhattan College, USA

ABSTRACT

Technoethics relates to the impact of ethics in technology and technological change in biological, medical, military, engineering, and other applications. Accordingly, new questions arise about the moral right and wrong of corresponding technological issues. These, in turn, generate novel trade-offs, many of them controversial, involving the desirable versus undesirable ethical aspects of the new invention or innovation from a moral viewpoint. The discussion in this chapter suggests that frequently much can be said on both sides of an ethical argument and that therefore, at times, agonizing decisions must be made about which side has the greater moral merit based on numerous variables. The minicases sprinkled throughout the text and the longer automobile engineering case at the end are used as illustrations.

INTRODUCTION

The term "technology" (or its abbreviated prefix "techno") can be defined in a variety of ways. An acceptable one is that it is the systematic knowledge and applications that can describe any current activity closely related to science and engineering and viewed as providing the means of doing useful work. Technology may be embodied in a physical reality or in a tool, a method, technique, skill, or know-how.

"Ethics" may be defined as a code or set of principles by which people live. Ethics is about what is considered to be morally right and wrong. When people make ethical judgments, they are voicing prescriptive or normative statements about what ought to be done, about moral duty and obligation, not descriptive statements about facts. In short, ethics involves concepts like good, right, just, duty, obligation, freedom, and responsibility and how these relate to what people should do.

DOI: 10.4018/978-1-4666-6122-6.ch006

TECHNOLOGY-RELATED ETHICS OF THE AMISH

But having said all this there is no gainsaying the fact that not everyone is sold or is willing to replace religion or any philosophy hook, line, or sinker with technology, considering it a world view. For instance, limits on technology are the signature mark of the Amish sect even in the 21st Century. Riding in horse-drawn buggies and living unplugged from the public grid clearly distinguish the various Amish communities totaling over 275,000 from mainstream Americans and their love affair with gadgets and "new and improved" processes.

Yet, the Amish do not categorically condemn technology but only selectively. Nor are they technologically illiterate; quite the opposite, despite their (generally) eighth or 10th grade education. Rather, various Amish communities selectively sort out what might help or harm them. For instance, they limit the use of telephones to one often housed under a shed in a community so as not to discourage face-to-face contact. More significantly, the Amish modify and adapt technology in creative ways to fit their cultural values, social goals, and ethics. Similarly, they discourage a refrigerator in every kitchen since their communal ones keep their milk, their primary produce, fresh. While they tolerate solar panels to generate sufficient electricity to power a saw or sewing machine, they are not all plugged into the electric grid. Accordingly, at the end of the day, Amish technologies are diverse and ever-changing and selective (Kraybill et al., 2013). Minimally, they are discriminating. They ask whether any particular technology provides tangible benefits or damage to the community (Fecht, 2013). Modern antibiotics or diagnostic and surgical techniques are acceptable. So is the use of mowers, hay balers, or generator-powered wood-working equipment. The latter are sometimes driven by compressed air. On the other hand, technology used primarily for entertainment or merely convenience is rejected as frivolous or unnecessary.

Too, the technology that changes the relationship to the community or causes pride or attracts attention such as plastic surgery, automobiles, or computers is unacceptable. So would any other equipment or procedure that would alter the nature of the community itself or reduce face-to-face contact such as e-mail or undermine harmonious community solidarity, cultural identity, or ethics.

Furthermore, any technology threatening the religious precepts of the Amish is shunned. For instance, an item that is available only to some and not all. In other words, the Amish ask whether a particular piece of technology imposes human dominion in a useful and responsible way or not. So, a particular technology that tends to the sick, relieves misery or pain, or makes the environment more sustainable and attractive is acceptable (Herzfeld, 2009). Technology that pushes community members apart is not.

But there are other contrarians of technology, scholars among them, outside the Amish community. A good example would be the late Neil Postman of New York University. In an address to a 1998 conference, he declared that "The human dilemma is, as it has always been, [that} it is a delusion to believe that technological changes of our era have rendered irrelevant the wisdom of the ages and the sages" (Postman, 1998). And yet, as Postman admits elsewhere, "...we discover, always to our surprise, that it has ideas of its own, that it is quite capable not only of changing our habits but .of changing our habits of mind" (Kelly, 2010).

WORK-RELATED ETHICS

This section is not about employees putting their hands in the till in their workplace or various kinds of fraud or corruption or even sabotage. Rather, it is about the fact highlighted in an Op-Ed article in *The New York Times* entitled "Where Have All the Jobs Gone?" (Bernstein, 2013). This is about how an economy in transition is trying to adapt to a changing job market and how the shortage of employment is raising questions of social justice and equity.

Indeed, despite what is generally considered the end of the recession following the 2008-2009 economic meltdown, stepped-up automation and robotics, globalization and outsourcing, and the greater efficiency of management all of which raise productivity, it appears that the economy in the long run can do with fewer workers than in the past to produce an equal output and even more. And while the introduction of new technology in the past—say, the automobile—resulted in a sharp loss of employment among those who grew fodder for horses, the blacksmiths who shoed them, the army of city workers who cleaned up after them, those who stabled them, the harness makers, the carriage and cart makers, and those who cared for them, eventually the opening up of jobs in the motor vehicle industry more than made up for the demise of the horse-and-buggy era. The newly employed came to include the automobile engineers and draftsmen, the assembly line workers, the auto salesmen and showroom workers, the mechanics, the used vehicle disposal operators, the garages and parking lot attendants, and many others. A similar transition happened when, say, human-operated elevators were replaced by automated ones. Too, agricultural innovation put nearly the entire farming population out of work in the last two centuries. Yet, over time, the displaced farmers (or at least their progeny) managed to relocate and engage in other lines of work.

The burning question now is whether the same historical pattern is bound to repeat itself and the same share of workers will be involved in the technological changes and innovations of the 21st century. Increasingly, observers are coming around to concluding that the job crisis is permanent and will not drastically improve with an even further uptick of the business cycle. Among other reasons, the growing income inequality of the last few decades, mostly because of the larger proportion of highly paid, creative, innovative, thinkers outside the box in the labor mix. Voluminous imports of cheaper foreign goods exacerbate the trends. From a macroeconomic viewpoint, this could be explained by the fact that among the various factors of production, labor as one of the aggregates has been losing its importance while capital, more streamlined entrepreneurial and managerial skills, and technology have been rising in value.

One of the social consequences of the continuing high unemployment rate is that women, working primarily in the services, have weathered the trend better as a group than men in production. For the American economy has become a more service-oriented economy rather than a manufacturing (let alone agricultural) one where men on the assembly line used to be in the majority.

And if this situation, in the United States and the developed world more generally, is indeed slated to last because it is not just a hiccup in the business cycle, then government, through its public policy, may have to decide job distribution issues authoritatively such as whether any workers are entitled to more than one full-time slot since so many have none or whether men, still traditionally though decreasingly the family breadwinners, should have priority over women, or whether the elderly qualifying for Social Security and/or employer retirement benefits should not cede their place to similarly qualified younger workers.

For indeed, with sharply rising inequality between socioeconomic classes, rewards seem to be concentrated in ever fewer hands in the upper echelons. However, such novel and drastic government intervention for those who cherish the capitalistic free-enterprise system based essentially on competitive demand and supply would run into sharp political opposition. Among other things and debate, it would intensify the one about merit, judging by the intense positions taken by politicians and the public relating to such issues as income, estate, and corporate tax rates and the philosophical and economic bases of the argument themselves. In other words, this controversy on social equity having to do with job distribution could open a much larger can of worms (Bazerman and Tenbrunsel, 2011).

MEDICAL ETHICS

Nowadays, on average, people are doing better in terms of their health but possibly also doing worse because they are more aware of their condition than in the old days of the many things that could go wrong with them, thanks to the stories and advertisements in the media to a large extent. Thus, individuals have become sensitized to a large number of variables flowing from metrics and terms yielded by all kinds of tests and correlations involving many medical procedures with the help of technological instruments, often high-tech. Hence, X-rays, CT (Computer Tomography) scans, PET (Positron Emission Tomography) scans, MRI's (Magnetic Resonance Imaging), and ultrasound machines not to mention blood tests and other routine ones are in widespread use. This happens because imaging the body with the results fed into a computer has transformed the patient, that is to say the real patient, into a record.

The current state of affairs is the result of both the urge to accuracy and additional detail and to the greater ease of diagnostics and treatment. The machine makers who manufacture and sell this hardware and software and the medical providers who use them both generate considerable income in their respective functions, and so do the hospitals and laboratories. The medical caregivers are also intimidated by the possibility of costly malpractice lawsuits (serious or frivolous) by patients, which increases their insurance premiums at times to six-figure amounts a year. Yet, at the end of the day, one may wonder whether such basic improvements as the availability of clean water, better nutrition, improved hygiene and sanitation are not as important as the miracles of modern medicine in improving life expectancy and morbidity rates in the past few decades (McKeown, 1979). This is all the more true as medical errors and negligence by caregivers are conservatively estimated to result in over 200,000 deaths a year in the United States.

Anyway, the resulting cost in the United States is the highest in the world, both per capita and in the aggregate. The pressure on government finances and the inflation that may result are some of the relative effects of the system. Thus, whereas in the old days physicians used to examine the patient with their fingers and ears or maximally use a stethoscope to detect any malfunctions of the subject's heart, circulatory, respiratory and other biological systems, currently the assembly-line medicine promoted and speeded up by various health insurance schemes and incentives, encourages "throughput." The latter involves getting the best results from blood tests to colonoscopies and everything in between.

Yet, diagnostic readings may be in error, and there is unnecessary testing and exposure to radiation and other side-effects of particular procedures such as the false-positives shown by the scans or biopsies (Trouble at the lab, 2013). Accordingly, on the basis of some tests, unnecessary mastectomies and radical prostate surgeries (among others) are being performed. As a physician opposed to all this high-tech medicine put it, the patient is lost amid the wiring. But there is something else. Thanks to the various devices for inputting information, the doctor often is more focused on that drill than in listening to the patient. Frequently, physicians engage in multitasking—for instance, hitting the keys of their computer while trying to, at the insistence of patients, have some kind of give-and-take with them. On these occasions the doctor, focused on the computer, barely hears the patient. In short, there is a distancing between caregiver and subject. A number of doctors are willing to go along with this flow of events but not all (*Letters to the editor*, February 2011; March 2011). Accordingly, the very concept of being a "patient" is now challenged. Many assert that the "patient" has in fact become the "consumer" of a commodity provided by medical practitioners and health insurance companies driven by profit-making and at times profiteering rather than in providing a service (*Letters to the editor,* April 2011; Cohn, 2013).

In fact, in the latter article, professionals predict that physicians will be seeing fewer patients in future as they are transitioning increasingly to the use of such technology as IBM's "Watson." The latter robot is being programmed to make diagnoses and recommendations for the treatment of patients. This is one in a series of developments suggesting that technology is about to disrupt health care in the same way it has derailed many other industries and services. In fact, in countries like Brazil and India, machines are already starting to diagnose and recommend treatment for primary care, mathematically following evidence without the medical practitioners' built-in bias that highlight some symptoms rather than others (Cohn, 2013).

Still, the opposite view in this conversation is that while robots can help doctors in making more accurate medical diagnoses, they are not likely to be the solution to all health care woes since human interaction and compassion also contribute to effective healing. This is all the more true as there is seemingly an increasing feeling of alienation between health caregiver—focusing on "billing events"--and patients (Dines, 2013; Rauch, 2013)

This was not always so. In the early 20th Century, after the blood pressure cuff (or sphygmomanometer) was invented, a number of physicians declined to use it on the ground that it would reduce the sensitivity in their fingers which they had applied to the patient's pulse to determine blood pressure manually. Eventually, all physicians came to accept this practical gadget. For, when technological innovation occurs, it is rare for it not to be adopted sooner or later since it is hard to obstruct progress, to put the genie back into the bottle.

But we are not there anymore. Rather, as mentioned above, in "Dr. Machine Will See You Now...," physical diagnosis has become a lost art (*Letters to the editor*, February 2011; March 2011). The word "art" is used advisedly inasmuch as until the 19th Century, medicine was much more about the art than the science of healing, as the medical practitioner was little informed of the specifics of cause and effect and was much more conditioned by human relationships. A physician would talk with a patient and make a diagnosis largely from the narrative of and readily visible signs and symptoms evidenced by the patient. But this is no more as machines intrude increasingly between the healer and the subject. This has been a long process, perhaps beginning in 1819 when René Laennec introduced the stethoscope. It has set into motion a series of changes in medicine and in the connections among doctor, patient, and instruments that completely transformed concepts of health, disease, and human relationships as a function of technology.

Is there an ethical dimension in this current love affair between health providers and high-tech for the imaging, monitoring, and measurement of pathology? The answer is that the excess of technology may leave little room for such human qualities as empathy, kindness, even serendipity or intuition, and the internalizing of the patient's emotions and even suffering by the doctor. What has been happening is that the ease of ordering any diagnostic test has caused the caregiver's most basic medical skills in examining the body hands-on and in relating to the patient to atrophy. And in the meantime, the physicians have acquired such blind faith in the machines and are so pressed for time (a lot of it self-inflicted) that they at times do not even ask the patient to disrobe during an auscultation, and so may miss symptoms visible to the naked eye. Or they may overlook the possible effects of mind over body, of psychosomatics, by not listening more attentively to the patient, and more importantly to his or her emotional vibrations.

None of this means that patients are not pleased with the use of diagnostic and treatment devices employed by the medical profession such as the stethoscope or the ophthalmoscope or X-rays machines (Roentgen rays) or the electrocardiogram or the thermometer invented long before then. The difference is that these manually handled gadgets did not erect a wall between the physician and patient that today's high-tech (even blood work) often does when the test is administered without the physician even having to look at the patient or even being in the same room by using remote control. But even more important than the consequences of the growing remoteness between caregiver and subject is the loss of the ritual of the bedside drill, which in the old days had inspired the patient (and relatives) with trust and confidence.

Not providing the above raises a question of not only values but also of ethics with reference to professional responsibility. The joker in the pack with the increasing use of technology in medicine and more aggressive treatment by specialists is that it does not necessarily lead to longer life expectancy per se than the holistic approach of primary care physicians (Wennberg et al., 2008). This does not mean that many individuals are not alive today thanks to, say, heart transplants, kidney dialysis, or antibiotics. But in less serious dysfunctions than these, something has been lost along the way.

Also, many of the procedures are very expensive. Thus, the time is fast approaching when, because of these high costs (and widespread government deficits), some hard choices will have to be made about whether the use of these expensive medical interventions is warranted, especially given that some may be unnecessary, unsafe, unkind, or unwise (Jennett, 1986). For instance, the British health care system sets an age limit on who can have dialysis for kidney failure or a time limit on Aricept pills for early-stage dementia.

In fact, even in the United States, there was great controversy when a number of years ago a committee tried to prioritize applicants for the relatively small number of dialysis machines available then in Washington State on the basis of criteria that reflected the committee members' own suburban middle class values (Fox and Swazey, 1978). The dilemma will be even greater when new and even more expensive medical technologies become available so that possibly more serious rationing by insurance companies or government or even self-rationing because of high costs will have to kick in. Indeed, a 2011 survey by a team of physicians has found that about a quarter of prescriptions to patients with serious morbidity such as cancer are not filled and used because of the patients' lack of adequate insurance or financial inability to purchase them (at least, in a recession). Too, advanced technology, far from marginalizing the rituals of the past, makes them even more crucial in multiplying the possibilities of hospital infections (Tenner, 1997).

As an example of how complicated an ethical question can be, take the case raised by a deaf medical student at the Creighton University School of Medicine in Omaha, Nebraska. The institution has refused to provide the student, Michael Argenyi, age 26, with all the interpreting facilities that would allow him to take full advantage of the clinical instruction provided by the school. In his lawsuit, Argenyi argues that because of his impairment and the university's refusal to provide him with the full range of interpreting services, he has been unable to benefit maximally from his medical training. Understandably, the relevant legislation on Americans with disabilities does not specifically mandate the provision of every conceivable facility by workplace employers or educational institutions, still leaving the handicapped at a disadvantage.

By the end of his second year at Creighton in 2011, the student had brought suit against the university about his alleged right to complete his medical studies with every known facility, including a human interpreter. Until then, the university had provided an FM sound amplifier, note- takers for lectures, priority seating, and other aids but fewer facilities than the audiologist had recommended. Creighton University contended that Mr. Argenyi, even though legally deaf, could speak well enough to communicate with patients, who could be more hesitant to share personal information in front of a third party (the interpreter). The medical school has also argued that doctors needed to focus on the patient, not a third party, and to rely on visual cues to make a proper diagnosis (Eligon, 2013).

END OF LIFE ETHICS

Should patients with terminal ailments or conditions be told about them and given the option to end aggressive leading-edge treatment and settle for palliative or hospice care? A few states such as California in 2009 and New York in 2010 have decided the matter legally by mandating doctors to provide information and counseling about measures that may reduce both suffering and the cost of care at the end of life. Physicians often object to such laws as interfering with how they choose to deal with their patients and on the grounds that these laws do not take into account the nuances of doctor-patient relationships. To the extent that such laws are intended to make the end, when it is near, easier, to spare terminally ill patients from futile and possibly painful medical interventions, one may view them as ethical (Brody, 2011).

But how about assisted suicide? Oregon's 1997 Death with Dignity Act made it legal in the state (and in a very few other cases). The legislation mandates that the patient be at the end stage of a terminal disease and not have psychiatric dysfunctions like depression or dementia. Also, that the patient take the drugs used in the terminal procedure without help to insure that the act is voluntary from start to finish. Dr. Jack Kevorkian, also known as Dr. Death and who passed away in June, 2011, had ignored a number of these requirements including screening patients to determine whether they were in fact close to the end and not merely depressed. Anyway, he had made his "suicide machines," known as Mercitron and Thanatron, readily available to those who sought the device. Even worse, he flaunted them on CBS's "60 Minutes" program in front of millions of viewers so that the authorities were then compelled to take note of Dr. Kevorkian's challenge and convict him.

But things are rarely that clear. Indeed, it seems that more people and institutions (non-profit organizations, government authorities) are wondering why the terminally ill should not be allowed to meet death on their own terms rather at the end of prolonged agonies. Why should the dying not leave life with dignity instead of under inexorable pain or loss of their physical or mental faculties? On these grounds the pathologist, Dr. Kevorkian, though an eccentric, may be regarded as a humanitarian seer rather than a convicted criminal. At the very least, he helped raise the issue of the moral (and political and legal) correctness of euthanasia as no other (Kevorkian, 2011).

The dilemma of who should live or not happened famously in the case of Terri Schiavo, a 34 year-old woman who, going into respiratory and cardiac arrest in 1990, was in a persistent vegetative state kept alive by a feeding tube, respirator, and other technology. She exhibited no awareness of her environment. In 1998, her husband and legal guardian, Michael Schiavo, petitioned the court to order the removal of her feeding tube on the grounds that there was no longer any hope of his wife's recovery. His petition thrust him into a protracted legal dispute with Terri's parents, who resisted the discontinuation of their daughter's feeding tube. Michael Schiavo's response was that Terri would never have wanted to be kept alive if she were in a vegetative state without a brain function for all these years with no hope of recovery. However, there was no living will to that effect.

In September 2003, after five years of hearings and back-and-forth court orders, the feeding tube was removed. But that did not end matters. For a few days later, Governor Jeb Bush of Florida, the brother of the-then U.S. President George W.

Bush, ordered the tube reinserted in accordance with the state's law granting the governor that authority. The Florida court struck down the law as unconstitutional and ordered the tube removed. President Bush's attempt to transfer the matter to the federal Court of Appeals for the Eleventh District was also unsuccessful as the latter refused to overturn the Florida court's ruling.

Terry Schiavo died on March 31, 2005, in a Florida hospice at age 41, 15 years after becoming comatose and after a large amount of money spent on medical and legal fees to keep her barely alive. With America's aging population and the ballooning number of baby boomers and lengthening lifespans in the 21st Century, such cases involving a legal and ethical dilemma are likely to increase. So will additional new technology. But in the meantime, the view that human life is sacred and must be prolonged at all cost remains the basis for legal and most ethical decisions, at least in the United States. It is the default position.

Needless to say, the issue of distributive justice is raised in other contexts as well. For instance, when it comes to heart or kidney transplants, what order of priorities should be assigned to the relatively few organs available on the basis of what criteria? For even though the highest duty of the medical profession is to preserve human life, choices often have to be made. For instance, about ninety percent of the average expenses on a patient is incurred in his or her last year of life. Thus, a point may come when such outsized costs may result in the denial of medical care to younger prospects who have a much higher possibility of survival and life expectancy than an older patient with a terminal or incurable ailment. The practice of triage, or sorting out the injured or the sick as on a battlefield, may still be alive and well when it comes to fair or appropriate resource distribution. In other words, categorizing the injured or the sick as on a battlefield according to their dysfunction and focusing care, not on those who would survive anyway without it or on those who would die anyway regardless of assistance, but on those for whom treatment would make a difference. Possibly the good news in all this is that advances in bionics have led to the creation of an increasing number of artificial organs and limbs (Brumfiel, 2013).

ETHICS OF CONCEPTIVE TECHNOLOGY

In 1978 in England, the first baby was conceived through in-vitro ("in glass") fertilization. Since that time, the procedure has become nearly commonplace. In-vitro fertilization involves bringing sperm and egg together in a (usually glass) Petri dish. The sperm is supplied by a male and the egg by the prospective mother or by a female donor, anonymous or otherwise. After the fertilized egg is cultured for a few days and develops into an embryo with a few cells, it is transplanted into the uterus of the egg donor or a surrogate. Because each procedure results in the fertilization of more than one egg, there are many surplus embryos. While these can be frozen for various lengths of time, many are discarded. Alternatively, the embryos can be a source of stem cells.

Since there is an absence of ethical or legal consensus regarding this practice, it is not clear whether in case of a dispute between the two, the biological mother or the surrogate mother is entitled to the baby. The tendency has been to favor parents, but there is no uniform regulation regarding this matter, or the practice may be banned outright. The point is that whether through in-vitro fertilization, artificial insemination, or surrogate motherhood and the possibility of bypassing the usual methods of procreating children, it could happen that a child can now have two fathers, namely, the sperm donor as well as the man taking on the conventional social role of a father. Also, three mothers, namely, an egg donor, an impregnated surrogate with another woman's egg and embryo carrying it through pregnancy, and the woman occupying the social role of mother.

A different set of issues involves babies, born prematurely or not, afflicted with major medical problems that result in a lack of function, constant pain, and early death. As with end-of-life issues mentioned above, the value of preserving infants born with major deficiencies, destined for a life of pain and sickness, is a serious ethical problem. In the animal world, a number of species destroy their defective offspring soon after birth, but of course, at this state of development, human ethics precludes such an extreme measure.

And so, as with those afflicted with incurable and terminal ailments, there is an ethical or humane question about preserving the life of severely dysfunctional infants who have serious medical or mental issues or genetic abnormalities, in spending the resources necessary for sustaining their frequently short lives in preference to using the funds on more promising alternatives. Thus, parents have an ethical and moral and religious issue on their hands when a fetus is about to be born with, say, Down's syndrome: Should the fetus be aborted or be allowed to survive? As technology makes possible the discovering of these potential dysfunctions earlier and earlier, additional ethical questions will arise. For many, mostly pro-life supporters, still believe that life begins at conception when a sperm impregnates and egg and not a few months into pregnancy when a fetus becomes viable outside the womb. As of now, the latter is the position of the U.S. Supreme Court which so ruled in the case of Roe versus Wade (1973) making abortions legal in all states up to that time limit.

ETHICS OF CONTRACEPTIVE TECHNOLOGY

A different focus involving distributive justice was raised in connection with the sale of two contraceptive methods a few years ago in the developing, emerging world (Heller, 1992). That case study involved the non-biodegradable soft plastic-like capsules that are inserted into a woman's blood stream to release hormones steadily and so to make her infertile for five years (unless removed beforehand). The study found that it was not only inappropriate technology but also unethical to distribute Norplant (as the device is called) in parts of the world with relatively low hygienic conditions. Indeed, the product calls for inserting the capsule into (or removing it from) a woman's arm. There are still scarce medical resources to perform such a surgical procedure and few clinics available in the emerging world.

In the case of the second product, the French-made RU 486, a steroidal derivative belonging to the antiprogestin class of drugs that inhibits the action of progesterone, one of the hormones necessary for sustaining pregnancy, is involved. The ethical question here arises because the drug can be used both as a contraceptive (a "morning after" pill) but also as an abortifacient within a time limit. This may make the pill objectionable to those who believe that abortion is inherently wrong. Furthermore, as in the case of the first birth control pill marketed under the trade name Enovid beginning in the 1960's, the use of RU 486 suffers from the difficulties of distributing the requisite daily doses and of educating illiterate Third World women where problems of overpopulation exist. Yet, this is where the need for contraceptives is most acute. Too, since both Norplant and RU 486 are non-barrier contraceptive technologies, they do not protect against sexually transmitted diseases (STDs) or immune dysfunctions such as acquired immunodeficiency syndrome (AIDS) or human immunodeficiency virus (HIV), giving rise to an ethical question in its own right. Yet, even among those advocating pro-choice, there are some who have an ethical problem with the use of therapeutic abortion as a form of birth control, while even some of those supporting pro-life may accept abortion in the case of rape or incest.

ETHICS OF HUMAN CLONING

Cloning is the practice of deriving one organism from another organism through asexual reproduction (Volti, 2014). Because all of the clone's genetic material originates from the "parent," the clone is a genetic duplicate or near identical genetic copy which may occur either by fission or fusion. Placing the nucleus in the egg reprograms the deoxyribonucleic acid (DNA) to replicate the whole individual from which the nucleus was derived. In February 1997, Ian Wilmut and Keith H.S. Campbell of the Roslin Institute, Edinburgh, Scotland, cloned a female sheep by transferring the gene-containing nucleus from a single cell of an adult sheep's mammary gland into an egg cell whose own nucleus had been removed and discarded. The resulting combination cell then developed into an embryo, and eventually a lamb was born in the same way a normal cell produces offspring after being fertilized with a sperm cell. The outcome was the birth of Dolly, a genetic duplicate of the ewe from which the udder cell's nucleus had been taken (Friedel, 2007).

A similar process had been undertaken years before with fish and frogs. Too, mammal embryos had previously been split to produce artificial twins. Then came the successful cloning of monkeys and other animals. The reaction of ethicists, politicians, the media, and the public at large has been largely negative. On ethical grounds, clergy have argued that humans cannot do what they want with nature. In 1994, the U.S. National Advisory Board on Ethics in Reproduction called the whole idea of cloning oneself "bizarre…, narcissistic, and ethically impoverished." Others have wondered at the ethical purpose of even trying. Conservative opinion maker George Will questioned whether humanity, which is supposed to be an endless chain, not a series of mirrors, is now endangered. Others have been even more categorical, claiming that human cloning is so repulsive that is should be banned entirely, that only God is entitled to create life.

President Bill Clinton asked the National Bioethics Advisory Board to opine on the ethics of human cloning. In May 1997, the board called for extending the prohibition to use government funds to support work on human cloning, calling the process "morally unacceptable." Many other countries followed suit, and prohibitions on cloning research were widely imposed because many say that humans should not try to play God. Opponents also argue that recent progress with adult stem cells makes it unnecessary. Still other objections against human reproductive cloning are that it is liable to abuse; that it involves a person's right to individual autonomy and selfhood; that is allows eugenic selection; that it uses humans as a means; that cloned humans may suffer psychologically; and that there are safety concerns, especially an increasing risk of genetic malformation, serious ailments like cancer, or shortened lifespans.

However, supporters hold that cloning could serve a great many useful purposes, and further development of the technology could lead to much less alarming procedures. Indeed, it has been demonstrated that single cells could be removed from human embryos and induced to grow into new ones. If permitted to develop normally, the cells would grow into genetically identical adults. These would be duplicates, but only of each other, like identical twins, not of some pre-existing adult. Cloning supporters also say that a sufficient number of useful applications could result from cloning and should thus not be forbidden because of unproven concerns, that human suffering could be relieved, and that therefore it is ethically justified to use human cloning to create embryos as a source of tissue for transplantation.

Among other justifications of human reproductive cloning, some of them exotic, are that people should be free to make personal reproductive choices, to engage in scientific inquiry, eugenic selection, social utility, treatment of infertility, as a source of human cells or tissue or replacement, to do research in stem differentiation to provide an understanding of aging and oncogenesis (tumor formation), or to prevent a genetic dysfunction among many others. Most arguments, perhaps the strongest, involve cloning as a way of treating or avoiding disease.

Now, as mentioned earlier, so great is the demand that only about five percent of the organs needed in the United States for therapeutic purposes are available. Furthermore, this discrepancy between the number of potential recipients and donor organs is increasing annually. This is an argument for raising the availability of organs and tissues.

Thus, the most justified use of human cloning is arguably to obtain stem cells for the treatment of disease so that, for instance, more kidneys would become available for transplants, reversing the need for lifelong dialysis for those suffering from kidney failure with all the physical disabilities and financial challenges that it imposes.

But arguments against cloning also persist. For instance, would it be prudent to differ from the normal process of reproduction, which makes for a healthy diversity among the population? Could cloning be turned into a business for profit so that only the wealthy or those with outsized egos could afford it? Finally, what kind of people would want to clone themselves, those most desirable for society or not necessarily?

All of the foregoing would tend to suggest that in the field of technoethics, specifically bioethics, there are more questions than answers. This is typical of most human problems, and technoethics is no exception.

COMPUTER-RELATED ETHICS

The central question facing the information technology (IT) industry is how to reward innovation without stifling creativity, but there is no obvious answer to this conundrum and no consensus as to what constitutes ethical practice. This question arises in the context of the social problems created by computer crime and the problem of computer security software theft, that is to say, of intellectual property rights. Thus, computers make possible hacking and the creation of viruses; the invasion of privacy; as well as the many problems associated with the monitoring of work performance by supervisors; or by employers' reading the electronic mail of their employees. Even the deleting of embarrassing e-mail messages may not be enough to insure privacy because there may be somewhere or other a backup copy of the original message. History has recorded many instances causing embarrassment to all those concerned as a result. Thus, society may be vulnerable as evidenced in 2010 and 2011 by the numerous confidential electronic messages disseminated for the entire world to see by WikiLeaks. Or consider the disclosure of highly classified government documents by the likes of Edward J. Snowden in 2013. Consequently, both perpetrators and the authorities compete to stay one step ahead of the game.

Lately, such websites as Facebook, Twitter, Linked-In and Skype video chats have made it easy to flirt with strangers and engage in sexual fantasy by, say, sexting suggestive text messages and photos like Anthony Weiner, a former U.S. Congressman who ran for mayor of New York City in 2013. That is, technology creates new rules for romance encouraging infidelity--just as the automobile had done in the 20th Century. This raises new questions of morality and ethics flowing from technology, the subject pertinent to technoethics (Parker-Pope, 2011; Heller, 2012a).

Computer users and software developers tend to have very different ethical positions on the issue of copying software, while government regulations are still inadequate to deal with these various problems. Several calls for improved network ethics have been just as ineffective In the meantime, critics like Julian Assange, editor-in-chief of WikiLeaks, are projecting a gloomy future for the new digital age. In his book, *Cyberpunks:*

Freedom and the Future of the Internet, Assange envisions a world in which Google and govern-

ments can intrude in the communications of every human being on the globe with few exceptions as even democracies, invidiously subverted by surveillance technologies, are not liberating but imprisoning the world. (Assange, 2012; Assange, 2013)

But in a more fundamental way, artificial intelligence (AI) critics have been asking whether AI is a proper goal for humanity or whether it is fair to replace wage-earners by technology in so many tasks when there is so much unemployment around. This is because employment for many different reasons, whether monetary, psychological, or social, is so central to the lives of so many people, especially men. Indeed, there has been steady concern about computers making outsourcing possible so that more and more types of office, manufacturing, and professional work can be performed more inexpensively or better abroad causing the loss of American jobs. Concerns about the constant use of computers with reference to such health hazards as repetitive strain injury (RSI) and the consequent lawsuits against employers and computer vendors have been frequent.

Such ethical dilemmas as to whether or not to copy software, and other issues of what is right or wrong, honesty, loyalty, responsibility, confidentiality, trust, accountability, security, and fairness are ongoing problems for which few answers – legal, social, ethical, or other – exist. Media such as e-mail, bulletin boards, faxes, mobile phones, and others generate ethical issues concerning user identity, authenticity, and other matters involving such forms of communications. Too, powerful capabilities such as monitoring, surveillance, database searching, and other means can be used for spying, as can scans of all kinds. IT transforms links between people, depersonalizing human contact and replacing it with instant, paperless and faceless communications.

Too, the unreliability of IT creates uncertainties and a whole series of ethical choices for those who operate complex systems or build them, often neglectful or unaware of the consequences of their system designs. As mentioned, copying software easily is more likely than not an infringement of the developer's intellectual property rights.

Also, is computer hacking harmless mischief or the equivalent of burglary, fraud, and theft? Are the brains behind the WikiLeaks disclosures online guilty of criminal behavior and/or merely unethical or neither? Or are they malicious? Or perhaps unpatriotic? Alternatively, consider the role of computers in military training. Thanks to computer games, soldiers can now be taught to "kill" a virtual enemy online. Having been desensitized in this way, it may be easier for them to dispose of the real enemy. This kind of training may be more effective than the old-fashioned way of demonizing "Fritz," "the Japs," or "the gooks" to motivate troops to do something that does not come naturally to most people given their religious or ethical upbringing or merely their basic human instincts of respecting life.

And yet, a psychologist at the Massachusetts Institute of Technology writes, "I think of how little resistance this generation will offer to the placement of robots in nursing homes" (Turkle, 2011). Translation: Constant contact with technology such as robots has dehumanized young people, especially with reference to an older generation to which they find it hard to relate.

In a sense, history is repeating itself. When repetitive assembly-line production was introduced in the early 20th hCentury by the likes of Frederick Winslow Taylor and Henry Ford, the question arose as to what the resulting numbing work during long shifts was doing to the workers, who disliked the new system of production because of the intense division of labor and possibly even what it was doing to their minds. Now, while computers in factories can be used to enhance the quality of workers' lives, to upgrade or re-skill the workforce, it can also make workers outright redundant or again transform them into deskilled, degraded machine-minders. human robots pressing buttons in a depersonalized environment. Anyone who has witnessed, say, the conversion of typewriters and teletype machines in offices to the computer electronic system with the speedups and repetitive stress injuries that followed will feel a resonant chord.

And while certain professional organizations such as the Association of Computing Machinery (ACM) and the Institute of Electrical and Electronics Engineers (IEEE) have issued and revised their codes of ethics, there is little enforcement machinery in place so that hacking, scans, fraud, virus creation, misappropriation of intellectual property, and other obvious legal and/or ethical violations are still commonplace and are likely to continue into the foreseeable future. People are people with all their virtues and vices, and codes rarely modify behavior (at least in the short run) any more than criminal law prevents all felonies. In both cases, greed and the possibility of profit or the enhancement of their power by governments are too hard to resist, especially in materialistically inclined societies constantly craving for the acquisition of more things and wealth.

AUTOMOBILE ENGINEERING ETHICS: THE FORD PINTO CASE

This case study is based to a large extent on four previous publications to which the current author acknowledges great indebtedness (Birsch and Fielder, 1994; Cullen, 1987; DeGeorge, 1990; Strobel, 1980). The topic, namely, being ethical, has to do with meeting standards of conduct concerning the welfare of others and their right to make their own choices rather than the producer making larger profit. The case will illustrate how ethical issues pose difficult questions about right and wrong for which existing norms provide no clear answer.

Thus, the Ford Pinto case raises several different matters sampled by Ruth Schwartz Cowan (Cowan, 1997). The most obvious ones concern choices of individuals within a corporation when faced with ethically significant decisions. Thus, do the actions of organizations or of their employees, policies, and institutional arrangements reflect an adequate respect for the rights and welfare of persons affected by their products, whether they are morally right or wrong? The focus is on evaluation that takes into account the complexity of all ethically relevant principles and circumstances and the controversy because of the lack of clear guidelines.

For instance, did the Ford Automobile Company design an adequately safe automobile? Were Ford's actions in responding to the design deficiencies ethical? Does the procedure through which automobiles are blueprinted and engineered pay sufficient attention to driver and passenger safety? Should more Ford engineers who believed that the Pinto's fuel system was unsafe under conditions of a rear-end collision have blown the whistle on the Ford Motor Company? Indeed, would it have been ethical for the designers and production engineers to go public to the media or to the authorities or to the consumers to warn them about a possibly lethally dangerous automobile? The answer seems to be yes, especially if the higher-ups, when apprised of the dangers, failed to take appropriate action. But the engineers may have worried about their professional future (say, avoidance of blacklisting) or their ability to survive if they made waves and antagonized their bosses. Is the profit-driven capitalist economic system as modified by government regulations adequate to insure the safety level that consumers have a right to expect? What conceptual tools can the ethical rules provide to answer these questions as in the Ford Pinto case?

In analyzing actual ethical problems, it is primarily a matter of articulating standards of right and wrong and applying them in light of particular circumstances. For instance, in the field of biomedical research involving human subjects, such guidelines as the informed consent of subjects used in experiments, the equitable distribution of burdens and benefits, the confidentiality of data, the compensation of, say, medical research subjects or their heirs for accidental death or injury, review procedures, and other criteria have been clearly laid out (Council for International Organizations of Medical Sciences/World Health Organization, 1993). Two relevant variables are obligations and rights.

Obligation is a concept that occurs in law, ethics, religion, and social life. It designates requirements of conduct in a particular tradition or practice. Thus, legal obligations arise from the framework of law governing human action. Religious traditions have specific requirements of conduct for their adherents. Membership in a community creates social obligations to neighbors and others. That is, ethical obligations arise primarily from one's participation in a community with shared norms of behavior.

Since any new technology is a social experiment, drivers and passengers in the case of an automobile are in a sense its subjects. For beyond the company's testing prior to release, only the varied conditions of actual daily driving under different conditions will demonstrate how viable a vehicle is. So, to what extent are drivers and passengers entitled to adequate safety provisions in the design of cars? That is, what are the ethical obligations of designers, organizations, and institutions with respect to the rights of automobile travelers while balancing safety with risk? How should managers trade off automobile safety with their obligation to generate a decent return on shareholders' investments in the company and their own desire to ascend the corporate ladder? Many ethical disagreements pit one moral claim against another and the question of which claim is more deserving.

The Ford Pinto was planned since the late 1960's to compete especially with the Japanese Honda and Toyota sub-compact cars but also with the German Volkswagen Beetle, the British Rover, and General Motors' Chevrolet Vega. The Pinto had problems from the start. The car was found to be unable to sustain a front-end collision without the windshield breaking—the so-called retention test. Thus, a quick-fix solution was adopted, namely, the drive train was moved backward. As a result, the differential was moved very close to the gas tank (Camps. 1981). This arrangement was supposed to solve the windshield problem in the following way: Some of the kinetic energy generated in a crash was channeled away from the windshield via the driveshaft to the differential housing. But this caused contact with the gas tank in this very small automobile. Now, whereas windshield retention tests were mandated by Federal Motor Vehicle Safety Standard No. 212, fuel system integrity was not at that time (1970).

The consequence was that gas tanks began exploding on impact in traffic. Too, to produce the car so cheaply (at $2,000 each), less than highest quality materials were used—again, all in the name of solving the major problem of all which, in the words of Henry Ford II, was "... making more money."

Furthermore, the new tests for certification by Ford were held without dummies in the driver's and passenger's seats and by stripping the car down to its minimum weight. That is how the Pinto was certified as being safe in front-end collisions by the car company.

Then, there were leaking fuel and fires after relatively low-speed, rear-end collisions. The major reason was that the car's gasoline tank had been mounted behind the rear axle so that the edge of the tank was located only six inches from the rear bumper, whose shock absorptive capacity in this case was minimal. The alternative location of the fuel tank would have been above the rear axle. But this site would have diminished the trunk space of the diminutive Pinto and raised the automobile's center of gravity, jeopardizing its stability. The latter might adversely have affected its handling. The-then Vice President and later Ford President Lee Iacocca, who had taken special interest in the Pinto from the start and had mandated that the vehicle be conceived, designed, and produced in a record time of some 25 months instead of the usual 43 for new automobile models (including the time to build a whole new set of tools and equipment to put the car together), had weighed in on the behind-the-rear-axle position for the gas tank.

But Ford's own report of a crash test of a 1971 Pinto two-door sedan performed in October 1970 showed that when the car was hit at 21.5 mph, the filler pipe was dislodged, causing fluid to leak out while a bolt on the differential housing (gearbox) punctured the dislocated gasoline tank. A spark from colliding metals could start a serious fire because several bolts on the differential housing were barely inches away from the gas tank. Indeed the rear-end impact in a collision pushed the tank in the direction of the bolts (and other sharp metal parts). Still, Ford decided not to place a rubber bladder inside the Pinto's tank, though such an addition had proved successful in fixed-barrier crash tests up to 26 mph. This was for reasons of economy. Indeed, Lee Iacocca had also set a cap of 2,000 lbs on the weight of the vehicle.

As the number of deaths by fire and burn injuries started to add up with the 1970 to 1976 Pinto models, appropriate changes based on the results of several on-going tests, whose results were usually not released to the general public, could have led to a much safer car. For instance, the addition of a 1-pound plastic shield to protect the gas tank from the rear or by lining the fuel tank with the aforementioned rubber bladder at a few dollars each would have improved the automobile's safety. Indeed, while production was still in progress, crash tests continued to show the high probability of explosion and fire and the ripping out of the gas filler pipe in rear-end collisions. Anyway, it was estimated that when 3 million 1970-1976 model Pintos were on the road, the total number of accidents would range from the low number of 23 a year given by Ford all the way to some 900 proposed by experts, consumerist groups, and some automobile writers.

The numerous test results provided to Ford engineers and management should have informed them of the car's potential danger and that the Pinto's safety could be significantly upgraded by a few minor changes. But these improvements would have increased the unit's total production cost by $11 each, Ford's figure considered to be highly inflated by the company's critics. This amount, multiplied by several million units, would have added substantially to the production cost of the car. The aggregate exceeded Ford's estimates of the monetary damages it would have had to pay to victims or survivors of crashes or their heirs. The computation was based on the federal government's NHTSA (National Highway Traffic Safety Administration) figures that in 1972 a human life was worth $200,000 each in lost working time, a serious injury $67,000, and a damaged vehicle $700 on the average. So, the charge has constantly been made that to Ford's management, even after they were notified of the Pinto prototype's poor test results preceding full-scale production, decisions were all about cost-benefit analysis, that is to say about profits before people. Indeed, in April 1971, Ford executives wrote a memorandum recommending that neither a bladder nor a flak vest be adopted till NHTSA's Standard 301 went into effect in September 1976 in order to save the company an estimated $20.9 million.

What is not clear is how much pressure the production engineers and up on the hierarchical ladder put on top management and, if necessary, why they did not blow the whistle outside the Ford Motor Company to inform the driver and passenger, the public, the media, and the authorities that the Pinto could be very dangerous if not deadly to people's health. And by the same token, why did the government's NHTSA wait for six years to finally issue its more stringent Standard 301 in 1976 (Federal Motor Vehicle Safety Standard 301) mandating that all 1977 cars had to be able to withstand a 30-mile-per hour rear moving-barrier crash with little gas leakage (1 oz/minute), a regulation which was considerably weaker than an earlier proposal which had aroused the strong opposition of the automobile industry and its allies. In other words, pressure group politics kicked in.

Ford then ordered the "voluntary" recall of half a million model 1971-1976 Pintos (and 30,000 Mercury Bobcats) to end public concern and debate over the matter (and do damage control to its declining reputation). A plastic shield was used for the gas tank to prevent it from being punctured by the differential housing, an improved sealing cap was placed on the tank, and a longer fuel tank filler pipe was substituted at an estimated total cost of $20 million to $40 million. As mentioned earlier, all of these changes were strenuously opposed by Ford, other automobile companies, and various big business lobbies like the U.S. Chamber of Commerce. Indeed, according to Ford's computations, the total compensation and damages payable to potential victims-survivors or their kin would have been about a third of the aggregate cost of recalls and retrofitting to build a safer car. These figures had been used by Ford with the NHTSA at the outset to enable it to roll out the Pinto in record time to meet the competition so that production was already under way when the crash tests subsequently configured that the car had a dangerously inflammable gas tank.

CONCLUSION

Synoptically, then, the ethical aspects of the Ford Pinto case are necessarily blurred. But here is a final word of caution: Ethical issues rarely present themselves in pure form. In other words, there may be mitigating circumstances to justify decisions in the application of technology, not just aggravating ones to condemn them. In short, there are rarely clear-cut good guys and bad guys, heroes or villains, in such matters.

Furthermore, as Ford's top executive at the time, Lee Iacocca, famously said: "Safety never sold any cars." For those buying them often look to other qualities such as inexpensive price, low gas consumption and maintenance costs, or frivolous ones such as style rather than safety. This fact has been known at least since the introduction of automobile seat belts, a safety feature that has saved many lives. While it was still optional, at first few passengers were willing to voluntarily acquire a seat belt for a few dollars each. Only when the government mandated it by law did seat belts become universal, especially with the built-in horns and whistles or automatic locks to make car drivers and passengers actually use them.

How about government's excessively slow performance (six years) before issuing safety directives to address the Pinto's rear-end collision tendency to catch fire and burn or maim those inside? Well, its reason (some would call it excuse) was that it needed further independent tests to prove the risks involved in riding the Pinto before issuing binding directives to automobile companies to spend millions for the recalls and retrofitting of their cars or redesign costs.

So, to reiterate, there are as many questions as there are answers. Be that as it may, here is the conclusion of the case. In September 1976, Standard 301 applicable to rear-end collisions went into effect. Cars were not to leak more than an ounce a minute after being impacted by a barrier moving at 30 mph. The 1977 Pinto, modified with a plastic shield between the differential housing and the tank, an improved filler pipe, and a sturdier bumper eventually met the standard after an earlier recall of some 300,000 1977 Pintos. In June 1978, Ford announced the "voluntary" recall of an additional 1.5 million Pintos and similar subcompacts (1970 to 1976 models) "to end public concern." Their retrofitting cost many millions. In July 1978, Lee Iacocca was dismissed by Ford for unspecified reasons. But in the meantime, a number of lawsuits were brought against the automobile company. The largest was in February

1978 when a mother-and-son twosome sustained fatal burns in the case of the mother and over 90 percent of his body in the case of the son. The damages award to the family came to $128.5 million (Camps, 1981).

Ultimately, after numerous accidents, fatalities and injuries, lawsuits, settlements for damages, and a public relations black-eye, Ford discontinued production of its Pinto in July 1980. In all fairness to automobile producers, however, it should be mentioned that variations on this profit-driven theme occur in other industries as well. Who has not heard the story of how a reputable baby formula producer at one point had its salesgirls dressed in white like nurses going around African villages trying to get new mothers to use baby powdered milk formula, ostensibly to free them from other chores instead of breast-feeding? It turned out that the mothers who went for the product scrimped on the dosage to make the powder last longer and thereby save money. In the process, many of the women used polluted water to "clean" the feeding bottles. The result was the undernourishment and sickness of many babies. For where there is only poverty and squalor, an artificial substitute for breast milk is profitable for the vendor but a bad choice for the users (Heller, 2012b),

REFERENCES

Assange, J. et al. (2012). *Cyberpunks: Freedom and the future of the internet.* Academic Press.

Bazerman, M. H., & Tenbrunsel, A. E. (2011, April). Ethical breakdowns. *Harvard Business Review, 89*(4), 58–65. PMID:21510519

Bernstein, A. J. (2013, May 4). Where have all the jobs gone? *The New York Times,* p. A21.

Birsch, D., & Fielder, J. H. (1994). *The Ford Pinto case: A study in applied ethics, business, and technology*. Albany, NY: University of New York Press.

Brody, J. E. (2011, June 7). Law on end-of-life care rankles doctors. *The New York Times*, p. D7.

Brumfiel, G. (2013). Replaceable you. *Smithsonian, 44*(5), 68–76.

Bryan, J. (1986). *High-technology medicine: Benefits and burdens*. Oxford, UK: Oxford University Press.

Camps, F. (1981). Warning an auto company about an unsafe design. In A. F. Westin (Ed.), *Whistle blowing: Loyalty and dissent in the corporation* (pp. 19–129). New York: McGraw-Hill Book Company.

Cohn, J. (2013). The robot will see you now. *Atlantic (Boston, Mass.), 311*(2), 58–67.

Council for International Organizations of Medical Sciences/World Health Organization (CIOMS/WHO). (1993). *International ethical guidelines for biomedical research involving human subjects.* Geneva, Switzerland: CIOMS/WHO.

Cowan, R. S. (1997). *A social history of American technology*. New York: Oxford University Press.

Cullen, F. T. et al. (1987). *Corporate crime under attack: The Ford Pinto case and beyond.* Cincinnati, OH: Anderson.

DeGeorge, R. T. (1990). *Business ethics* (3rd ed.). New York: Macmillan.

Dines, M. (2013). The conversation: Responses and reverberations. *Atlantic (Boston, Mass.), 311*(4), 10–11.

Eligon, J. (2013, August 20). Deaf student, denied interpreter by medical school, draws focus of advocates. *The New York Times*, p. A11.

Fecht, S. (2013). Where the Amish and technology meet. *Popular Mechanics, 190*(2), 17.

Fox, R. C., & Swazey, J. P. (2002). *The courage to fail: A social view of organ transplants and dialysis*. Piscataway, NJ: Transaction Publishers.

Friedel, S. D. (2007). *A culture of improvement: Technology and the Western millennium*. Cambridge, MA: The MIT Press.

Heller, P. B. (1992). Ethical considerations in the international distribution of new contraceptive technology: The case of Norplant and RU 486. In J. R. Wilcox (Ed.), *The internationalization of American business: Ethical issues and cases* (pp. 57–75). New York: McGraw-Hill.

Heller, P. B. (2012a). Technoethics. *International Journal of Technoethics*, *3*(1), 14–27. doi:10.4018/jte.2012010102

Heller, P. B. (2012b). *Technology and Society Reader: Case Studies* (rev. ed.). Lanham, MD: University Press of America.

Herzfeld, N. (2009). *Technology and religion: Remaining human in a co-created world*. West Conshohocken, PA: Templeton Press.

Jennett, B. (1986). *High-technology medicine: Burdens and benefits* (p. 174). Oxford: Oxford Universty Press.

Kelly, K. (2010). *What technology wants* (p. 194). New York: Penguin Books.

Kevorkian, J. Obituary. (2011). *The Economist, 399*(8737), 88.

Kevorkian, J. Obituary (2011), *The Economist*, 399(8737), 88.

Kraybill, D. B., Johnson-Weiner, K. M., & Nolt, S. M. (2013). *The Amish*. Baltimore, MD: The Johns Hopkins University Press.

Letters to the editor, Dr. machine will see you now.... (2011, March 2). *The New York Times*, p. A24.

Letters to the editor, The special doctor-patient relationship (2011, April 26). *The New York Times,* p. A24.

Letters to the editor, Treat the patient not the CT scan (2011, February 28). *The New York Times,* p. WK10.

McKeown, T. (1979). *The role of medicine: Dream, mirage, or nemesis?* Princeton, NJ: Princeton University Press.

New York Times. (2013, June 2). The banality of don't be evil. *The New York Times*, p. SR4.

Parker-Pope, T. (2011, June 14). Digital flirting: Easy to gct caught. *The New York Times,* p. D5.

Postman, N. (2013). Five things we need to know about technological change. In *Technologies, social media, and society* (18th ed., pp. 3–6). New York: McGraw-Hill.

Rauch, J. (2013). How not to die. *Atlantic (Boston, Mass.), 311*(4), 64–69.

Strobel, L. P. (1980). *Reckless homicide? Ford's Pinto trial*. South Bend, IN: And Books.

Tenner, E. (1997). *Why things bite back: Technology and the revenge of unintended consequences*. New York: Random House Vintage.

Trouble at the Lab. (2013). *The Economist, 409*(8858), 26-30.

Turkle, S. O. (2011). *Alone together: Why we expect more from technology and less from each other*. New York: Basic Books.

Volti, R. (2014). *Society and technological change* (7th ed.). New York: Worth Publishers.

Wennberg, J. E., Fisher, E. S., Goodman, D. C., & Skinner, E. S. (2008). *Tracking the care of patients with severe chronic illness*. NH, Lebanon: The Dartmouth Institute for Health Policy and Clinical Practice.

KEY TERMS AND DEFINITIONS

Cloning: The practice of deriving one organism from another organism through asexual reproduction from a single sexually produced individual.

Computer Ethics: Moral questions having to do with abuse, fraud, hacking, whistle-blowing, ownership, and privacy among others.

Engineering Ethics: Moral questions having to do with the engineer's loyalty, whether to the standards of his/her profession, to the employer, the consuming public, the authorities, the engineer's professional judgment, or existential needs (see the Ford Pinto case).

High Technology (or High Tech): A form of advanced technology that is contrasted with low or intermediate technology tending to generate specific moral issues as in the case of dangerous products or weapons of mass destruction.

Moral: Having to do with right and wrong.

Robot: A labor-saving automaton, often a computer-driven efficient device with seemingly human intelligence.

Science: Verifiable knowledge often based on experiments.

Technoethics: Considerations of moral right and wrong applied to the industrial arts.

Technology: Applied or industrial science, systematic knowledge of the industrial arts responding to practical problems.

Section 2
Un/Ethical Waves in Cyberspace

Chapter 7
Sacrificing Credibility for Sleaze: Mainstream Media's Use of Tabloidization

Jenn Burleson Mackay
Virginia Tech, USA

Erica Bailey
Pennsylvania State University, USA

ABSTRACT

This chapter uses an experiment to analyze how mainstream journalists' use of sensationalized or tabloid-style writing techniques affect the credibility of online news. Participants read four news stories and rated their credibility using McCroskey's Source Credibility Scale. Participants found stories written with a tabloid style less credible than more traditional stories. Soft news stories written with a tabloidized style were rated more credible than hard news stories that also had a tabloidized style. Results suggest that online news media may damage their credibility by using tabloidized writing techniques to increase readership. Furthermore, participants were less likely to enjoy stories written in a tabloidized style. The authors conclude by utilizing act utilitarianism to argue that tabloidized writing is an unethical journalistic technique.

INTRODUCTION

There's a journalistic world where old-school objectivity fights for existence against dramatic disasters and fuzzy features. It is a place where journalistic ethics might take a backseat, while reporters or editors douse newspapers with sleaze and entertainment. With little more than a creative selection of verbs and a thirst for a sizzling story, American journalists can venture into tabloid territory.

This technique of spicing up mainstream media news often is called tabloidization. The exact definition of the term varies from one scholar to the next, but it is viewed as a method for attaining audiences in an ever-competitive media

DOI: 10.4018/978-1-4666-6122-6.ch007

environment. Tabloidization has been described as dumbing down the news by giving consumers the stories that they want rather than providing useful public service information (Nice, 2007). The writing tone in these tabloidized stories is designed to be stimulating and exciting (McLachlan & Golding, 2000). Tabloization results in lower journalistic standards, less hard news, and more soft, sensational, or entertaining stories (Kurtz, 1993). It is far too simplistic a notion to assume that tabloidization is a completely negative practice, however. The method also might be utilized to increase the audience of the media and to increase their knowledge of news and information (Gans, 2009).

This chapter is an effort to understand the effects of online news tabloidization on credibility. The study will examine how readers evaluate the credibility of stories written with a tabloidized format compared to how they rate stories written with a more traditional journalistic style. A more traditional reporter's story would stick to the facts and get to the point of the story, whereas a tabloidized story might include sleazy wording or unnecessary intimate details designed to grab the reader's attention rather than inform him or her. In addition, the researchers will consider whether tabloidization is more accepted in certain types of stories, such as feature pieces, as compared to hard news stories. We predict that participants will report a higher level of enjoyment of tabloidized content than non-tabloidized content.

The study also asks how the media should respond to tabloidization pressures. In addition to studying participant responses to tabloidized content, this study will apply normative ethical theory to the tabloidization of the media. By using act utilititarianism, the researchers will examine how journalists should address the challenges of the new media climate and whether utilizing tabloidization for media survival is an acceptable ethical practice.

BACKGROUND

Tabloidization can result from competition, technology, and the desire for circulation. News organizations essentially have restructured, redesigned, and degraded their content in an effort to survive. Tabloidization can be viewed as a way of appealing to advertisers above other competing interests (Conboy, 2006). The deregulation of the media is one reason that current affairs programs have become increasingly commercialized. The programs have reverted to a hybrid format that is a combination of news and reality television (Baker, 2006). Not all countries are experiencing the same level of tabloidization, however. Research suggests that the increase of democracy in Brazil resulted in a less tabloidized, and less politically affiliated media (Porto, 2007). An increase in media privatization and deregulation in India, on the other hand, has led to more entertainment news and fewer public service-oriented stories (Rao & Johal, 2006).

Signs of tabloidization can be found in some of the earliest mass media (Tulloch, 2000). Scholars have cited several characteristics as signs of tabloidization. It has been described as an increase in entertainment coverage, a decrease in long stories, an increase in shorter stories with illustrations, and an increase in informal language within news stories. Frank Esser (1999) says the concept "implies a 'contamination' of the so-called serious media by adopting the 'tabloid agenda'" (p. 293). Howard Kurtz (1993) argues that tabloidization results in lower journalistic standards, an increase in sleazy tales in place of thoughtful political pieces, and a transition as to what journalists feel audiences need to know about a politician's capabilities for office. An overall increase in visual elements such as photographs and large headlines are another sign of the tabloidization process (Rooney, 2000).

While many news organizations are developing a tabloid style, mainstream news organizations tend to avoid using the term "tabloid". Journalists have cited the complexities of trying to maintain a serious journalism tradition while reverting to shorter, less complex news stories (Rowe, 2011). Although males and females do not acknowledge it to the same degree, audiences say they enjoy reading tabloid stories. Both sports and celebrity gossip pieces are considered particularly entertaining (Johansson, 2008).

Some scholars argue that those decrying the tabloidization of the media should consider that the media is a complex entity featuring multiple types of journalism. One should not try to distinguish merely between tabloid media and traditional media (Peters, 2011; Harrington, 2008). Furthermore, traditional news stories share some of the same characteristics as tabloid stories. For example, both types of media embrace emotion (Peters, 2011). Tabloidized stories also do not necessarily contain more emotional elements than more traditional news stories (Uribe & Gumter, 2007). While much scholarship criticizes the tabloidization of the media, there is some indication that tabloidization may have positive effects, such as giving the media a way to reach the readers of teen magazines (Nice, 2007). Using journalistic methods to reach out to large groups of people may not be at the heart of the journalist's professional interest, but in a sense, tabloidization or "popularization" could, and perhaps should, be used to increase the audience for news (Gans, 2000, p. 21). Ensuring that the public consumes plenty of news is increasingly important as we operate in a growing global society. It is important for the media to find a way to reach everyone. As Gans (2000) explains: "Although the news media cannot chase away real and imagined demons, and there are other limits to what they can do and whom they can reach, they can try harder to get the news out to the people who may unknowingly need it most" (p. 27). Nonetheless, there are scholars who suggest that sensationalized news has negative effects on audiences. Meijer (2013) conducted an ethnographic inquiry into the relationship between sensationalized news coverage and residents' views of reality using content analyses. Her research revealed that sensationalized and one-dimensional news coverage led readers to feel as though they had lost touch of reality.

Credibility

Information is more believable when it comes from a highly credible source. People are more likely to experience a greater belief change when they receive a message from a highly credible source (Hovland & Weiss, 1951).. Articles from more traditional newspapers such as the *Washington Post* are perceived as more credible than articles from tabloid publications, such as the *National Enquirer.* Highly credible sources also are more believable and accurate (Kaufman, Stasson & Hart, 1999). Audience judgment of credibility may depend on the situation to which credibility is accessed (Kim & Pasadeos, 2007). Media consumers who are liberal and trust the government are more like to trust the media (Lee, 2010).

Credibility often is measured in terms of either the source or the medium. Source credibility emphasizes the reputation of a specific individual who relays a media message (Hovland & Weiss, 1951). Medium credibility emphasizes the reputation of the medium as a whole (Gaziano & McGrath, 1996). This article focuses primarily on source credibility. Source credibility can be measured in terms of multiple dimensions: competence, care, and trustworthiness (McCroskey & Teven, 1999).

Online Credibility

Criteria such as credibility, trustworthiness, and accuracy are among the most important content criteria for online news editors (Gladney, Shapiro, & Castalodo, 2007). Internet credibility research often compares online publications to traditional media outlets. People who trust mainstream media are more likely to use mainstream media websites.

Those who are more skeptical of the media are more likely to visit nonmainstream media websites (Tsfati, 2007). Some research suggests that the Internet is as credible as most or all other media (Flanagan & Metzger, 2000; Johnson, Kaye, Bichard & Wong, 2007). However, the credibility of online news websites has dropped during the past decade. One possible explanation for that drop is that audiences are becoming more web savvy (Johnson & Kaye, 2010).

Several factors have been connected to credibility. A higher level of religiosity is associated with a greater trust in the news media than a lower religiosity (Golan & Kiousis, 2010). Age also affects how credibility is rated. College students rated television news more credible than online sources, whereas older participants found online news more credible than television formats (Bucy, 2003). In Westerwick's (2013) investigation of the effects of Web site design appeal, website sponsorship, and search engine ranking on viewers' judgment of online information credibility, website sponsorship was shown to be a significant predictor of credibility ratings. Specifically, a credible website sponsor was shown to be more important for overall information credibility than the website design and the website's ranking within a Google search.

Credibility is important to bloggers (Perlmutter & Schoen, 2007). Heavy blog users find blogs more credible than those who rely less on the medium and individuals interested in politics find blogs somewhat credible (Johnson et. al., 2007; Trammell, Porter, Chung, & Kim, 2006). Blogs are not considered to be fair, but blog users view bias as a strength of the medium (Johnson et al., 2007).

News Type

Research frequently refers to the differences in hard and soft news, but scholarship presents multiple definitions for the terms. Hard news has been defined as stories that need to be reported immediately, whereas soft news refers to stories that do not require such timely publication (Shoemaker & Cohen, 2006). Hard news also has been defined as stories that are important to the audience's understanding of public affairs, while soft news has been classified as "news that typically is more personality-centered, less time-bound, more practical, and more incident-based than other news" (Patterson, 2000, 3-4).

Research indicates that the media are increasing their soft news coverage and decreasing hard news coverage. This increase in soft, entertaining coverage often is associated with newspaper conglomerate and *USA Today* owner Gannett. The *Arkansas Gazette*'s feature coverage expanded after Gannett purchased the paper (Plopper, 1991). *USA Today*'s style includes an over-emphasis on unimportant soft news stories (Logan, 1985-1996). Media practitioners may rely on this soft news coverage to appeal to larger audiences (Scott & Gobertz, 1992). Soft news attracts viewers who might not otherwise watch the news (Baum, 2002). Individuals who prefer hard news to soft news are better informed about several news issues (Prior, 2003). Some research has suggested that audiences can learn from soft news, but that learning might be limited (Baum, 2002). Boczkowski and Peer (2011) propose that there is a 'choice gap' in what journalists' choose to display as top news stories, and what consumers choose to read. They found that journalists choices in stories tended to be soft in terms of subject matter, and that consumers tended to prefer more hard news stories.

Utilitarianism

John Stuart Mill typically is associated with the normative theory utilitarianism, which proposes "actions are right in proportion as they tend to promote happiness; wrong as they tend to produce the reverse of happiness" (Mill, 2009, p. 55). There are two types of utilitarianism. Act utilitarianism is considered a direct moral theory that emphasizes analyzing actions on the basis of each individual act or situation (Crisp, 1997, p. 102).

Rule utilitarianism is an indirect moral theory in that actions are judged on how they conform to something else, namely a rule or norm. The two approaches can resolve ethical issues differently. For example, consider a situation in which a journalist has promised a source – a drug dealer who sells cocaine to youth – confidentiality in exchange for information about the city's drug crisis. In terms of act utilitarianism, breaking that promise might yield the most utility because it allows the authorities to arrest someone who is hurting the city's youth. Act utilitarianism, however, would not suggest that the journalist should always reveal the name of the source. The decision would be based on the circumstances of each act. The act utilitarian approach to decision making asks about the balance of harm versus good that will result from a specific action and endorses the action from which the most utility will result. Rule utilitarianism would consider what would happen if the decision to break a promise to the source became a rule that was consistently followed every time the journalist faced the same set of circumstances. Typically, breaking promises is thought to be morally impermissible. Rule utilitarianism requires me to abide by that rule and keep that promise. Therefore, rule utilitarianism would suggest that the journalists should not break the promise of confidentiality to the source.

For this study, act utilitarianism, rather than rule utilitarianism, is used as a guide for evaluating study results, as ethical decisions in journalism are seldom answered by simply adhering to a rule. We believe that the complexities of practicing journalism require each situation to be evaluated by its specific circumstances.

Utilitarianism requires behavior that promotes happiness. When measuring which action will yield maximum utility, everyone's pleasure is of equal value, including the one doing the act. This begs the question of what is to be considered pleasurable. According to Mill, ''pleasures of the intellect, of the feeling and imagination, and of the moral sentiments,'' should be given ''a much higher value as pleasures than those of mere sensation'' (Mill, 2009, p. 56). Put another way, "it is better to be a dissatisfied Einstein than a blissfully happy ignoramus" (Christians, 2007, p. 114). This clarification of what is pleasurable makes it easy to see how utilitarianism is applicable to journalism studies as the aim of journalists is to disseminate information in order to educate the public about important matters.

In journalism ethics research, some scholars argue that utilitarianism is not an adequate method for evaluating ethical decisions in media studies. They argue that other ethical schools of thought are better suited for studying the complexity of media and journalism ethics (Christians, 2007; Quinn, 2007; Ward, 2007). Others argue that utilitarianism often is simplified and therefore misunderstood (Elliot, 2007; Peck, 2006). Nevertheless, utilitarianism is commonly used in journalism ethics research (Christians, 2007; Elliot, 2007; Peck, 2006; Ward, 2007).

Utilitarianism has been used to examine the practice of unnamed sourcing in journalism. Utilitarianism supports the use of unnamed sourcing in particular instances where a greater aggregate good would be achieved. The journalist is required to evaluate whether using the unnamed source is a greater benefit than the harm that can be caused by anonymous sources (Duffy & Freeman, 2011). Utilitarianism has been used to justify the documentary filmmakers' responsibility to prevent harm to his or her subjects. Since documentary filmmaking involves human interaction and is a practice of social institution, filmmakers have a moral obligation to avoid harming subjects (Maccarone, 2010). Some scholars also have argued that utilitarianism is practiced in newsrooms. Editors can justify using gory and controversial images in the news with editors and journalists arguing that gruesome images from car accidents will encourage most people to drive more carefully (Lester, 1999).

While audiences may enjoy reading tabloizied content, this chapter will use act utilitarianism to demonstrate that tabloidization generally is not an ethical journalistic technique. There may be instances where it can be justified, but generally speaking, its usage will not serve the greatest good for the greatest number of people.

MAIN FOCUS OF THE CHAPTER

Issues, Controversies, Problems

Scholars have suggested that sensationalized, tabloid-style writing has cluttered the content of the mainstream media. An experiment was designed to test how audiences respond to these writing techniques. Our goal was to evaluate how tabloidized writing influences the credibility and the enjoyment of news content. We initially posed two research questions and one hypothesis to address these issues. Next, we examined tabloidization through the lens of act utilitarianism.

The media value their own credibility (Gladney et al., 2007). People are more likely to believe information that comes from a credible source (Hovland & Weiss, 1951). Furthermore, audiences find traditional news media more credible that tabloid media (Kaufman, Stasson, & Hart, 1999). Scholarship has not explained what relationship credibility may have to tabloidization.

There also is some question as to whether the news media's use of tabloidization techniques has a positive or negative effect on audiences. On one hand, the increase in personalization and sleaze can be viewed as corruption of professional journalism (Esser, 1999; Kurtz, 1993). On the other hand, tabloidization may help audiences want to consume news (Gans, 2000; Nice 2007).

To explore the relationship between the growing use of tabloidization in the media and credibility, the following research question was proposed:

RQ1: Will participants exposed to tabloidized stories rate the sources of those stories as having lower levels of credibility than participants exposed to traditional news stories?

The media appears to be shifting emphasis from hard news to soft news, which is a characteristic of tabloidization (Esser, 1999). Some scholars believe that the use of soft news lowers the quality of information (Logan, 1985-1986). This type of news, however, may help the media to attract new audiences (Baum, 2002). If credibility is important to audiences and both tabloidization and the increase of soft news are attempts to garner audiences, then the news media would want audiences to find those soft news stories credible. Yet, the relationship between credibility and different types of news has not been adequately explored. Thus, the following research question was posed:

RQ2: Will readers perceive tabloidized hard news stories as more or less credible than soft news stories?

Scholarship suggests that audiences enjoy reading tabloidized content (Johansson, 2008). Research also suggests that editors have increased the tabloidization of their content because they are responding to audience demands for that type of content. This combined research suggests that audiences enjoy tabloidized material. This leads to the following hypothesis:

H1: Participants who read tabloidized stories will report a higher level of enjoyment than participants that read traditional, untabloidized stories.

Participants for the study came from undergraduate classes at one researcher's institution. Of the 74 participants, about 34% were male and 66% were female. About 53% of the participants were in the traditional group and 47% were in the tabloid group. In regards to media consumption, 33.8% of participants reported they got their news from the newspaper, 35.1% the Internet, 8.1% radio, 36.5% network news, and 31.1% cable news.

Participants viewed a series of mock news stories. The stimuli that participants viewed depended on whether they were randomly assigned to the traditional group or the tabloidized group.

Stimuli

Four fictional news stories were written by one of the researchers. Each story was written in two formats. The first format followed more traditional news writing techniques, such as what is commonly called the inverted pyramid style, a writing format in which the most important information is located at the beginning of the story and the least important details are at the end of the story. The other version kept the same basic facts, but was written in a sensational style. The guidelines for the tabloidized writing style came from Kurtz (1993) and Esser (1999). They included: a decrease in conventional hard news coverage and an increase in soft news and sleaze; a broadened view of the information readers need to determine if a candidate is fit for a political office; lower journalistic standards overall.

Before the experiment began, five current/former journalists were asked to read all eight stories and to evaluate that the stories accurately reflected writing techniques in the way that the researcher intended. The journalists generally agreed that the tabloidized stories were more sensationalized than the traditional stories.

Two mock Web pages were designed for a publication called "The Daily News." The pages were identical, including the story headlines. The only difference was the content of the stories.

The first treatment group stories, which were written in the conventional journalistic style, were placed on one website. The second treatment group's stories were placed on another site. The files for the websites were loaded onto the desktops of several computers.

Both groups of stories were based on the same topics and general information. They also had the same headlines. Two stories used a hard news format and were considered timeline. In other words, these stories contained information that would become outdated if the stories were not published immediately. The other two stories were soft, feature-style stories, the sort of stories that fulfill an entertainment value and do not need to be published immediately.

The first story on the Web pages focused on a kidnapping incident, where authorities suspected was related to an Internet stalking. The first version of the story was presented in a traditional inverted pyramid format with the most recent information at the beginning of the story, such as the fact that a teen is missing, and her age. In the second version of the story, the same facts are presented, but they are interwoven into a narrative depicting what the stalker and teen might have been thinking and doing just before the disappearance. The second version is highly sensationalized and uses sordid details to grab the reader's attention.

The second story focused on a city council meeting in which officials voted to ban cell phones in downtown businesses. The first version of the story begins with what is traditionally considered the most important information: the fact that the ban was put into action. The least important information, which is related to a heated discussion that occurred after the meeting, is left for the end of the story. In the tabloidized version of the story, the order is changed, allowing the story to immediately emphasize the fight that occurred after the meeting rather than the information about the new law.

The third story was a profile of a candidate for sheriff. Both stories were written as profiles, with the emphasis on informing the reader who the candidate is and what he hopes to do if elected. The tabloidized version highlights details that readers do not necessarily need in order to judge whether the candidate is qualified for the office, such as a physical description of the candidate. Several lines in the story emphasize the candidate's popularity among women. The information is kept to a minimum in the traditional group version of the story.

The final story was a feature about a new teen dance club. The traditional story mostly gave the bare-bones information necessary for the story, whereas the second story was more detailed and sensationalized. Unlike the traditional story, the tabloidized story is more of a narrative that explains how events evolved throughout the night, whereas the traditional group story just gives the essential facts of the story, along with a sprinkling of quotes.

The Instrument

Students were asked to evaluate the source of each news story using James McCroskey's Source Credibility scale (McCroskey & Teven, 1999). The scale measures three dimensions of credibility: competence; caring and goodwill; trustworthiness. The scale has been used in several studies. Reliability of the scale was tested using Cronbach's Alpha. The overall competence alpha was .88; caring, .77; trust, .90.

Participants were asked several demographic questions at the end of the survey. They also were asked to state how enjoyable the stories were and how likely they were to seek out stories similar to the ones that they just read.

Procedure

Participants were recruited from communication classes and randomly assigned to the traditional or the tabloidized group. Participants were asked to read the first story and then answer a series of questions. Then, they were asked to read the next story and answer a series of questions, and so on.

Results

On any of the credibility measures, such as competence, a higher score indicated a higher level of credibility. A probability of .05 was used for all statistical results.

RQ1: Will participants exposed to tabloidized stories rate the sources of those stories as having lower levels of credibility than participants exposed to traditional news stories?

There were statistically significant differences in how the traditional group and the tabloid group rated all three types of credibility. The means on each of the credibility scales are shown in Table 1. The participants reading the traditional stories found the news sources to be more competent with a higher competence factor score than the participants reading the tabloidized stories, t(72)=3.41, p. = .00. The sources of the traditional stories also demonstrated more care, as shown with the caring and goodwill scores, t(72)=2.3, p.=.02, and more trust with a trustworthiness score, t(72)=2.28, p. =.03.

There also was statistical significance in how the traditional participants and the tabloidized group evaluated the credibility of the hard news and the soft news stories. Participants in the traditional group found the sources of the hard news stories more competent than the tabloidized group, t(72)=4.45, p. =.00. The traditional group also found the sources of the soft news stories more competent than the tabloidized group, but the relationship was not statistically significant. Traditional participants found the sources of the hard news stories more trustworthy than the tabloidized group, t(72)=2.61, p.=.01. The traditional group also found the sources of the soft news stories more trustworthy than the tabloidized group, but the relationship was not statistically significant. Traditional participants found the sources of the hard news stories more caring than the tabloidized group, t(72)=2.47, p.=.016. The traditional group found the soft news stories more credible than the tabloidized group, but the relationship was not statistically significant.

RQ2: Will readers perceive tabloidized hard news stories as more or less credible than soft news stories?

There was a statistically significant difference in how the tabloidized group rated the credibility

Table 1. Credibility scores for traditional versus tabloidized participants

Credibility Type	Traditional	Tabloidized
Competence overall	165.21*	145.14*
Competence hard news	57.64*	48.97*
Competence soft news	57.49	54.97
Caring/goodwill overall	111.38*	102.69*
Caring/goodwill hard news	55.85*	50.86*
Caring/goodwill soft news	55.54	51.83
Trustworthiness overall	120.77*	111.54*
Trustworthiness hard news	60.77*	54.74*
Trustworthiness soft news	60.00	56.80

*Denotes statistical significance

of the hard and soft news stories, $t(35) = -3.04$, p.=01. Participants found the sources of the soft news stories (54.97) more competent than the authors of the hard news stories (48.97). There was no statistical significance on the caring or the trust factor.

H1: Participants who read tabloidized stories will report a higher level of enjoyment than participants that read traditional, untabloidized stories.

This hypothesis was tested by analyzing participant responses to two questions. One question asked how much the participant "enjoyed reading the style of writing" used on the website. The other question asked how likely the participant would be to "seek out stories that were written in a style similar" to the stories that they viewed. The relationships were tested with Chi-square tests.

There was statistical significance in how participants responded to both questions. Participants in the traditional group were more likely to report a higher level of enjoyment in reading the stories than the tabloidized participants, $\chi^2(3)=9.776$, $p=.02$. The results are shown in Table 2.

As shown in Table 3, participants in the traditional group also were more likely to report that they would seek out stories written in a similar style than tabloidized participants, $\chi^2(3)=10.739$, $p=.01$.

The hypothesis was not supported.

Discussion

Tabloidization has become a tool for mainstream journalists (Conboy, 2006). It has been interpreted as a method for attracting younger audiences and as an agent for increasing the audience size overall (Nice, 2007; Gans, 2000). The participants in this study suggested, however, that tabloidized material posted online is less credible than content written in a more traditional journalistic style. Participants were more accepting of sleazy and lurid details when the story was soft news-oriented. While the news media may offer tabloidized content because they believe that audiences want that writing style, the participants in the tabloidized group were less likely to report that they enjoyed reading the stories in this study than participants who read stories written in a more traditional style. While it certainly is possible that participants did not want to admit that they enjoyed reading trashy content and two questions are a limited means for accessing enjoyment, this result raises a question as to whether journalists truly are providing audiences with the content that readers

Table 2. Enjoyment responses for traditional versus tabloidized participants

Enjoyment level	Traditional (%)	Tabloidized (%)
Very enjoyable	38.5	8.6
Somewhat enjoyable	51.3	68.6
Not very enjoyable	7.7	20
Not enjoyable	2.6	2.9

want the most. Other scholars have attempted to tackle that question, but more research is needed in this important area. The study did not attempt to ascertain the reasons as to why participants enjoyed a particular style of writing more than another. That is another important question that could be addressed in future research.

Utilitarianism

In addition to studying how audiences responded to tabloidized content, we sought to understand how journalists should address the tabloidization temptation in terms of normative ethical theory. We have chosen to address that issue by applying act utilitarianism to our study results.

Utilitarianism is interested in maximizing utility. The theory suggests that the highest pleasure is that of intellect (Mill, 2009). When applying act utilitarianism, specific actions are evaluated based on the utility that will result from said action. Actual outcomes are difficult to predict, so utilitarians determine the best action based on expected outcomes. If a utilitarian approach is used, journalists should be concerned with providing the highest pleasure for the majority of people. Traditional news stories offer important details and relevant information that might be buried in tabloidized stories. Utilitarianism would suggest that news readers experience more intellectual pleasure by reading stories that stick to the facts rather than sleazy content.

The participants in this study reported that they were more likely to enjoy and seek out stories written in a traditional style rather than tabloidized content. From the perspective of utilitarianism, these participants were more likely to seek out a higher pleasure rather than a lower pleasure. The participants were more interested in seeking out the material that would maximize utility. Therefore, the best way for journalists to accommodate the enjoyment and overall pleasure needs of the audience is to supply readers with more traditional content more so than tabloidized material. There is some question as to why tabloid content is popular even though participants reported that they were less likely to seek out this type of story. One explanation for the popularity of tabloidized content may be that some people do not have enthusiasm for higher pleasures. Mill argues that people who chose not to enjoy higher

Table 3. Seek similar content responses for traditional versus tabloidized participants

Seek Likelihood	Traditional (%)	Tabloidized (%)
Very likely	30.8	17.1
Somewhat likely	48.7	28.6
Not very likely	20.5	42.9
Would not	0	11.4

pleasures have become incapable of experiencing such pleasures (Crisp, 1997).

Research suggests that people value credibility (Kaufman et al., 1999). Journalists also value credibility. The concept is found throughout journalism textbooks and news organization ethics codes. Participants in this study were more likely to find the journalists who wrote with a more traditional style to be more credible on the basis of competence, trustworthiness, and good will. Utilitarianism calls for one (the journalist) to be impartial to his own happiness when considering actions that yield the most utility (Mill, 2009). Utilitarianism would suggest, therefore, that it would be in the best interest of both the journalist and the audience if content is presented in a more traditional style.

When considering utilitarianism, the news writing styles are relevant not just to journalists, but news consumers as well. Mill suggested that people have a duty "to form the truest opinions they can" (Mill, 1859, p. 102). Being informed on matters of concern to society is the best way to do that. Elliot (2007) furthers this point by emphasizing the importance of public discussion in providing the opportunity to compare one's beliefs against others. As such, in most cases, utilitarianism would suggest that journalists should use a more traditional writing style that would appeal to the highest possible intellectual pleasure. There may be instances, however, when the only way to reach the masses may be to use a more tabloidized writing style that will grab the attention of audiences and ensure that they pay attention to a story. Act utilitarianism recognizes that there is no single one-size-fits-all answer to all complex ethical issues. It suggests that decisions should be made on the basis of specific circumstances that the journalist faces. Perhaps the tabloidized style should be utilized when evidence suggests that audiences who have failed to take note of a similar story that was important. Participants here suggested that they were more comfortable with the credibility of writers when they read tabloidized soft news stories than they were with tabloidized hard news stories. This also might suggest that the journalist can have more flexibility in choosing to use a more tabloidized writing style for those feature stories than with breaking news stories, such as homocides and car accidents.

Solutions and Recommendations

The study reconfirms what Mill (2009) suggests, individuals prefer higher intellectual pleasures. Sleazy, sensationalized writing styles appeal to lower pleasures. The participants in this study reported less enjoyment of tabloidized stories. They also tended to find the sources of tabloidized stories less credibile than the sources of more traditional stories.

This study suggests that journalists typically should not rely on tabloidized writing techniques. That writing style may help journalists to grab reader attention, but it interferes with the credibility of the news story and the intellectual pleasures. Journalists should revert to more traditional, tried and true writing styles. That will make their work more ethical and it will satisfy audiences. Futhermore, it will improve the overall credibility of journalists.

FUTURE RESEARCH DIRECTIONS

Future research should consider how tabloidized writing techniques have seeped into mainstream journalism through social media. Certainly, social media outlets such as Twitter have sped up the news consumption process, with information released in some cases before journalists have had time to confirm the facts. Might journalists also reveal more sleazy details through social media than they do in more traditional media? Is tabloidized information, which never makes it into mainstream news stories, revealed in social media or have mainstream news stories become more sleazy as social media has become a significant journalistic tool? Scholars also should consider the type of stories that are being shared through social media. Are audiences more likely to share tabloidized stories through social media sites than they are to share more traditional stories? Are tabloidized stories more frequently shared through some social media sites than they are on other social media sites?

CONCLUSION

New technology gives audiences more access to information than ever before, forcing journalists to compete with one another on a global stage. It can be tempting to add sleazy details to a story in the hopes of garnering some repeat posts on Facebook and other social networking sites. Nonetheless, the participants in this study suggested that they were not fond of tabloidized content and they were less likely to trust the writers of lurid stories. Futhermore, utilitarianism suggests that tabloidization generally is not an ethical way to reach audiences. As journalists continue to ponder their survival in a market that is over-saturated with free content, perhaps they should take a moment to consider not just what they write about, but how they write stories. Perhaps this will help them to appease the ever-technologically savvy audience.

ACKNOWLEDGMENT

The material in this chapter was previously published as a journal article. The original citation is: Mackay, J.B and Bailey, E. (2012). Succulent Sins, Personalized Politics, and Mainstream Media's Tabloidization Temptation. *International Journal of Technoethics*, 3(4), 41-53.

REFERENCES

Baker, S. (2006). The changing face of current affairs programmes in New Zealand, the United States and Britain 1984-2004. *Communication Journal of New Zealand*, *7*, 1–22.

Baum, M. A. (2003). Soft news and political knowledge: Evidence of absence or absence of evidence. *Political Communication*, *20*, 173–190. doi:10.1080/10584600390211181

Boczkowski, P. J., & Peer, L. (2011). The choice gap: The divergent online news preferences of journalists and consumers. *The Journal of Communication*, *61*(5), 857–876. doi:10.1111/j.1460-2466.2011.01582.x

Bucy, E. P. (2003). Media credibility reconsidered: Synergy effects between on-air and online news. *Journalism & Mass Communication Quarterly*, *80*, 247–264. doi:10.1177/107769900308000202

Christians, C. G. (2007). Utilitarianism in media ethics and its discontents. *Journal of Mass Media Ethics*, *22*, 113–131. doi:10.1080/08900520701315640

Conboy, M. (2006). *Tabloid Britain: Constructing a community through language*. New York, NY: Routledge.

Costera Meijer, I. (2013). When news hurts. *Journalism Studies*, *14*(1), 13–28. doi:10.1080/1461670X.2012.662398

Crisp, R. (1997). *Mill on Utilitarianism*. London, UK: Routledge.

Duffy, M. J., & Freeman, C. P. (2011). *Unnamed sources: A utilitarian exploration of their justification and guidelines for limited use*. Retrieved from http://digitalarchive.gsu.edu/cgi/viewcontent.cgi?filename=0&article=1012&context=communication_facpub&type=additional

Elliott, D. (2007). Getting Mill right. *Journal of Mass Media Ethics*, *22*, 100–112. doi:10.1080/08900520701315806

Esser, F. (1999). Tabloidization' of news. *European Journal of Communication, 14*, 291–324. doi:10.1177/026732319901400300l

Flanagin, A. J., & Metzer, M. J. (2000). Perceptions of Internet information credibility. *Journalism & Mass Communication Quarterly, 77*, 515–540. doi:10.1177/107769900007700304

Gans, H. J. (2009). Can popularization help the news media. In B. Zeilzer (Ed.), *The changing faces of journalism tabloidization, technology, and truthiness* (pp. 17–28). New York, NY: Routledge.

Gaziano, C., & McGrath, K. (1986). Measuring the concept of credibility. *The Journalism Quarterly, 63*, 451–462. doi:10.1177/107769908606300301

Gladney, G. A., Shaprio, I., & Castaldo, J. (2007). Online editors rate web news quality criteria. *Newspaper Research Journal, 28*, 55–69.

Golan, G. J., & Kiousis, S. K. (2010). Religion, media credibility, and support for democracy in the Arab world. *Journal of Media and Religion, 9*, 84–98. doi:10.1080/15348421003738793

Harrington, S. (2008). Popular news in the 21st century. *Journalism, 9*, 266–284. doi:10.1177/1464884907089008

Hovland, C. I., & Weiss, W. (1951). The influence of source credibility on communication effectiveness. *Public Opinion Quarterly, 15*, 635–650. doi:10.1086/266350

Johansson, S. (2008). Gossip, sport and pretty girls. *Journalism Practice, 2*, 402–413. doi:10.1080/17512780802281131

Johnson, T. J., & Kaye, B. K. (2010). Still cruising and believing? An analysis of online credibility across three presidential campaigns. *The American Behavioral Scientist, 54*, 57–77. doi:10.1177/0002764210376311

Johnson, T. J., Kaye, B. K., Bichard, S. L., & Wong, W. J. (2007). Every blog has its day: Politically-interested internet users' perceptions of blog credibility. *Journal of Computer-Mediated Communication, 13*, 100–122. doi:10.1111/j.1083-6101.2007.00388.x

Kaufman, D., Stasson, M. F., & Hart, J. W. (1999). Are the tabloids always wrong or is that just what we think? Need for cognition and perceptions of articles in print media. *Journal of Applied Social Psychology, 29*, 1984–1997. doi:10.1111/j.1559-1816.1999.tb00160.x

Kim, K. S., & Pasadeos, Y. (2007). Study of partisan news readers reveals hostile media perceptions of balanced stories. *Newspaper Research Journal, 28*, 99–106.

Kurtz, H. (1993). *Media circus: The trouble with America's newspapers*. New York, NY: Times Books.

Lee, T. (2010). Why they don't trust the media: An examination of factors predicting trust. *The American Behavioral Scientist, 54*, 8–21. doi:10.1177/0002764210376308

Logan, R. A. (1985-1986). USA Today's innovations and their impact on journalism ethics. *Journal of Mass Media Ethics, 1*, 74-87.

Maccarone, E. M. (2010). Ethical responsibilities to subjects and documentary filmmaking. *Journal of Mass Media Ethics, 25*, 192–206. doi:10.1080/08900523.2010.497025

McCroskey, J. C., & Teven, J. J. (1999). Goodwill: A reexamination of the construct and its measurement. *Communication Monographs, 66*, 90–103. doi:10.1080/03637759909376464

McLachlan, S., & Golding, P. (2000). Tabloidization in the British press: A quantitative investigation into changes in British Newspapers. In C. Sparks, & J. Tulloch (Eds.), *Tabloid tales: Global debates over media standards* (pp. 76–90). Lanham, MD: Rowman and Littlefield.

Melican, D. B., & Dixon, T. L. (2008). News on the net: Credibility, selective exposure, and racial prejudice. *Communication Research*, *35*, 151–168. doi:10.1177/0093650207313157

Mill, J. S. (2009). Utilitarianism. R. Crisp (Ed.), J.S. Mill Utilitarianism. New York, NY: Oxford.

Nice, L. (2007). Tabloidization and the teen market. *Journalism Studies*, *8*, 117–136. doi:10.1080/14616700601056882

Patterson, T. E. (2000). *Doing well and doing good: How soft news are shrinking the news audience and weakening democracy*. Cambridge, MA: Harvard University Press.

Peck, L. A. (2006). A ''fool satisfied''? Journalists and Mill's principle of utility. *Journalism and Mass Communication Educator*, *61*, 205–213. doi:10.1177/107769580606100207

Perlmutter, D. D., & Schoen, M. (2007). If I break a rule, what do I do, fire myself? Ethics codes of independent blogs. *Journal of Mass Media Ethics*, *22*, 37–48. doi:10.1080/08900520701315269

Peters, C. (2011). Emotion aside or emotional side? Crafting an 'experience of involvement' in the news. *Journalism*, *12*, 297–316. doi:10.1177/1464884910388224

Porto, M. (2007). TV news and political change in Brazil: The impact of democratization on TV Globo's journalism. *Journalism*, *8*, 263–284. doi:10.1177/1464884907078656

Prior, M. (2003). Any good news in soft news? The impact of soft news preference on political knowledge. *Political Communication*, *20*, 149–171. doi:10.1080/10584600390211172

Rao, S., & Johal, N. S. (2006). Ethics and news making in the changing Indian mediascape. *Journal of Mass Media Ethics*, *21*, 286–303. doi:10.1207/s15327728jmme2104_5

Rooney, D. (2000). Thirty years of competition in British tabloid press: *The Mirror* and *The Sun* 1968-1998. In C. Sparks, & J. Tulloch (Eds.), *Tabloid tales: Global debates about media standards* (pp. 91–109). Lanham, MD: Rowman & Littlefield.

Rowe, D. (2011). Obituary for the newspaper? Tracking the tabloid. *Journalism*, *12*, 449–466. doi:10.1177/1464884910388232

Scott, D. K., & Gobetz, R. H. (1992). Hard News/Soft News Content of the National Broadcast Networks, 1972-1987. *The Journalism Quarterly*, *69*(2), 406–412. doi:10.1177/107769909206900214

Shoemaker, P. J., & Cohen, A. A. (2006). *News around the world: Practitioners, content and the public*. Oxford, UK: Routledge.

Sparks, C., & Tulloch, J. (2000). *Tabloid tales: Global debates over media standards*. New York, NY: Rowman & Littlefield Publishers, Inc.

Trammell, K., Porter, L., Chung, D., & Kim, E. (2006). *Credibility and the uses of blogs among professionals in the communication industry*. Paper presented to the Credibility divide Communication Technology Division at the 2006 Association for Education in Journalism and Mass Communication. San Francisco, CA.

Tsfati, Y. (2010). Online news exposure and trust in the mainstream media: Exploring possible associations. *The American Behavioral Scientist*, *54*, 22–42. doi:10.1177/0002764210376309

Tulloch, J. (2000). The eternal recurrence of new journalism. In C. Sparks, & J. Tulloch (Eds.), *Tabloid tales: Global debates about media standards* (pp. 13–46). Lanham, MD: Rowman & Littlefield.

Uribe, R., & Gunter, B. (2007). Are 'sensational' news stories more likely to trigger viewers' emotions than non-sensational news stories? *European Journal of Communication*, *22*, 207–228. doi:10.1177/0267323107076770

Ward, S. J. A. (2007). Utility and impartiality: Being impartial in a partial world. *Journal of Mass Media Ethics*, *22*, 151–167. doi:10.1080/08900520701315913

Westerwick, A. (2013). Effects of sponsorship, web site design, and google ranking on the credibility of online information. *Journal of Computer-Mediated Communication*, *18*(2), 80–97. doi:10.1111/jcc4.12006

ADDITIONAL READING

Arendt, F. (2013). News Stereotypes, Time, and Fading Priming Effects. *Journalism & Mass Communication Quarterly*, *90*(2), 347–362. doi:10.1177/1077699013482907

Burgers, C., & de Graaf, A. (2013). Language intensity as a sensationalistic news feature: The influence of style on sensationalism perceptions and effects. *Communications: The European Journal of Communication Research*, *38*(2), 167–188. doi:10.1515/commun-2013-0010

Cassidy, W. P. (2007). Online News Credibility: An Examination of the Perceptions of Newspaper Journalists. *Journal of Computer-Mediated Communication*, *12*(2), 144–164. doi:10.1111/j.1083-6101.2007.00334.x

Chung, C. J., Nam, Y., & Stefanone, M. A. (2012). Exploring Online News Credibility: The Relative Influence of Traditional and Technological Factors. *Journal of Computer-Mediated Communication*, *17*(2), 171–186. doi:10.1111/j.1083-6101.2011.01565.x

Duffy, M. J., & Freeman, C. P. (2011). Unnamed Sources: A Utilitarian Exploration of their Justification and Guidelines for Limited Use. *Journal of Mass Media Ethics*, *26*(4), 297–315. doi:10.1080/08900523.2011.606006

Elliott, D. (2007a). [), Taylor & Francis Ltd.]. *Getting Mill Right*, *22*, 100–112.

Hendriks Vettehen, P., Beentjes, J., Nuijten, K., & Peeters, A. (2010). Arousing news characteristics in Dutch television news 1990–2004: An exploration of competitive strategies. *Mass Communication & Society*, *14*(1), 93–112. doi:10.1080/15205431003615893

Hofstetter, C. R. (1986). Useful news, sensational news: Quality, sensationalism and local TV news. *The Journalism Quarterly*, *63*(4), 815. doi:10.1177/107769908606300421

Karlsson, M., & Clerwall, C. (2011). Patterns and origins in the evolution of multimedia on broadsheet and tabloid news sites: Swedish online news 2005–2010. *Journalism Studies*, 1–16.

Lee, S., Stavrositu, C., Yang, H., & Kim, J. (2004). Effects of Multimedia and Sensationalism on Processing and Perceptions of Online News. *Conference Papers -- International Communication Association*, 1.

McChesney, R. W. (2012). Farewell to journalism? *Journalism Studies*, *13*(5–6), 682–694. doi:10.1080/1461670X.2012.679868

Molek-Kozakowska, K. (2013). Towards a pragma-linguistic framework for the study of sensationalism in news headlines. *Discourse & Communication*, *7*(2), 173–197. doi:10.1177/1750481312471668

Nah, S., & Chung, D. S. (2012). When citizens meet both professional and citizen journalists: Social trust, media credibility, and perceived journalistic roles among online community news readers. *Journalism*, *13*(6), 714–730. doi:10.1177/1464884911431381

Netzley, S. B., & Hemmer, M. (2012). Citizen Journalism Just as Credible As Stories by Pros, Students Say. *Newspaper Research Journal*, *33*(3), 49–61.

Online News Sensationalism: The Effects of Sensational Levels of Online News Stories and Photographs on Viewers' Attention, Arousal, and Information Recall. (2012). *Conference Papers -- International Communication Association*, 1-29.

Örnebring, H., & Jönsson, A. M. (2004). Tabloid journalism and the public sphere: a historical perspective on tabloid journalism. *Journalism Studies*, *5*(3), 283–295. doi:10.1080/1461670042000246052

Parmelee, J. H., & Perkins, S. C. (2012). Exploring social and psychological factors that influence the gathering of political information online. *Telematics and Informatics*, *29*(1), 90–98. doi:10.1016/j.tele.2010.12.001

Peck, L. A. (2006). A Fool Satisfied? Journalists and Mill's Principle of Utility. *Journalism & Mass Communication Educator*, *61*(2), 205–213. doi:10.1177/107769580606100207

Plassner, F. (2005). From hard to soft news standards? How political journalists in different media systems evaluate the shifting quality of news. *The Harvard International Journal of Press/Politics*, *10*(2), 47–68. doi:10.1177/1081180X05277746

Skorpen, E. (1989). Are Journalistic Ethics Self-Generated? *Journal of Mass Media Ethics*, *4*(2), 157–173. doi:10.1080/08900528909358341

Sparks, C., & Tulloch, J. (2000). *Tabloid tales: Global debates over media standards*. Lanham, Md: Rowman & Littlefield Publishers.

Sundar, S. S., Knobloch, S., & Hastall, M. (2005/05/26/2005 Annual Meeting, New York, NY). *Clicking News: Impacts of Newsworthiness, Source Credibility, and Timeliness as Online News Features on News Consumption*.

Thorson, K., Vraga, E., & Ekdale, B. (2010). Credibility in Context: How Uncivil Online Commentary Affects News Credibility. *Mass Communication & Society*, *13*(3), 289–313. doi:10.1080/15205430903225571

Vettehen, P. H., Nuijten, K., & Peeters, A. (2008). Explaining Effects of Sensationalism on Liking of Television News Stories: The Role of Emotional Arousal. *Communication Research*, *35*(3), 319–338. doi:10.1177/0093650208315960

Wang, T.-L. (2012). Presentation and impact of market-driven journalism on sensationalism in global TV news. *International Communication Gazette*, *74*(8), 711–727. doi:10.1177/1748048512459143

KEY TERMS AND DEFINITIONS

Act Utilitarianism: Normative ethical theory that considers the consequences that a decision will have when a specific situation is faced. The theory suggests that ethical decisions should be made on a case-by-case basis.

Credibility: The believability of information.

Hard news: Stories that require immediate coverage, such as crime news or stories that assist in an individual's ability to vote intelligently.

Soft News: Stories that generally serve an entertainment function rather than a public affairs function.

Source Credibility: The degree to which an individual source, such as the author of a news story, is considered believable.

Tabloidization: The contamination of tabloid writing techniques into the mainstream news media. It often is associated with less political information and an increase in sleazy details into news stories.

Utilitarianism: Normative ethical theory which suggests that decisions should promote happiness, providing the greatest good to the greatest number of people.

Chapter 8
Internet Companies and the Great Firewall of China:
Google's Choices

Richard A. Spinello
Boston College, USA

ABSTRACT

This chapter, focusing primarily on the search engine company Google, considers the problems Internet companies have confronted in adapting to the strict censorship regime in China, which is executed with the help of its "great firewall." Google initially decided to comply with that regime but later changed its mind to the detriment of its whole China strategy. Companies like Yahoo and Microsoft have encountered similar problems in China, while social media firms like Twitter have avoided the Chinese market because of these issues. The moral analysis concludes that a socially responsible company must not cooperate with the implementation of the censorship regimes of these authoritarian sovereignties. This conclusion is based on natural law reasoning and on the moral salience that must be given to the ideal of universal human rights, including the natural right of free expression.

INTRODUCTION

Since the 1990s the technology landscape has been dominated by Internet gatekeepers which provide tools like search engines and portals that help users access and navigate the Internet. As Yahoo, Microsoft, and Google have expanded into markets like China or Saudi Arabia they have been asked to support various censorship laws and other online restrictions. Yahoo, for example, signed the "Public Pledge on Self-Discipline for the Chinese Internet Industry" in 2002. This pledge required Yahoo to "inspect and monitor" any information on domestic and foreign websites (Goldsmith and Yu, 2008, p.9).

Despite the Internet's great promise as a borderless global technology and a free marketplace of ideas, there has been considerable friction between the speech enabled by Internet technologies and the laws of authoritarian countries

DOI: 10.4018/978-1-4666-6122-6.ch008

which define their culture in a more paternalistic fashion. Cyberspace was supposed to be an open environment where anyone could express their opinions, start a new business, or create a web site. Its end-to-end design created an environment conducive to liberty and democracy, with unfettered access to information. As the U.S. Supreme Court eloquently wrote in its *Reno v. ACLU* (1997, p. 857) decision, the Internet enables an ordinary citizen to become "a pamphleteer, . . .a town crier with a voice that resonates farther than it could from any soapbox." But a lethal combination of stringent censorship laws and software code in the form of filtering programs has enabled totalitarian societies to effectively undermine the Internet's libertarian ethos. As a result, Internet freedoms have been replaced by "network authoritarianism" (MacKinnon, 2012, p. 32).

Many Western companies like Yahoo have been forced by these foreign governments to cooperate in the regulation of cyberspace activities as a condition of doing business within that country. This regulation or control most often comes in the form of code, filtering software which allows a sovereign nation to restrict its citizens from accessing or disseminating certain information on the Internet. Consider the case of China. There are eight large gateways coming into China from the global network that provides Internet access for Chinese citizens. The Chinese Government requires the Chinese telecomm companies that control these gateways to configure their routers in order to screen and filter out objectionable content. The Chinese government controls these telecom companies (such as China Telcom), which depend on simple routers for the "backbone" of the Chinese network. Those routers are equipped with packet filtering capability that enables the filtering out of unwanted content. While the primary purpose of routers is to direct or "route" Internet traffic to its correct destination, they can also be easily configured to block content and thereby *prevent* information from getting to its destination. These specially modified gateway routers, can block an entire web site (such as taiwandemocracy.com) based on an access control list, or process web content through "deep packet inspection." With deep packet inspection software, the router can examine the specific content of data packets. Any message or "packet" of information that contains forbidden language will be blocked. The router itself, therefore, becomes the censor (Goldsmith and Wu, 2008, pp. 93-94).

Thus, China has transformed the Internet into a "giant cage," and the cage metaphor suggests the extent to which the Chinese government has infused regulatory controls into the Internet's architectures and processes. As Schmidt and Cohen (2013, p. 85) point out, "China is the world's most active and enthusiastic filterer of information." MacKinnon (2010) describes additional methods used by Chinese government to control online speech. These methods include cyber-attacks that undermine web sites or infect the email accounts of dissidents. In extreme cases the government relies on "localized disconnection" to deter Chinese citizens from using the Internet altogether (p. 5). When ethnic riots broke out in Xinjiang province, the government shut down the Internet in the entire province for six months.

However, with hundreds of millions of Internet users, China also represents an attractive investment opportunity for multinational high tech companies like Google, Yahoo, Twitter, and Facebook. But the Chinese government demands that companies investing in China play by their rules, and their rules include cooperating with its strict censorship and surveillance laws. As a consequence, multinationals like Yahoo and Google have had to navigate an ethical minefield in China, and many would argue that their navigation has been quite poor. Other companies like Twitter, whose services are completely blocked in China, have refused to comply with China's censorship laws. "We are not going to make the kinds of sacrifices... [necessary] to be unblocked in China," declares Twitter CEO Dick Costolo (Ovide, 2013, p. B2). Although many authoritarian regimes besides China censor the Internet aggressively, our focus of attention will be primarily on Chinese policies.

Internet gatekeepers, which provide a service such as online access or search engine capability, cannot subvert this firewall if they expect to do business in China. Countries like Iran and Saudi Arabia deploy similar techniques. The gatekeepers along with other technology and social media companies are caught in a vice between countries exercising their legitimate sovereignty and individual citizens seeking to exercise their basic speech rights. It was once thought that states would have a difficult time enforcing their sovereignty in cyberspace but, thanks to code such as filtering software, freedom of expression is threatened by state power often assisted by private companies. But states are re-asserting their authority and demanding compliance with local law. As a result, the Internet loses some of its "generative" potential as a viable force for semiotic democracy (Zittrain, 2003).

The Internet gatekeepers are especially vulnerable and must find ways to responsibly navigate this perilous virtual terrain. Their corporate strategies, oriented to rapid global expansion, cannot ignore the question of the Internet's role in authoritarian societies like China, Iran, and Egypt. The problem is exacerbated by the lack of international laws that govern cyberspace along with the policy disputes that prevent the dissemination of anti-censorship technologies. Without the guidance of law, companies must determine whether to side with the host government or with many of their citizens who have a different conception about free speech.

Google's unfortunate experience in China will be the main springboard for our discussion, but we will also take into account the practices of companies like Microsoft and Yahoo. After briefly reviewing some background on Google, which is attracted to foreign markets by the need to sustain its economic growth, we will turn to the legal issues and the prospects that there may be some legal resolution on the horizon. We conclude that those prospects are dim and that corporate self-regulation is essential in the face of this policy vacuum. *Ethical self-regulation* subordinates rational self interest to the legitimate needs and rights of others and, above all, respect for the common good of the Internet community.

We then turn to a moral analysis of Google's strategy, which revolves around an apparently irresolvable polarity: either the company can initiate cultural and normative changes in China *or* compromise its core values and adapt to China's norms and law. This analysis pursues several key questions. If it chooses the latter alternative, can Google's conduct in China be morally justified according to the pragmatic and utilitarian reasoning adopted by the company to defend its decision? Or should companies like Google and Yahoo refrain from participating in online censorship or surveillance as a cost of doing business in certain states? We argue that Google cannot responsibly cooperate with China's systematic repression of free speech rights. The right to free expression is universal and cannot be relativized away despite the persuasive claims of some ethicists to the contrary. In reaching this conclusion we must explore the philosophical grounding for this right and address the valid concerns of the pluralists. We begin all this with a look at Google and the search engine business.

BACKGROUND ISSUES: THE TROUBLED HISTORY OF GOOGLE IN CHINA

Google, the ubiquitous U.S. Internet search engine company, was founded in 1998 by two Stanford graduate students, Sergey Brin and Larry Page. Their ambitious goal was to create software that facilitated the searching and organizing of the world's information. Thanks to its PageRank algorithm the Google search engine delivered more reliable search results than its rivals by giving priority to web pages that were referenced or "linked to" by other web pages. The company continues to refine its search algorithm so that it can better respond to obscure and complicated queries. Google monetized its technology by licensing its search engine and by paid listings, or "sponsored links," that appeared next to web search results.

Google is the most popular search engine on the Web and still powers the search technology of most major portals and related sites. In 2013 Google enjoyed a 67% share of the global search engine market with rival search engines like Microsoft's Bing falling far behind with only an 18% share; 86% of all Internet searches in the U.S. are now done on Google. Searches on Google's site generated about $60 billion in economic activity in the United States, giving the search engine giant considerable power over companies that rely on Google for their traffic (Goodwin, 2013).

Early in Google's history its founders insisted that the company must be centered upon several distinctive corporate values: "technology matters," "we make our own rules," and "don't be evil." In accord with its first two principles, Google is committed to technology innovation and to sustaining a creative leadership role in the industry. The "don't be evil" principle is realized primarily through the company's commitment not to compromise the integrity of its search results, but it has been given a broader meaning over the years. This core value, however, has been a source of continual controversy for Google whenever it is perceived to have violated social norms or transgressed ethical boundaries.

Google introduced a version of its search engine for the Chinese market in early 2006, google.cn. Google's biggest rival in China is Baidu.com, Inc., which has a 60% share of the China Internet search-engine market. Given the size of the market and its future potential, the company admitted at the time that China was "strategically important" (Dean, 2005, p. A12). In order to comply with China's strict censorship laws, Google agreed to purge its search engine results of any links to politically sensitive web sites and other content disapproved by the Chinese government. These include web sites supporting the Falun Gong cult or the independence movements in Tibet and Taiwan. As one reporter indicated:

If you search for 'Tibet' or 'Falun Gong' most anywhere in the world on google.com, you'll find thousands of blog entries, news items and chat rooms on Chinese repression. Do the same search inside China on google.cn and most, if not all, of these links will be gone. Google will have erased them completely (Thompson, 2006, p. 37).

In order to avoid further complications, the company does not host user-generated content such as blogs or e-mail on its computer servers in China for fear of the government's role in restricting their content. In this way it avoided the plight of companies like Yahoo who were compelled by Chinese law to hand over information about dissidents using Yahoo's email. Unlike its competitors, Google alerted users to censored material by putting a disclaimer at the top of the search results indicating that certain links have been removed in accordance with Chinese law. Also, Chinese users could still access Google.com with its uncensored search results (though links to controversial sites would not work thanks to the firewall).

Google's cooperation with the Chinese government was met with considerable consternation by a plethora of human rights groups such as Human Rights Watch. These groups and many other critics chastised Google for so blatantly violating its high-minded corporate ethos. The company was reprimanded by several U.S. Congressmen who sought to craft legislation aimed at stopping Google and other companies from abetting the Chinese government and forcing them to comply with the United Nation's Universal Declaration of Human Rights.

In defense of its corporate policies, the company argued that its presence in China created abundant opportunities for Chinese citizens to have greater access to information. Most technical experts agree that Baidu's search results are not nearly as comprehensive or unbiased as Google's. According to a Google spokesperson, "While removing search results is inconsistent with Google's mission, providing no information, (or a heavily degraded user experience that amounts to no information) is more inconsistent with our mission" ("Google in China," 2006, p. A23). Google sincerely believed that its presence in China contributes to the country's modernization, and that this consideration must be balanced with the legal requirements imposed by the Chinese government.

The corporate hierarchy has been ambivalent about the moral ramifications of the company's controversial foray into the China market. One of the Google's founders, Sergey Brin, explained that his company was grappling with difficult questions and challenges: "Sometimes the 'Don't be evil' policy leads to many discussions about what exactly is evil" (Dean, 2005, p. A12). Google has apparently assumed that by improving access to information in a repressive country like China the company is bringing about sufficient benefits that outweigh the costs of abetting censorship. Despite its censorship of some information sources, Google still provides Chinese citizens with an opportunity to learn about AIDS and other health related issues, environmental concerns, world economic markets, and political developments in other parts of the world.

After a few tumultuous years and several notorious security breaches, Google decided to stop censoring its Web search and news services in China. In March, 2010 the company announced that it would redirect people who come to google.cn to an uncensored site hosted in Hong Kong. Of course, the government still has the ability to block Google searches by mainland Chinese on the Hong Kong site. Google's defiance of the Chinese government will undoubtedly damage its efforts to compete successfully with Baidu, China's own Internet search engine. While moral issues were part of the company's overall calculus, its decision to confront Chinese authorities reflected a more realistic view about its "limited business prospects" in this country (Waters, 2010, p. B1).

Google must decide about the status of its future in China. Its share of the Web-search market fell to 17% in 2012 largely to the benefit of Baidu. However, Google has not completely abandoned the Chinese market nor has it given up on the idea of a stronger presence there in the future. It hopes to capitalize on the growth of Android, the operating system that powers many of the mobile phones used in China. One of its tentative goals is to introduce Android Market which offers thousands of apps for cell phones into the Chinese market. It has also launched a product called Shihui which allows people to search Chinese sites that offer discounts at local stores (Efrati and Chao, 2012). The problem is that Google's insubordination has upset many Chinese government officials and so its future success in this huge Chinese market is far from guaranteed.

Google is not the only company accused of yielding to the pressures of authoritarian governments. In 2006 Microsoft shut down a popular Chinese language blog hosted on MSN, worried that some of the content was offensive to the Chinese government. The site criticized the firing of editors at a progressive Beijing newspaper called the Beijing News. In a statement defending its actions, Microsoft simply said that "MSN is committed to ensuring that products and services comply with global and local laws, norms, and industry practices in China" (Chen, 2006, p. A9). Yahoo has also run into problems in China. In 2001 Wang Xiaoning posted comments calling for democratic reform in China to a listserv called Yahoo Group where users could send and receive emails with their identities kept anonymous. In one posting Wang said that "we should never forget that China is still a totalitarian and despotic country." The Chinese government forced Yahoo to remove the content and sought the culprit's identity which it was able to do thanks to information Yahoo Hong Kong provided to the Chinese police. Wang was convicted of "incitement to subvert state power" and sentenced to ten years in state prison where he was allegedly tortured into signing a confession about his "seditious" activities (Bryne, 2008, pp. 159-60).

There have been similar incidents in other countries, but these examples will suffice to illustrate the nature of this problem, which has vexed American technology companies for almost a decade. How can this dilemma be effectively resolved? Does the resolution lie in law and policy implemented at the national or international level, or does it lie deep within the recesses of corporate conscience?

LEGAL ISSUES

U.S. law does not directly address Google's indirect participation in China's censorship regime nor does it deal with Yahoo's apparent willingness to aid the Chinese government in tracking down dissident journalists. There are, however, several laws under which U.S. companies abetting censorship or surveillance might be held accountable, including the Alien Tort Claims Act (ATCA), enacted as part of the Judiciary Act in 1789. The ATCA (2000) gives US federal courts jurisdiction over torts filed by aliens against US companies which have acted "in violation of the law of nations or a treaty of the United States." There has been a contentious debate about the scope of this Act. Some have argued that the scope is quite narrow, confined to violations of law of nations at the time the law was written such as piracy. But in *Filartiga v. Pena-Irala* (1980, p.878) the Second Circuit ruled the ATCA provides a broad jurisdiction for violations of "universally accepted norms of the international law of human rights, regardless of the nationality of the parties." This and related rulings have opened the door for corporations to be found liable for specific violations of international law by direct actions or by simply aiding and abetting the violations of foreign governments. Courts have not specifically delineated what constitutes the law of nations. Nonetheless they have affirmed that "certain forms of conduct violate the law of nations whether undertaken by those acting under the auspices of the state or only as private individuals." (*Kadic v. Karadzic*, 1995, p. 239)

In the case of Google, a Chinese citizen could file a claim under ATCA alleging that the search engine company was complicit in the infringement of his or her free speech rights. In most cases, however, it would be difficult to argue that this infringement constituted a tort violation. Violations actionable under the ATCA include crimes against humanity such as torture, racial discrimination, or enslavement. Hence it is far from clear that the ATCA would be an effective mechanism for Chinese citizens looking for legal redress for Google's complicity in censoring free speech unless Google's actions abetted other crimes on the government's part.

On the other hand, it would appear that Wang Xiaoning has a stronger case, given his imprisonment and treatment in Chinese jails. Wang filed suit against Yahoo in 2007 under the auspices of ATCA. Wang's lawyer contended that Yahoo aided and abetted China as it violated the law of nations through its torture of Wang and its infliction of cruel and inhuman punishment. Yahoo settled with Wang for an undisclosed amount so this case does not tell us anything about the applicability of ATCA in other cases involving Internet gatekeepers nor does it set any precedent for future litigation.

However, a recent Supreme Court decision will likely close off this avenue. In a case involving Royal Dutch Shell and Nigeria the Court decided by a narrow margin that the statute only covers violations of international law in the United States. But the Court left open the possibility that some acts abroad could so severely "touch and concern the territory of the United States," to thereby "displace the presumption" against the statute's use. (*Kiobel v. Royal Dutch Petroleum Co.*, 2013)

It's remotely possible that the US will develop new laws to cover the complicity of companies like Google and Yahoo. In 2007 several members of Congress proposed legislation called the Global Online Freedom Act that was designed to prescribe minimum standards for online freedom of expression. Title II of this act would eliminate jurisdiction of foreign countries for information housed on their servers and it would prohibit businesses from modifying the functionality of their search engines to produce different results compatible with a censorship regime like China's. Other federal legislators have called for a global first amendment that would constrain censorship activities in Internet restricting countries.

It's not evident that the U.S. Congress has the wherewithal to pass such legislation. Efforts have received some impetus from the State Department which now says that Internet freedom should be a cornerstone of American foreign policy. There is a spirit of "techno-utopianism" in Washington inspired by the sentiment that democratization throughout the world will happen more quickly once the Internet becomes liberated from the bonds of online censorship. The U.S. State Department, however, offered no help for Google claiming that the issue was strictly between Google and China.

But is such legislation aimed at US companies a feasible solution to this problem or will it simply put gatekeepers like Google and Microsoft at a competitive disadvantage around the globe? Will these laws really enable Chinese or Iranian citizens greater access to the Internet or will they just complicate the efforts of US companies to compete in these markets? There are few legal issues as complex and sensitive as state sovereignty and so any legislation must be carefully crafted to avoid unintended consequences. While the intentions of the U.S. government are laudable, there are at least three problems with such legislation. While these laws apply to US firms, some companies like Yahoo operate through local ventures in which they have an ownership stake, and those companies would be immune from the legislation. Yahoo now runs its Chinese operations through Alibaba.com, which it controls but does not own. In this era of global cooperative capitalism, businesses are no longer centralized vertically-integrated entities, but decentralized networks. If a new law is to be effective it must somehow deal with complicated jurisdictional and ownership issues where accountability is often murky. Second, it's hard to conceive that China would terminate its filtering system and tear down its great firewall just because companies like Google or Microsoft were forced out. Filtering technologies are widely available, and China has its own search engine company, Baidu, whose services are improving as it gains experience with this technology. Third, this legislation has been characterized as "an arrogant attempt for the United States to serve as a world police," and there is some merit to this claim, since the U.S. would be seeking to impose its will on the Internet (Eastwood 2008, p. 310). Fresh efforts to promote free speech throughout the world will likely be treated in the same way. Those efforts, however well-intended, could be counterproductive and only risk a backlash against the United States.

AN ETHICAL PERSPECTIVE AND RESOLUTION

If the law offers little direction, and there are scant prospects for a significant change, we have a policy vacuum which means that a responsible company can only discern the right course of action through moral and social analysis. Philosophers have been constructing the field of Information and Computer Ethics (ICE) for decades, so what does it have to say about Google's conundrum? While much has been written about the Western ideal of speech and the need to curb digital content controls (Spinello, 2014, pp. 63-102), little has been said about the more complex problem of intercultural ethical disputes with free speech as the focal point. Analysis has been focused on the problem of pornography and what some see as misguided efforts to limit its diffusion in cyberspace either through federal law or through filtering programs. Many ICE scholars are wary of restricting Internet speech, and especially concerned about the deployment of filtering technologies given that these technologies are so imprecise and opaque. The trend has been to argue for broad free speech rights on the Internet even for children. Yet at the same time these scholars embrace the notion of cultural pluralism, and, in this case, these two divergent objectives cannot be easily reconciled.

Hence the need for some original analysis which must begin with whether or not China is doing anything wrong when it engages in systematic censorship of political speech on the Internet. One argument in China's favor centers on the cultural moral imperialism of its critics who are intolerant of China's different standard for free speech. Accordingly, in response to calls for "Internet freedom," Chinese officials have accused the United States of "information imperialism." They have consistently maintained that China's Internet regulations are compatible with the country's "national conditions and cultural traditions" (Buruma, 2010, p. W1).

Moreover, computer ethicists like Hausmanniger (2007) have staunchly defended the ethical obligation to respect different moral belief systems, however discordant they are with traditional Western norms. Correlative with the turn to subjectivity, which began with Descartes' grounding of certitude in the *cogito*, is the post-Cartesian "turn to contingency" which gives primacy to difference and plurality, instead of cultural or social uniformity (p. 45). As a logical consequence, there must be respect towards "the free actions that create difference and plurality" (p. 56). Similarly, Ess and Thorseth (2010) have argued that we must be more sensitive to the reality of ethical pluralism. They claim that many Western philosophers such as Plato and Aristotle are really pluralists at heart because they support general or formal moral norms (such as community well-being or privacy) that are applied differently depending on the cultural context. Discrepant free speech norms represent another instance of ethical pluralism, and should not really surprise us. Despite the reality of moral and cultural diversity, Ess and Thorseth hope for some sort of global ICE and a harmonization of policies.

Defenders of China's policy also point out that it has a different conception of the person and a more collectivist view of human rights which may justify the country's overall approach to censorship. The Chinese government respects in broad terms the value of free expression but interprets the scope of that value differently than its counterparts in the West. The Confucian tradition sees the purpose of law as the protection of social harmony, which is inconsistent with the normative individualism of the Western liberal tradition. Confucianism stresses obedience to authority which is part of Chinese culture. Deference to authority is an essential aspect of being Chinese and implies the need for severe limits on political dissent (Buruma, 2010, W1).

Since the days of Chairman Mao's cultural revolution, the country has sought to control knowledge and restrict expression in order to propagate the state's uniform message unencumbered by the dissonant voices of dissenters. This restriction is consistent with China's nationalist strategy which sees unequivocal support for the state as the only way for China to regain its long lost greatness and avoid the humiliation the country has repeatedly suffered at the hands of the West. According to Vincent (1988, p. 42), "The fundamental rights and duties of citizens are to support the leadership of the Communist Party of China, support the Socialist system, and abide by the Constitution and the laws of the People's Republic of China."

Given these cultural anomalies and different human rights standards, it is no surprise that China adopts a divergent view of free speech "rights" which are construed with such a narrow scope. The general norm of free expression is being interpreted according to a different set of particularities and cultural imperatives. The Chinese standard, which heavily limits freedom of information for the sake of the collective good, represents the concrete reality of ethical pluralism, which must be factored into moral decision making by those doing business in China. In their defense of pluralism, Ess and Thorseth (2010) argue that "As global citizens...we must learn and respect the values practices, beliefs, communication styles and language of 'the Other'" (p. 168). Failure to do so, they contend, "makes us complicit in forms of computer-mediated imperialism and colonialism" (p. 168). Pluralism, therefore, amounts to a strong version of cultural moral relativism where each culture's moral values are equally valid and deserving of our respect.

Rawls, who is also sympathetic with ethical pluralism, offers a more nuanced perspective that calls for recognition of a few basic rights that set a few limits to pluralism. According to this "thin" theory of rights, all persons do not have equal basic rights or entitlements grounded in a philosophical or moral conception of the person. Rather, there is a special class of "urgent rights" such as freedom from slavery and serfdom and security against genocide, whose violation should be condemned by all peoples (Rawls, 2001, p. 79). Free expression, however, does not qualify as one of these "urgent" rights.

This pluralistic understanding of ethics is implied in the public responses of Google, Yahoo, and Microsoft to criticism about their policies. All three companies have argued at one point that their policies reflect the moral flexibility mandated by cultural moral relativism. According to Microsoft, "Like other global organizations we must abide by the laws, regulations, and norms of each country in which we operate" (BBC News 2005). Implicit in this argument defending a "when in Rome" approach to morality is the notion that ethical norms, like customs, have only local validity because they are prescriptive or action-guiding. The social norms and civil liberties in China and Iran are simply different from the norms and liberties enjoyed by U.S. citizens, and it's imperialistic to maintain that U.S. norms are superior. This moral perspective, if tenable, would seem to validate the behavior of the Internet gatekeepers like Google. In this context, despite the vociferousness of their critics, the companies are doing nothing morally wrong when they cooperate in China's extensive censorship regime. In an age of ethical relativism, where the definition of right and wrong is so indeterminate, and where cultural differences demand their due, how can we criticize Google's malleable ethical policy or Yahoo's cooperation with the Chinese government?

The problem with cultural moral relativism, however, is that it offers virtually no standards for judging, evaluating, or ranking particular cultures. It assumes that all culture are equal and all cultural differences normatively acceptable. Also, how does the cultural relativist account for the source of a culture's moral norms? If not some transcendent standard, then it must be consensus of the majority or the will of the sovereign. But how can non-democratic countries determine the will of the people? Do the majority of citizens in China still accept the dogma of Confucianism as interpreted by the state autocracy? Grounding rights in sheer consensus or sovereign preference ignores the possibility that those rights are intrinsic to the human condition. Basic human rights become contingent on the whim of cultural consensus, which is often shaped by those in power. If pluralism is pressed too far, values like free expression, due process, or even human life itself could become feeble and shallow and lose any semblance of objectivity.

On the other hand, many philosophers ranging from Aristotle and Aquinas to contemporary thinkers like Phillipa Foot and Elizabeth Anscombe accept the thesis that we possess a common human nature. Despite our ethnic diversity, there are essential features we all share in common such as rationality and free will. According to Foot (1979, p. 6):

Granted that it may be wrong to assume identity of aim between people of different cultures; nevertheless there is a great deal all men have in common.

All need affection, the cooperation of others, a place in community, and help in trouble. It isn't true to suppose that human beings can flourish without these things -- being isolated, despised or embattled, or without courage or hope. We are not, therefore, simply expressing values that we happen to have if we think of some moral systems as good moral systems and others as bad.

Rejection of a common human nature is also a rejection of the equality of persons which is the foundation of human rights and justice. If we assume that all humans share in some essential common features, it follows that there must be intrinsic human values expressed in *transcultural norms* that are more specific and less culturally contingent than Ess and Thorseth are willing to admit. To be sure, even these specific norms must be applied with some cultural flexibility and sensitivity. Reasonable people can discern whether or not a particular norm is being flouted in a given culture or properly applied in a way that accounts for a culture's particular circumstances.

What is the basis for discovering these transcultural norms? As Plato first theorized, they must be grounded in a coherent notion of the Good (τό ἀγάθόν), the ultimate source of all efficacy and normativity. Plato (1935, VII 516b) believed that our actions must be in conformity with the Good, "cause of all that is correct and beautiful in anything." Almost every moral philosophy since Plato endorses some notion of the good. Even deontologists such as Rawls, who give priority to the concept of right over the concept of good, concede the need for a "thin" theory of the good. Rawls (1971) contends that there are certain primary goods, "liberty and opportunity, income and wealth, and above all self-respect," which are necessary for the "framing and execution of a rational plan of life" (pp. 433-34).

More robust or "thicker" theories of the good are offered in teleological ethical frameworks, such as the new natural law, an updated and secularized version of the natural law theory found in the writings of St. Thomas Aquinas. According to this framework, practical reasoning about morality begins with the intelligible reasons people have for their choices and actions. Some of those reasons are based on ends that are intelligible only as a means to other ends. Tangible goods like money have only instrumental value, since they possess no intrinsic worth. Other goods are intrinsic, that is, they are valued and sought after for their own sake. These intrinsic or "basic" human goods provide reasons to consider some possibilities as worthwhile and choiceworthy for their own sake. These goods are "basic" not for survival but for human flourishing.

What then are these intrinsic human goods valued for their own sake as constitutive aspects of human flourishing? According to natural law theorists like Finnis (1980) and George (2007) it is reasonable to conclude that a list of basic goods should include the following: bodily life and health; knowledge of the truth; aesthetic appreciation; sociability or harmony with others; skillful performance in work and play. These intelligible goods are all distinct aspects of basic well-being, intrinsically valuable and sought after for their own sake because they are perfective of the human person. Knowledge of what is true, for example, is a beneficial possibility, because its possession represents a mode of existence superior to the mode of ignorance. Hence, knowledge is not just instrumentally valuable but is worthy of being pursued as an end-in-itself. These substantive goods constitute the foundation of normativity and provide a secure grounding for moral judgments about justice and human rights.

Where does free expression fit into this paradigm? Free expression is not a basic good, since it does not directly contribute to human flourishing. It is difficult to see how free speech would be valued in itself apart from the relational goods it supports. On its own, it doesn't really fulfill or perfect us, but it does allow us to pursue other forms of personal fulfillment. Thus, it is desired as a means to another end, that is, as instrumental to certain intrinsic goods that directly provide personal fulfillment. For example, speech or communication is essential for the harmonious cooperation necessary to build community and create bonds of fellowship. Miscommunication or misunderstanding among people is common, but this reality means that communication efforts must be refined or revised, but certainly not suppressed. Dissenting political speech often brings to light problems and conflicts that must be resolved if a political community is to overcome differences and evolve into a more authentic communion of persons based on the common good. Thus, a strong case can be made that the right to free expression is justified as an instrumental good promoting our intrinsic sociability.

Speech is also essential to support and preserve the intrinsic goods of knowledge and reflective understanding. People cannot be coerced in matters of speech and communication in a way that interferes with their capacity to inform others of the truth. The acquisition of objective, true knowledge by people in a community is contingent on the ability of teachers and others within that community to disseminate that truth without fear of retribution of punishment. Censorship and suppression of certain information is typically motivated by a desire to keep the truth from citizens and to prevent them from overcoming ignorance and error. Censors aim to achieve conformity of thought rather than propagation of the truth.

From this analysis we can deduce that there is at least a moral presumption in favor of free speech rights because free speech is a vital instrumental good (or value). These rights are necessary to protect individuals in their efforts to appropriate and share several intrinsic human goods that constitute the basis of human flourishing. These goods (objective knowledge and sociability) are not confined to the West or to liberal societies, but are sought by all rational human persons as intrinsic to their personal fulfillment, and therefore it is plausible to argue for the universality of this right. Basic human rights are not grounded on the basis of utility. Rather, they are grounded in necessity, in what human persons need and rationally desire "for the exercise and development of distinctive human powers" (Hart, 1983, p.17). The right to free expression, therefore, is necessary for the pursuit of several intrinsic goods common to all persons. Human beings need free expression to build authentic community and grow in fellowship and to advance in knowledge of the truth. Like all rights, the right to free speech must be limited in different ways that are consistent with the common good. Most sovereignties do not protect perverted forms of speech such as obscenity nor hate speech that incites violence.

Further support for the universality and intrinsic value of this right to free expression is its endorsement by the United Nations in its famous Universal Declaration of Human Rights first promulgated in 1948. This declaration, which represents a moral and political consensus among United Nations member nations, still carries great weight throughout the world. The U.N. has never disavowed this declaration nor qualified its support for these rights. Article 19 of the U.N.'s Universal Declaration (2007) states: "Everyone has the right to freedom of opinion and expression; this right includes freedom to hold opinions without interference and to seek, receive and impart information and ideas through any media and regardless of frontiers." The United Nations is certainly sensitive to cultural issues but it also recognizes that some rights transcend those cultural differences. The U.N document clearly favors the universalism suggested by Plato and natural law theorists instead of the radical ethical pluralism of more contemporary philosophers. Implicit in its endorsement of these rights is the assumption that people possess them not by some government's fiat or cultural consensus but as a matter of natural justice. This declaration also assumes that some cultures can be morally deficient and blind to the truth about particular rights such as free speech.

If we assume that there is such a natural, universal right to free speech, properly configured to protect morally justified privacy, secrecy, and security concerns, the Chinese government infringes on this right by ruthlessly imposing orthodox political beliefs on the entire community and providing no basis of dissent, good faith disagreement, or attempts to correct the historical record so that future generations will know the truth about events such as Tiananmen Square. As we have seen, China relies on culture and its history as an oppressed state as a pretext to support the government's authoritarian regime. However, China is guilty of a moral failing by not respecting this right to free political expression, which many of its citizens have demanded for decades. China's nationalism is excessive and unhealthy, since it involves the pursuit of the nation's social welfare without proper regard for the needs and natural rights of all its citizens. If China is willfully infringing on the rights of its citizens by blocking speech and keeping important information from its people, its actions are unjust and immoral.

If this analysis is sound, Google's moral culpability logically follows, since the company willingly cooperated in perpetuating restrictions on this basic right. For a sincere company committed to corporate responsibility, the "don't be evil" principle must preclude cooperation in evil. Instead of defying the Chinese government, Google facilitated and supported the perpetuation of its censorship regime until its change of policy in 2010.

The basic moral imperative at stake in the Google case is that a moral agent should not cooperate in or become involved in the wrongdoing initiated by another. This simple moral principle seems axiomatic. If someone intentionally helps another individual carry out an objectively wrong choice, that person shares in the wrong intention and bad will of the person who is executing such a choice and is guilty of formal cooperation. In this context, what one chooses to do coincides with or includes what is objectively wrong in the other's choice. For example, a scientist provides his laboratory and thereby willingly assists in harmful medical experiments conducted on human beings by a group of unscrupulous medical doctors because he is interested in the results for his own research. Although this scientist did not conduct the experiments, he shares in the wrongful intentions and actions of the doctors who did. In the domain of criminal law, if a person helps his friend commit or conceal a crime, that person can be charged as an accessory. Hence, it is part of moral common sense that a person who willingly helps or cooperates with the actions initiated by a wrongdoer deserves part of the blame for the evil that has been perpetrated. There are also various forms of material cooperation whereby the acts of the cooperator and wrongdoer are distinct and share no bad intention (Grisez, 1991, pp. 871-74).

Google is accountable for its formal (not merely material) cooperation, since it intentionally participated in the wrongdoing (censorship) initiated by Chinese government officials. Google does not share completely in the bad will of the Chinese censors since it disagrees with their ends. Nonetheless, it intended the chosen means of abetting those censors in order to have a presence in the world's second largest market. However reluctantly, Google intended to censor its search results and further the aims of China's censorship regime as a condition of doing business in China. Neither this ulterior motive nor any extenuating factors mitigate its responsibility. It makes no difference that Google disapproves of China's policy and its reasons or motives for censoring search results does not coincide with the reasons of the Chinese government. It was guilty of formal cooperation by virtue of choosing the wrong means to achieve its valid business objectives. Responsible companies, which function as moral agents in society, must eschew both formal cooperation and those forms of material cooperation deemed ethically unacceptable according to traditional standards.

To sum up: in the absence of clear international laws, corporations like Google must rely on ethical self-regulation. We maintain that proper self-regulation should lead companies to preserve their core values by respect for universal rights and avoidance of formal cooperation with host governments willing to infringe those rights.

FUTURE RESEARCH DIRECTIONS

Given the other priorities of the U.S. government it is unlikely that there will be major legislation to address the issues delineated here. It's now highly unlikely that companies will be sued under ATCA for their alleged misdeeds abroad, so current law will not work as a mechanism for forcing companies to confront their moral responsibilities. We may hear more rhetoric from the West about "Internet freedom" but it will probably fall on deaf ears within the halls of the Chinese government. This will leave the corporate Internet gatekeepers and social media sites still caught in the crosshairs, perplexed about how to navigate this difficult terrain. Should companies like Google be forced into the *de facto* role of "policy maker" dictating the terms of its engagement in totalitarian countries like China? Should they resist helping China or Iran to enforce their laws in cyberspace? Finally, what would be the long term social and policy implications of helping these countries erect borders in cyberspace?

What also needs consideration and further in depth research is the normative issue of the scope and nature of free speech rights. Much has been written about problems like cyberporn, spam, and hate speech along with the use of content controls (see Spinello and Tavani, 2004; Spinello, 2014). But very little has been written about free speech and communication as an intercultural issue. Those who advocate for ethical pluralism typically avoid addressing specific thorny issues like speech let alone review the moral liability of gatekeepers who directly cooperate with authoritarian governments. On the other hand, they write at length about the need to bridge cultural differences to create more harmony in the infoshpere. One might contend, however, that the proliferation of censorship regimes and the intractability of many sovereign states on this issue does not augur well for "fostering a *shared* ICE that 'works' across the globe" (Ess and Torseth, 2010, p. 164). While working to promote this shared global ethos, ethicists must consider whether this right to free expression is contingent on culture and history or is a universal right. Is there a philosophical grounding for the United Nations claim that *everyone* has the right to freedom of opinion and expression? We have argued that this is so but recognize that this debate is far from settled. There is no doubt that public and corporate policy should turn on how that vital question is resolved.

CONCLUSION

Many governments have reasserted their sovereignty in cyberspace, and as a result, the Web's openness and universal character has been diminished. This has created problems for corporations like Google attempting to balance their core values with restrictions imposed by China and other countries. We have looked at the related issues of free expression and corporate liability through the prism of law and ethics. Thanks to jurisdictional constraints, the law offers little guidance for corporations who aspire to be morally responsible and to cooperate at least implicitly with the U.S. policy to maximize the free flow of information over the Internet. Given this policy vacuum, these corporations must fall back on moral reasoning as Google did, grappling with what "do no evil" really means in a context of moral and cultural diversity. We have argued in this paper that despite this diversity and even the so-called "turn to contingency," free expression is a fixed and universal value rather than a merely contingent one. We have based this analysis on natural law reasoning along with the argument that speech is an instrumental good necessary for instantiating intrinsic goods such as community and knowledge. Because of its status as a necessary instrumental good, it follows that people have a right to free speech since rights are based on what persons need and rationally desire. It also follows that companies like Google behave irresponsibly if they actively cooperate in the undermining of this right. In making this case we have suggested the propriety of a thick theory of rights as a matter of universal justice. This position is compatible with the UN's 1948 consensus on universal rights and stands in opposition to the thin theory proposed by Rawls who recognizes only a small set of "urgent" rights.

As a practical matter, this ethical analysis will probably not make corporate decision making any easier. Private companies must still struggle with how to respond when they are asked to participate in a censorship or surveillance regime like those in Iran or Saudi Arabia. As this paper has demonstrated, companies are caught in a vice between universal moral values like free expression and pressure to conform to local norms. The problem of online censorship will probably intensify as some countries continue to resist the Internet's open technology. As a result, gatekeepers and other Internet companies will be forced to give this issue the cogent moral reflection it deserves.

REFERENCES

Alien Tort Claims Act (2000) 28 U.S.C. § 1350.

Buruma, I. (2010, January 30). Battling the information barbarians. *The Wall Street Journal*, p. W1-2.

Byrne, M. (2008). When in Rome: Aiding and abetting in Wang Xiaoning v. Yahoo. *Brooklyn Journal of International Law*, *34*, 151.

Chen, K. (2006, January 6). Microsoft defends censoring a dissident's blog in China. *The Wall Street Journal*, p. A9.

Dean, J. (2005, December 16). As Google pushes into China, it faces clashes with censors. *The Wall Street Journal*, p. A12.

Eastwood, L. (2008). Google faces the Chinese internet market and the Global Online Freedom Act. *Minnesota Journal of Law, Science, &. Technology (Elmsford, N.Y.)*, *9*, 287.

Efrati, A., & Chao, L. (2012, January 12). Google softens China stance. *The Wall Street Journal*, pp. B1-B2.

Ess, C., & Thorseth, M. (2010). Global information and computer ethics. In L. Floridi (Ed.), *The Cambridge handbook of information and computer ethics* (pp. 163–180). Cambridge, UK: Cambridge University Press. doi:10.1017/CBO9780511845239.011

Filartiga v. Pena-Irala 630 F. 2d 876, 2d Cir. (1980).

Finnis, J. (1980). *Natural law and natural rights*. Oxford, UK: Oxford University Press.

Foot, P. (1979, June). *Moral relativism*. Paper presented for Lindley Lecture, Department of Philosophy, University of Kansas. Lawrence, KS.

George, R. P. (2007). Natural law. *The American Journal of Jurisprudence*, *52*, 55. doi:10.1093/ajj/52.1.55

Goldsmith, J., & Wu, T. (2008). *Who controls the internet?* Oxford, UK: Oxford University Press.

Goodwin, D. (2013, November 14). Google fails to gain search market share. *Search Engine Watch*. Retrieved December 28, 2013, from http://searchenginewatch.com/article/2307115/Google-Fails-to-Gain-Search-Market-Share

Google in China Editorial. (2006, January 30). *The Wall Street Journal*, p. A18.

Grisez, G. (1991). *Difficult moral questions*. Chicago: Franciscan Herald Press.

Hart, H. L. (1983). *Essays in jurisprudence and philosophy*. Oxford, UK: Oxford University Press.

Hausmanninger, T. (2007). Allowing for difference: Some preliminary remarks concerning intercultural information ethics. In *Localizing the Internet: Ethical issues in intercultural perspective* (pp. 39–56). Munich: Wilhelm Fink Verlag.

Kadic v. Karadzic 70 F.3d 232, 2d Cir. (1995).

Kiobel v. Royal Dutch Petroleum Co. 133 U.S. 1659 (2013).

MacKinnon, R. (2010, March 24). Testimony for the congressional-executive commission on China. *Congressional Hearing on Google and Internet Control in China*. Retrieved January 2, 2014 from http://rconversation.blogs.com/MacKinnonCECC_Mar24.pdf

MacKinnon, R. (2012). *Consent of the networked*. New York: Basic Books.

News, B. B. C. (2005, September 7). Yahoo helped jail China writer. *BBC Online*. Retrieved July 31, 2010 from http://www.news.bbc.co.uk/l/hi/world/4221538.stm

Ovide, S. (2013, August 5). Free speech a test for Twitter. *The Wall Street Journal*, pp. B1-2.

Plato, . (1935). *The republic*. Cambridge, MA: Harvard University Press.

Rawls, J. (1971). *A theory of justice*. Cambridge, MA: Harvard University Press.

Rawls, J. (2001). *The law of peoples*. Cambridge, MA: Harvard University Press.

Reno v. ACLU. 521 U.S. 844. (1997).

Schmidt, E., & Cohen, J. (2013). *The new digital age*. New York: Knopf.

Spinello, R. A. (2014). *Cyberethics: Morality and law in cyberspace* (5th ed.). Sudbury, MA: Jones & Bartlett.

Spinello, R. A., & Tavani, H. (2004). *Readings in cyberethics* (2nd ed.). Sudbury, MA: Jones & Bartlett.

Thompson, C. (2006, April 23). China's Google problem. *New York Times Magazine*, 36-41, 73-76.

United Nations Charter. (2007). The Universal Declaration of Human Rights. In R. A. Spinello (Ed.), *Moral philosophy for managers* (5th ed., pp. 293–297). New York: McGraw-Hill.

Vincent, R. J. (1988). *Human rights and international relations*. New York: Cambridge University Press.

Waters, R. (2010, March 24). Realism lies behind decision to quit. *Financial Times*, p. B1.

Zittrain, J. (2003). Internet points of control. *Boston College Law Review. Boston College. Law School*, *44*, 653.

KEY TERMS AND DEFINITIONS

Censorship: The intentional suppression or regulation of expression based on its content.

Code: Hardware and software applications that use Internet protocols and can function as a regulatory constraint.

Content Filtering: Software that restricts access to Internet content by scanning that content based on keyword searches.

Firewall: An electronic barrier restricting communications between two points of control on the Internet.

Formal Cooperation: Intentionally sharing in another person's or group's wrongdoing.

Internet Filtering: Technologies that prevent users from accessing or disseminating information on the Internet.

Pluralism: Different ethical responses to a moral problem, usually based on cultural diversity.

Router: Directs or "routes" Internet traffic to its correct destination.

Search Engine: Navigation tool for searching web sites usually based on proprietary algorithms.

Chapter 9
The Legal Challenges of the Information Revolution and the Principle of "Privacy by Design"

Ugo Pagallo
University of Turin, Italy

ABSTRACT

The chapter examines how the information revolution impacts the field of data protection in a twofold way. On the one hand, the scale and amount of cross-border interaction taking place in cyberspace illustrate how the information revolution affects basic tenets of current legal frameworks, such as the idea of the law as a set of rules enforced through the menace of physical sanctions and matters of jurisdiction on the Internet. On the other hand, many impasses of today's legal systems on data protection, liability, and jurisdiction can properly be tackled by embedding normative constraints into information and communication technologies, as shown by the principle of privacy by design in such cases as information systems in hospitals, video surveillance networks in public transports, or smart cards for biometric identifiers. Normative safeguards and constitutional constraints can indeed be embedded in places and spaces, products and processes, so as to strengthen the rights of the individuals and widen the range of their choices. Although it is unlikely that "privacy by design" can offer the one-size-fits-all solution to the problems emerging in the field, it is plausible that the principle will be the key to understanding how today's data protection issues are being handled.

INTRODUCTION

In order to understand how the information revolution challenges current tenets of the law, focus should be on the ways in which the use of information and communication technologies (ICTs) has changed over the past decades: whereas human societies have been ICTs-related for hundreds of years, but mainly dependent on technologies that revolve around energy and basic resources, today's societies are increasingly dependent on ICTs and, moreover, on information as a vital re-

DOI: 10.4018/978-1-4666-6122-6.ch009

source (Floridi 2014). Several contributions to both *Information Technology Law* (Bainbridge, 2008; Lloyd, 2011; etc.), and *Artificial Intelligence and the Law* (Casanovas et al. 2011; Palmirani et al. 2013; etc.) make this point clear: the information revolution has affected not only the substantial and procedural sides of the law, but its cognitive features as well. The impact of technology on today's legal systems can be fully appreciated through a threefold perspective.

First, technology has engendered new types of lawsuits or modified old ones. As the next generation of offences arose within the field of computer crimes in the early 1990s, technology also impacted on traditional rights such as copyright (1709) and privacy (1890), turning them into a matter of access, control, and protection over information in digital environments (Heide, 2001; Tavani and Moor, 2001; Ginsburg, 2003; Floridi, 2006).

Secondly, technology has blurred traditional national boundaries as information on the internet tends to have a ubiquitous nature. This challenges the very conception of the law as enforced through physical sanctions in the nation-state. Spamming, for instance, offers a good example: it is transnational par excellence and does not diminish despite harshening criminal laws (like the *CAN-SPAM Act* passed by the U.S. Congress in 2003). No threat of sanctions, in other words, seems to limit spamming.

Finally, technology has deeply transformed the approach of experts to legal information. As Herbert A. Simon pointed out in his seminal book on *The Sciences of Artificial*, this transformation is conveniently illustrated by research in design theory, which "is aimed at broadening the capabilities of computers to aid design, drawing upon the tools of artificial intelligence and operations research" (Simon, 1996). While scholars increasingly insist on the specific impact of design or "architecture" and "code" on legal systems (Lessig, 1999; Katyal, 2002; Zittrain, 2008; van Schewick, 2010), both artificial intelligence and operations research not only further design but, in doing so, affect the structure and evolution of legal systems (Pagallo, 2010).

These three levels of impact have, nonetheless, led some scholars to adopt a sort of techno-deterministic approach, leaving no way open to shape or, at least, to influence the evolution of technology. It is enough to mention that some have announced "The End of Privacy" (Sykes, 1999), "The Death of Privacy in the 21st Century" (Jarfinkel, 2000), or "Privacy Lost" (Holtzmann, 2006). On this reading, technology would allow these scholars to unveil an already written future: while, in digital environments, spyware, root-kits, profiling techniques, or data mining would erase data protection, further means like RFID, GPS, CCTV, AmI, or satellites, much as FBI programs like Carnivore or the NSA's 2013 scandal of the Prism project would lead to the same effect in everyday (or analog) life. Strongly decentralized and encrypted architectures providing anonymity to their users, as well as systems that permit plausible deniability and a high degree of confidentiality in communications, suggest however that rumours of the death of privacy have been greatly exaggerated. Techno-deterministic approaches are liable to the same criticism that John Kenneth Galbraith put forward in his own field: "The only function of economic forecasting is to make astrology look respectable." On this basis, taking leave from all sorts of techno-deterministic drifts, the chapter aims to provide a more balanced picture of the current state-of-art, by examining two of the hottest legal topics in data protection, namely, online responsibility and jurisdiction, which are then analyzed in connection with today's debate on the idea of embedding data protection safeguards in ICT and other types of technologies, that is, the principle of "privacy by design." The overall goal is to shed further light on the aforementioned threefold level-impact of technology on contemporary legal systems. Accordingly, the chapter is presented in five sections.

First, the *background of the analysis* sums up the claims of "unexceptionalism." In its substantial form, it vindicates the analogy between cyberspace and the "real world," that is, between digital and traditional boundaries of legal systems. In the phrasing of Allan R. Stein, "*The Internet is a medium*. It connects people in different places. The injuries inflicted over the Internet are inflicted by people on people. In this sense, the Internet is no different from the myriad of ways that people from one place injure people in other places" (Stein, 1998). In its procedural form, "unexceptionalism" argues that traditional tools and principles of international law can find a solution to the regulatory issues of the digital era. These ideas have been adopted by the European authorities on data protection, *i.e.*, the EU Working Party art. 29 D-95/46/EC. Notwithstanding the ubiquity of information on the internet, since the early 2000s the EU WP29 has proposed to solve conflicts of law on the international level through the use of "several alternative criteria for determining extensively the scope of application of national law" (see the WP29 Opinion 5035/01/WP56 from 2002).

There is a paramount difference, however, between cross-border regulations of the internet and the traditional criterion of territoriality, grounded upon the Westphalian paradigm (1648). As remarked in the *section on international conflicts of law*, the right of the states to control events within their territory was originally conceived in a world where cross-border regulations were the exception and not the rule. Vice versa, in a world where virtually all events and transactions have transnational consequences, "unexceptionalism" would result in a form of pan-national jurisdiction covering the entire world (see for instance the aforementioned WP29 Opinion on the regulatory framework for transnational cookies). As I refer in the *section on privacy and design*, such evident drawbacks have pushed scholars and policy makers alike to address issues of responsibility and jurisdiction from another perspective, namely, by embedding data protection safeguards as a default setting in information and communication technologies (ICTs). Since the mid 1990s, the overall idea is to bypass otherwise unsolvable issues of transnational jurisdiction and cross-border liability, by totally or partially preventing the impact of harm-generating behaviour in digital environments.

As the *section on future research directions* illustrates, additional work is required. Taking into account current investigations in the field of artificial intelligence (AI) & Law and, more particularly, in the realm of legal ontologies, we should further examine the modelling of highly context-dependent normative concepts like personal data, security measures, or data controllers, as well as the decomposition of the complete design project into its functional components (Pagallo, 2011 and 2012). Yet, we need not rely either on prophetic powers or on divinatory commitments, to reach a workable conclusion: the principle of "privacy by design" ought to be a default mode of operation for both private companies and public institutions, if, and only if, it strengthens individual rights by widening the range of their choices. Whilst the principle is not likely to offer the one-size-fits-all solution to the problems we are dealing with in relation to data protection, it is nonetheless plausible that suggestions coming from the "privacy by design"-debate will be the key to understand most crucial issues of today's legal framework.

BACKGROUND

Legal debate between advocates of the "unexceptionalist" theses and advocates of the uniqueness of the information revolution is not new. The same problem, after all, arose within the field of computer ethics in 1985. On the side of unexceptionalism, Deborah Johnson's idea was that ICT provides for new ways to instrument human actions which raise specific questions that, nonetheless, would only be a "new species of old moral issues"

(Johnson, 1985). On the other side, Walter Maner insisted on the new generation of problems which are *unique* to computer ethics: "For all of these issues, there was an essential involvement of computing technology. Except for this technology, these issues would not have arisen, or would not have arisen in their highly altered form. The failure to find satisfactory non-computer analogies testifies to the uniqueness of these issues. (…) Lack of an effective analogy forces us to discover new moral values, formulate new moral principles, develop new policies, and find new ways to think about the issues presented to us" (Maner, 1996; and, previously, Moor, 1985).

A decade later, the information revolution forced legal scholars into the debate. In the early 1990s, law-makers introduced the first provisions on computer crimes. In 1995, the European Community approved its first general directive on the protection and processing of personal data, *i.e.*, the aforementioned directive 95/46/EC. A year later, in 1996, it was the turn of the exclusivity rights over public communication pursuant to art. 20 of the Berne Convention (1886) to be complemented by two international copyright treaties, namely, the WIPO's *Copyright Treaty* (WCT) and the *Performances and Phonograms Treaty* (WPPT). In 1998, the U.S. Congress amended the *Digital Performance Rights in Sound Recordings Act* from 1995 with both the *Digital Millennium Copyright Act* (DMCA) and the *Sonny Bono Act* on the extension of exclusivity rights. From the standard fourteen-year term of protection granted by the U.S. *Copyright Act* in 1790, copyright was extended to cover twenty-eight years in 1831, a further twenty-eight years of renewal in 1909, fifty years after the author's death in 1976, down to the current seventy years of protection set up in 1998.

In the unexceptionalism vs. uniqueness debate, it is crucial to distinguish the substantial from the procedural side. What the unexceptionalists have indeed been claiming is that we can (and should) handle the new IT law-cases on computer crimes, data protection, or digital copyright, with the settled principles and traditional tools of international law. In the wording of Jack Goldsmith's *Against Cybernarchy*, IT law-problems are "no more complex and challenging than similar issues presented by increasingly prevalent real-space events such as airplane crashes, mass torts, multistate insurance coverage, or multinational commercial transactions, all of which form the bread and butter of modern conflict of law" (Goldsmith, 1998, p. 1234). On this view, traditional legal tools would be capable to resolve the regulatory problems of the information revolution because "activity in cyberspace is functionally identical to transnational activity mediated by other means, such as mail or telephone or smoke signal" (*op. cit.*, p. 1240).

On the side of the uniqueness advocates, both the scale and amount of cross-border transactions taking place in cyberspace question this "functional identity." According to David Post's criticism of the unexceptionalist ideas, "border-crossing events and transactions, previously at the margins of the legal system and of sufficient rarity to be cabined off into a small corner of the legal universe (…) have migrated, in cyberspace, to the core of that system" (Post, 2002, p. 1380). Like in other fields of scientific research such as physics, biology, or engineering, scale does matter: "A world in which virtually *all* events and transactions have border-crossing effects is surely not 'functionally identical' to a world in which most do not, at least not with respect to the application of a principle that necessarily requires consideration of the distribution of those effects." (*ibid.*)

To further clarify the terms of the debate, let me go back to the 1998 provisions of the DMCA and, more specifically, to the "safe harbour"-clauses set up by section 512 of the US Code. They define responsibility regimes for copyright liability, in a way that corresponds to the European provisions established by articles 12-15 of directive 2000/31/EC on e-commerce. Theoretically speaking, law-givers could have chosen one of three different situations in which individuals and corporations find themselves confronted to copyright liability: (i) legal irresponsibility, (ii) strict liability, and (iii) personal responsibility which depends on "fault."

Following the good old idea that "all which is not prohibited is allowed," the first hypothesis of legal irresponsibility is properly illustrated by the immunity provisions for online speech approved by the U.S. Congress in 1996, pursuant to section 230 of the *Communications Decency Act*: "No provider or user of an interactive computer service shall be treated as the publisher or speaker of any information provided by another content provider." The reason hinges on the fact that intermediaries should not be considered responsible for what users do or say through their network services, so as to foster people's freedom of speech and the flow of information on the internet. Vice versa, the responsibility of traditional publishers clarifies the hypothesis of liability without fault or strict liability. Notwithstanding eventual illicit or culpable behaviour, editors, publishers, and media owners (newspapers, TV channels, etc.), are liable for damages caused by their employees, *e.g.*, pre-digital media's journalists and writers. This mechanism applies to many other cases in which law imposes liability regardless of the person's intention or use of ordinary care, as it occurs with people's responsibility for the behaviour of their animals and, in most legal systems, of their children. Whereas the rationale of the liability involving traditional publishers and editors hinges on the "one-to-many" architecture of pre-digital medias, it seems inappropriate to apply this mechanism of distributing risk to current internet service providers (ISPs), because the architecture of the internet is par excellence "many-to-many."

Still, people are liable mostly for what they voluntarily agree upon through strict contractual obligations and, moreover, for obligations that are imposed by the government to compensate damage done by wrongdoing. There is liability for intentional torts when a person has voluntarily performed the wrongful action prohibited by the law; but legal systems also provide for liability based on lack of due care when the "reasonable" person fails to guard against "foreseeable" harm. This kind of responsibility is neither excluded nor established *a priori*: it is instead grounded on the circumstances of the case. It therefore fits particularly well in this context, as shown by the decision of the EU Court of Justice in the *Google v. Louis Vuitton case* from March 23rd, 2010. Although the judgement concerns issues of trade marks, keyword advertising, and search engines, it allows to clarify the third kind of responsibility which depends on personal fault. "In order to establish whether the liability of a referencing service provider may be limited under Article 14 of Directive 2000/31, it is necessary to examine whether the role played by that service provider is neutral, in the sense that its conduct is merely technical, automatic and passive, pointing to a lack of knowledge or control of the data which it stores" (§ 114 of the decision). The responsibility is thus neither excluded *a priori* (legal irresponsibility), nor established *a priori* (liability without fault), because it depends on "the actual terms on which the service in the cases in the main proceedings is supplied." As a consequence, the Court of Paris should "assess whether the role thus played by Google corresponds to that described in paragraph 114 of the present judgment" (*ibid.*, § 117).

The EU WP29 already remarked this latter point in its 2009 Opinion on social networks. Suggesting some convergences with the legal framework established by section 230(c) of the U.S. *Communications Decency Act*, the EU WP29 affirms that ISPs as well as social network services (SNS) should only be obliged to provide information and adequate warning to users about privacy risks when uploading data: "users should be advised by SNS that pictures or information about other individuals, should only be uploaded with the individual's consent" (WP's Opinion 5/2009). As a result, users are personally responsible for what they do online via social networks, P2P systems, cloud computing, and the like. This has been confirmed by cases of defamation and privacy or copyright infringements. Conversely, ISPs and SNS should be held responsible only when they fail to remove illegitimate content after having been asked to do so by a judicial or administrative authority. The reason for these novel forms of legal irresponsibility for ISPs, which covers activities that provide for, among other things, connectivity, cache content, information location tools and search engines, is summed up by the Judiciary Report of the U.S. Senate Committee from 1998 (S. Rep. No. 105-190, p. 8): "Due to the ease with which digital works can be copied and distributed worldwide virtually instantaneously, copyright owners will hesitate to make their works readily available on the Internet without reasonable assurance that they will be protected against massive piracy. (…) At the same time, without clarification of their liability, service providers may hesitate to make the necessary investment in the expansion of the speed and capacity on the Internet. (…) Many service providers engage in directing users to sites in response to inquiries by users or they volunteer sites that users may find attractive. Some of these sites might contain infringing material. In short, by limiting the liability of service providers, the DMCA ensures that the efficiency of the Internet will continue to improve and that the variety and quality of services on the Internet will continue to expand."

Although this general framework has usually improved the efficiency of the internet, these legal provisions on ISP immunity present some limits of their own. For instance, the notice and takedown-procedure in section 512 of DMCA has now and then been used to censor legitimate criticism or to silence adversarial political speech. Besides, an empirical study on intermediary immunity under section 230 of the *Communications Decency Act* argues that, while section 230 largely protects intermediaries from liability for third-party speech, this provision does not represent the free pass that many of its critics lament. In a nutshell, "judges have been haphazard in their approach to its application" (Ardia, 2010). Moreover, we should not forget the procedural, rather than substantial, side of the debate between unexceptionalism and the uniqueness of the information revolution. Here, attention should be again drawn to the ubiquitous nature of information on the internet and the unexceptionalist claim that every nation state has both the "right to control events within its territory and to protect the citizens [which] permits it to regulate the local effects of extraterritorial acts" (Goldsmith, 1998). Even admitting the uniqueness of the information revolution and, thereby, the novelty of its impact on contemporary legal systems, it does not follow that the settled principles and traditional tools of international law are unable to address the new IT law-cases arisen in digital environments. After all, this mechanism was applied to a well-known American company, when an Italian court admitted the responsibility of the internet provider by sentencing some of Google's executives in the *Vividown case*. The executives were, in fact, held liable for "illicit treatment of personal data" pursuant to article 167 of the Italian Data Protection Code, that is, for allowing a video to be posted showing an autistic youth being abused (Tribunal of Milan, decision 1972 from February 24th, 2010). Although the Court of Appeals in Milan overruled this decision two years later, both courts have rejected the defendants' idea that data processing performed by Google's servers in California is not governed by the Italian law, not even in the case of data transmitted by Italian users. Otherwise – so goes the argument of the courts in Milan – it would be easy to avoid being subject to Italian and European rules by simply locating the company (and its servers) outside EU borders.

In spite of the noble aim to provide global protection of people's fundamental rights, determining the applicability of national law to cross-borders interaction on the internet may end up in a paradox. In order to understand why scale matters not only in physics or biology, but in legal science as well, let us examine how lawyers deal with issues of jurisdiction and criteria for the international resolution of legal conflicts.

INTERNATIONAL CONFLICTS OF LAW

Jack Goldsmith is probably right when affirming that preliminary issues concerning the applicable law in the international arena represent "the bread and butter" of contemporary lawyers (Goldsmith, 1998). However, what the American scholar apparently missed is that the settled principles and traditional tools of both public and private international law are unable to fully meet the challenges set by the current transnational relationships. The conventional representation of the international legal order as grounded upon the principle of national sovereignty – so that "in the absence of consensual international solutions, prevailing concepts of territorial sovereignty permit a nation to regulate the local effects of extraterritorial conduct" (Goldsmith, 1998, p. 1212) – often falls short in tackling a world with no clear boundaries, such as the cyberspace and, even less so, today's computer cloud-environments. The ubiquity of information on the internet leads to the illegitimate situation where a state claims to regulate extraterritorial conduct by imposing norms on individuals who have no say in the decisions affecting them (thus jeopardizing the legitimacy of democratic rule of law). Moreover, this situation determines the ineffectiveness of state action within the realm of cyberspace for citizens are often affected by conduct that the state is simply unable to regulate (Post, 2002).

In order to further illustrate the drawbacks of unexceptionalism, let me go back to the 2002 Opinion of the European authorities on data protection (5035/01/EN/Final WP 56). On that occasion, the WP29 examined "alternative criteria for determining extensively the scope of application of national law," such as the doctrine of selective enforcement, or techniques of the private international law tradition, as the effects-doctrine and the presence of persons, or property, in territory as criteria for solving international conflicts of law, *e.g.*, Articles 3 and 4 of the EU e-commerce directive. Among such criteria, the WP29 found a decisive element in today's cookies, *i.e.*, the file-texts put on the computer's hard disk when people access a web site, in order to determine whether or not EU provisions on data protection should be enforced. According to the aforementioned WP29 Opinion 5/2002, cookies should in fact be considered as "equipment" pursuant to art. 4 (1)c of D-95/46/EC: "Each Member State shall apply the national provisions it adopts pursuant to this Directive to the processing of personal data when (…) the controller is not established on Community territory and, for purposes of processing personal data, makes use of equipment, automated or otherwise, situated on the territory of the said Member State." In addition, the WP29 argued that the aim of the extensive interpretation of the directive was not only to broaden the range of applicability of EU law. By considering cookies to be a sort of equipment pursuant to the European directive on data protection, the WP29 claimed that the goal is to ensure the protection of people's fundamental rights: "The objective of this provision in Article 4 paragraph 1 lit. c) of Directive 95/46/EC is that an individual should not be without protection as regards processing taking place within his country, solely because the controller is not established on Community territory. This could be simply, because the controller has, in principle, nothing to do with the Community. But it is also imaginable that controllers locate their establishment outside the EU in order to bypass the application of EU law."

Later, on December 1st, 2009, this viewpoint was partially confirmed in a joint contribution by the WP29 and the EU Working Party on Police and Justice (WPPJ). In the document on "The Future of Privacy" (02356/09/EN – WP168), both the WP29 and WPPJ admitted that "article

4 of the directive, determining when the directive is applicable to data processing, leaves room for different interpretation." Nevertheless, in accordance with the previous Opinion from May 30th, 2002, the European working parties insisted on the idea that the protection of people's fundamental rights "means that individuals can claim protection also if their data are processed outside the European Union." Whereas the EU cookie directive 2009/136/EC from November 25th, 2009, establishes that users should be provided with clear and comprehensive information about such practices, *e.g.*, third parties storing information on the equipment of the user, the traditional approach to issues of international conflicts of law was confirmed on January 25th, 2012, by the EU Commission's proposal for a new regulation on data protection. In the wording of Article 3 on the "territorial scope" of the new regulation, the latter should be applied "to the processing of personal data of data subjects residing in the Union by a controller not established in the Union, where the processing activities are related to (a) the offering of goods or services to such data subjects in the Union; or (b) the monitoring of their behaviour."

This one-size-fits-all approach to issues of jurisdiction in the field of data protection suggests three main concerns. First, since many EU provisions apply to non-European companies doing business in Europe, this framework involves issues in the field of consumer law as well. For instance, many of these companies have trouble in excluding EU users from their services because, in order to avoid matters of data protection, they would need to know the residence and the name of the users, clearly entailing potential infringements on data protection law and other issues of jurisdiction. Ultimately, this leads to a vicious circle.

Second, the proposed regulation does not consider further ways for "determining extensively the scope of application of national law," in the phrasing of WP29. Consider such techniques of the private international law tradition, as specifying a list of countries for (dis)applying the law, or determining the presence of persons or property in the territory as a condition of enforceability. Whereas such criteria of selective enforcement obviously restrict the "territoriality scope" of the laws, *e.g.*, the "principle of comity," they make the law more effective, and even clearer, in a number of cases. For example, "if a cyberspace actor takes reasonable precautions to avoid collecting data about EU residents then it is not targeting the EU, thus permitting the actor to demonstrate its actual mental state so far as targeting is concerned" (Reed 2012, p. 230).

Third, the traditional approaches to current conflicts of law, whether selective, or generally grounded on the principle of sovereignty, are often ineffective, because the state is simply unable to enforce its own rules on the internet. Significantly, in the Opinion from July 25th, 2007 (2007/C 255/01), the European Data Protection Supervisor (EDPS), Peter Hustinx, emphasized these limits of traditional approaches to current conflicts of law, since "this system, a logical and necessary consequence of the territorial limitations of the European Union, will not provide full protection to the European data subject in a networked society where physical borders lose importance (...): the information on the Internet has an ubiquitous nature, but the jurisdiction of the European legislator is not ubiquitous." Like other cases put forward by contemporary *lex mercatoria*, corporate governance, or human rights litigation, cyberspace issues show the shortcomings of international approaches based upon the principle of sovereignty and nations' right to unilaterally control events within their territories. As Peter Hustinx stressed in the aforementioned Opinion from 2007, the challenge of protecting personal data at the international level "will be to find practical solutions" through typical transnational measures such as "the use of binding corporate rules by multinational companies." Furthermore, we need to promote "private enforcement of data protection principles through self-regulation and competition," while "accepted standards such as the OECD-guidelines for data protection (1980) and UN-Guidelines could be used as basis" for international agreements on jurisdiction and enforceability.

However, such international forms of cooperation and integration do not offer the magic bullet. As confirmed by the passenger name record (PNR)-agreements between the U.S. and Europe, along with the hot debate followed within the European institutions, traditionally close allies may engage in often problematic covenants and conventions (Pagallo, 2008; Brouwer, 2009). In addition, practices of censorship, corruption, bribery and filtering, which are unfortunately spread throughout the world, confirm the difficulties to settle most of today's data protection issues through standard international solutions (Deibert *et al.*, 2008). Consequently, in the 2009 document on "The Future of Privacy," it is telling that both the WP29 and the WPPJ have illustrated a different thesis. Indeed, global issues of data protection should be approached by "incorporating technological protection safeguards in information and communication technologies," according to the principle of privacy by design, which "should be binding for technology designers and producers as well as for data controllers who have to decide on the acquisition and use of ICT." In the next section I will examine how far this idea goes.

PRIVACY AND DESIGN

More than a decade ago, in *Code and Other Laws of Cyberspace*, Lawrence Lessig lamented the poverty of research involving the impact of design on both social relationships and the functioning of legal systems (Lessig, 1999). In a few years, however, this gap has been filled by work on privacy (Ackerman and Cranor, 1999); universal usability (Shneiderman, 2000); informed consent (Friedman, Howe, and Felten, 2002); crime control and architecture (Katyal, 2002, 2003); social justice (Borning, Friedman, and Kahn, 2004); allegedly perfect self-enforcement technologies on the internet (Zittrain, 2007); "design-based instruments for implementing social policy that will aid our understanding of their ethical, legal and public policy complexities" (Yeung, 2007); and so forth. In particular, Karen Yeung has proposed a theory of legal design, by distinguishing between the subjects in which the design is embedded and the underlying design mechanisms or "modalities of design." On one side, it is feasible to design not only places and spaces, products and processes, but biological organisms as well. This is the case of plants grown through OGM technology, or of genetically modified animals like Norwegian salmons, down to the current debate on humans, post-humans, and cyborgs. On the other side, the modalities of design may aim to encourage the change of social behaviour, to decrease the impact of harm-generating conducts, or to prevent that those harm-generating conducts may even occur. As an illustration of the first kind of design mechanisms, consider the installation of speed bumps in roads as a means to reduce the velocity of cars (lest drivers opt to destroy their own vehicles). As an example of the second modality of design, think about the introduction of air-bags to reduce the impact of harm-generating conduct. Finally, as an instance of total prevention, it is enough to mention current projects on "smart cars" able to stop or to limit their own speed according to the driver's health conditions and the inputs of the surrounding environment. In light of Karen Yeung's taxonomy and its nine possible combinations between subjects (*i.e.*, places, products, organisms) and modalities (*i.e.*, behavioural change and reduction or prevention of harm-generating conducts), what are the relevant scenarios in the field of privacy and design? Leaving aside cases of genetically modified salmons and of OGM plants, what about the most interesting hypotheses for data protection?

In the aforementioned document on "The Future of Privacy," the WP29 and the WPPJ pointed out the goals that should be reached by embedding appropriate technical and organizational measures "both at the time of the design of the processing system and at the time of the processing itself, particularly in order to maintain security and thereby to prevent any unauthorized processing" of personal data (as the recital 46 of directive 95/46/EC establishes). Specifically, the European authorities on data protection claim that the principle of privacy by design "should be binding for technology designers and producers as well as for data controllers who have to decide on the acquisition and use of ICT," so that data minimization and quality of the data should be ensured together with its controllability, transparency, confidentiality, and user friendliness of information interfaces. Among the examples of how the new principle can contribute to better data protection, the EU Working Parties recommend that biometric identifiers "should be stored in devices under control of the data subjects (*i.e.*, smart cards) rather than in external data bases." In addition, the EU authorities suggest that making personal data anonymous both in public transportation systems and in hospitals should be considered a priority. While, in the first case, video surveillance must be designed in such a way that faces of individuals cannot be recognizable, in the case of hospitals' information systems patient names should be kept separated from data on medical treatments and health status (Pagallo and Bassi, 2010).

The soft law-proposal according to which the principle of privacy by design should be applied "as early as possible," namely, at the time of the design of the processing system, has been deepened by the EU Commission's proposal for a new regulation on data protection (2012/0011 (COD) from January 2012). More particularly, the principle sets two new obligations for data controllers on "data protection by design and by default," in accordance with Article 23 of the proposed regulation. On one side, "having regard to the state of the art and the cost of implementation, the controller shall, both at the time of the determination of the means for processing and at the time of the processing itself, implement appropriate technical and organisational measures and procedures in such a way that the processing will meet the requirements of this Regulation and ensure the protection of the rights of the data subject" (Art. 23(1)). On the other side, "the controller shall implement mechanisms for ensuring that, by default, only those personal data are processed which are necessary for each specific purpose of the processing and... in particular, those mechanisms shall ensure that by default personal data are not made accessible to an indefinite number of individuals" (Art. 23(2)).

These and further provisions of the EU Commission's proposal have sparked much controversy, *e.g.*, Article 17 on "the right to be forgotten" (Pagallo and Durante, 2014). Yet, it is still unclear how we should interpret the aim to enforce data protection by design. Should we conceive the latter as a means for total prevention? Could privacy by design constitute an infallible self-enforcing technology preventing harm generating-conducts overall? Is this aim either achievable, or desirable?

First, contemplate the technical difficulties of modelling concepts traditionally employed by lawyers, through the formalization of norms, rights, or duties, to fit the processing of a machine. As a matter of fact, "a rich body of scholarship concerning the theory and practice of 'traditional' rule-based regulation bears witness to the impossibility of designing regulatory standards in the form of legal rules that will hit their target with perfect accuracy" (Yeung, 2007). As Eugene Spafford warns, legal scholars should understand that "the only truly secure system is one that is powered off, cast in a block of concrete and sealed in a lead-lined room with armed guards – and even then I have my doubts" (Garfinkel and Spafford, 1997).

Secondly, privacy is not a zero-sum game between multiple instances of access, control, and protection over information in digital environments. Personal choices indeed play the main role when individuals modulate different levels of access and control, depending on the context and its circumstances (Nissenbaum, 2004). After all, people may enjoy privacy in the midst of a crowd and without having total control over their personal data, whereas total control over that data does not necessarily entail any guarantee of privacy (Tavani, 2007).

Finally, there are ethical issues behind the use of self-enforcing technologies, since people's behaviour would unilaterally be determined on the basis of automatic techniques rather than by choices of the relevant political institutions. A kind of infallible self-enforcing technology, in other words, not only "collapses the public understanding of law with its application eliminating a useful interface between the law's terms and its application" (Zittrain, 2007). What is more, there are instances of self-enforcing technology, *e.g.*, Digital Rights Management (DRM), which enable copyright holders to monitor and regulate the use of their protected artefacts, thus impinging on people's privacy again (Pagallo, 2008).

These various possible applications do not imply that technology is simply "neutral," a bare means to obtain whatever end. Rather, the idea of design, responsibility, and jurisdiction brings us back to the fundamentally political aspects of data protection. As stressed by Flanagan, Howe and Nissenbaum (2008), "some technical artefacts bear directly and systematically on the realization, or suppression, of particular configurations of social, ethical, and political values." As a result, we might end up in a vicious circle by embedding values in technology, *e.g.*, through design policies. Consider, for instance, how conflicting values or interpretations thereof may impact on the very design of an artefact. Likewise, striking a balance between different goals design can aim at is often necessary, so that multiple choices of design can result in (another) conflict of values. Therefore, is the principle of "privacy by design" replicating the substantial and procedural divergences we find in today's debate among lawyers and policy makers? Was not the principle intended to overcome possible conflicts among jurisdictions, and between jurisdiction and enforcement, due to a nation state's refusal to enforce the judgements of foreign courts, by embedding data protection safeguards in ICT and other types of technology?

My view is that such a paradoxical conclusion can be rejected, if, and only if, we assume the use of self-enforcing technologies as the exception, or last resort option, for coping with the challenges of the information revolution. In other words, issues on data protection should be grasped 'by' design, not 'as' design: to conceive data protection 'as design' would mean to aim at some kind of self-enforcing technology eliminating Zittrain's "useful interface" between legal rules and their application (Zittrain, 2008). What is at stake concerns the opportunity of reducing the impact of harm-generating conducts by strengthening people's rights and widening their choices in digital environments. Otherwise, compliance with regulatory frameworks through design policies would be grounded either on a techno-deterministic viewpoint proposing to solve data protection issues 'as' simply matter of design, or on a paternalistic approach planning to change individual behaviour. Consequently, we may claim that "privacy assurance must ideally become an organization's default mode of operation" (Cavoukian, 2010), if, and only if, the principle of privacy by design is devoted to broaden the options of the individuals by letting people take security measures by themselves. Moreover, the principle of privacy by design suggests we prevent some of the conflicts among values and interpretations by adopting the bottom-up approach put forward by the European working parties on data protection, that is, self-regulation and competition among private organizations, within the parameters established by the explicit and often detailed guides of the national authorities. Further conflicts among values and divergent aims of design are mitigated by a stricter (but more effective) version of the principle, according to which the goal is to reinforce people's pre-existing autonomy, rather than building it from scratch (Pagallo, 2011).

Open issues persist concerning the technical feasibility of replacing standard international agreements with design patterns so as to prevent online related conflicts over jurisdiction, substantial divergences on the role of people's consent, and the opt-in vs. opt-out diatribe involving the decisions of data subjects. In addition, scholars are confronted with recent developments in artificial intelligence, which are disclosing new perspectives in how we can deal with flows of information in digital environments. A section on future research directions is thus required.

FUTURE RESEARCH DIRECTIONS

Over the last decade and a half privacy commissioners and national authorities have discussed the idea of embedding data protection safeguards in ICTs. While the obligation of data controllers to implement appropriate technical and organizational measures was laid down in the first European directive on data protection, namely, in art. 17 of D-95/46/EC, the concept of "Privacy by Design" was further developed by the Ontario's Privacy Commissioner, Ann Cavoukian, in the late 1990s, up to Article 23 of the aforementioned proposal for a new regulation on data protection, presented by the EU Commission on January 2012. Besides national advisors and working parties on data protection, scholars have dwelled on the topic as well. There has been seminal work on the ethics of design (Friedman, 1986; Mitcham, 1995; Whitbeck, 1996), and privacy (Agre, 1997), and several publications have focused on data protection issues involved in the design of ICT by the means of value-sensitive design (Friedman and Kahn, 2003; Friedman *et al.*, 2006), and of legal ontologies (Abou-Tair and Berlik, 2006; Mitre *et al.*, 2006; Lioukadis *et al.*, 2007; etc.). The topic being very popular, is there any particular reason why the principle of "privacy by design" is cutting the edge among scholars today? All in all, this growing interest depends on three reasons: they are political, ethical, and pretty technical.

First, most of the provisions on data protection and design have been disappointing. As frankly stated by the EU WP29 and the WPPJ in their joint document on "The Future of Privacy," a new legal framework is indispensable and it "has to include a provision translating the currently punctual requirements into a broader and consistent principle of privacy by design." However, as mentioned in the previous section, the proposal of the EU Commission does not seem to have attained this end, since many provisions on the right to be forgotten, on new obligations for ISPs, and new powers for the Commission on the technical measures and design mechanisms that should be reckoned as compulsory, have sparked much controversy.

Second, the call for "a broader and consistent principle of privacy by design" depends on the asymmetry between the ubiquity of information on the internet and the fact that national provisions and jurisdictions are not. In addition to traditional approaches to international conflicts of law, a viable solution could be to implement privacy safeguards in ICT as default settings, by promoting the enforcement of data protection standards through a bottom-up, rather than top-down, approach, that is, via self-regulation, competition, and the use of binding corporate rules by multinational companies.

Third, research on the principle of privacy by design is continuously stimulated by developments in the field of artificial intelligence and operations research, which may not only improve the science of design, but cast further light on the structure and evolution of legal systems. As an interesting example, consider the "Neurona Ontology"-project developed by Pompeu Casanovas and his research team in Barcelona, Spain (Casellas, 2010). Here, the field of "legal ontologies" is the key to implement new technological advances in managing personal data and providing organizations and citizens "with better guarantees of proper access, storage, management and sharing of files." The explicit goal of the project is to help company officers and citizens "who may have little or no legal knowledge whatsoever," when processing personal data in accordance with mandatory frameworks in force.

The aim of legal ontologies is in fact to model concepts traditionally employed by lawyers through the formalization of norms, rights, and duties, in fields like criminal law, administrative law, civil law, etc. The objective being that even a machine should comprehend and process this very information, it is necessary to distinguish between the part of the ontology containing all the relevant concepts of the problem domain through the use of taxonomies, and the ontology which includes both the set of rules and constraints that belong to that problem domain (Breuker *et al.*, 2008). An expert system should therefore process the information in compliance with regulatory frameworks on data protection through the conceptualization of classes, relations, properties, and instances pertaining to a given problem domain.

However, several provisions of data protection show how such regulations not only include "top normative concepts," as notions of validity, obligation, prohibition, and the like. These rules also present highly context-dependent normative concepts as in the case of personal data, security measures, or data controllers. These notions raise a number of relevant issues when reducing the informational complexity of a legal system where concepts and relations are subject to evolution (Pagallo, 2010). For example, we already met some hermeneutical issues in data protection law, *e.g.*, matters of jurisdiction and definitions of equipment, which can hardly be reduced to an automation process. These technical difficulties make it clear why several projects of legal ontologies have adopted a bottom-up rather than a top-down approach, "starting from smaller parts and sub-solutions to end up with global" answers (Casellas, 2010). While splitting the work into several tasks and assigning each to a working team, the evaluation phase consists in testing the internal consistency of the project and, according to Herbert A. Simon's "generator test-cycle," involves the decomposition of the complete design into functional components. By generating alternatives and examining them against a set of requirements and constraints, "the test guarantees that important indirect consequences will be noticed and weighed. Alternative decompositions correspond to different ways of dividing the responsibilities for the final design between generators and tests" (Simon, 1996).

Leaving aside criteria such as the functional efficiency, robustness, reliability, elegance, and usability of design projects, the ability to deal with our own ignorance helps us striking a fair balance between the know-how of current research on privacy by design and its limits. The unfeasible dream of automatizing all data protection does not imply that expert systems and friendly interfaces cannot be developed.to achieve such goals as enabling businesses and individuals to take relevant security measures by themselves, while enhancing people's rights and encouraging their behavioural change, so as to restrict the discretion of company officers and public bureaucrats. Moreover, in the field of police and criminal justice data protection, an ongoing EU project hinges on the formalization of legal ontologies, so as to enforce the principle of privacy by design and strike a balance between the flow of information between Law Enforcement Agencies (LEAs) and the protection of people's fundamental rights (CAPER, 2013). Waiting for further applications of the principle in, for example, smart environments, online social lending, data loss prevention, wrap contracts, business, digital forensics, or the civilian use of drones (Pagallo, 2013), it is highly likely that "privacy by design" will represent the privileged understanding of our data protection abilities.

CONCLUSION

This chapter focused on the three level-impact of technology on current legal systems, considering the substantial, procedural, and cognitive features of the subject.

First, I dwelled on the substantial impact taking into account matters of responsibility on the internet. The main reason why, in the U.S. as in Europe, law-makers finally opted for a new generation of "safe harbour"-clauses and immunity provisions for copyright liability and freedom of speech, depends on the crucial difference between the "one-to-many" structure of pre-digital medias and the "many-to-many" architecture of the internet.

I then examined the procedural features of this technological change including issues of jurisdiction as well as different ways to work out traditional international conflicts of law. What used to be the exception has now turned into the rule, in that virtually all events and transactions have transnational effects on the internet. The consequence is a fundamental asymmetry between the ubiquity of information in digital environments and circumscribed territoriality of national jurisdictions, so that tools and settled principles of international law fall often short when meeting the challenge of this novel kind of contrast.

Finally, I considered the cognitive implications of technology and how artificial intelligence and operations research help us addressing the new legal issues in the field of privacy by design. Work on legal ontologies and the development of expert systems illustrated some of the automated ways in which it is feasible to process and control personal data in compliance with regulatory frameworks, so as to advise company officers and citizens "who may have little or no legal knowledge whatsoever" (Casellas, 2010).

Emphasis on this three level-impact of technology, however, does not ignore the mutual interaction through which political decisions influence developments in technology, while technology is reshaping key legal concepts and their environmental framework. Ultimately, the introduction of a new generation of "safe harbour"-clauses and of immunity provisions for copyright liability and freedom of speech makes evident the role of political decisions in, for example, "the future of the internet" (Zittrain, 2008), the improvement of P2P file sharing-applications systems (Pagallo and Durante, 2009), and the like. Still, political decisions have their own limits when it comes to problems of responsibility and jurisdiction concerning data protection. Leaving behind the pitfalls of "unexceptionalism" as well as the panacea of standard international agreements, it is noteworthy that privacy authorities, commissioners, and law makers have suggested to insert data protection safeguards in ICT at the time of design of these processing systems. As an ideal default setting, the principle allows us to remark the most relevant cases of the aforementioned taxonomy on subjects and modalities of design (Yeung, 2007), in the field of data protection.

On the one hand, regarding the application of the principle, we should focus on places and spaces, products and processes, rather than other human fellows. (Apart from Sci-Fi hypotheses – remember the scene from *Minority Report* where Tom Cruise acquires eye bulbs at the black market – it would be illegal, and even morally dubious, to redesign individuals so as to protect their personal data. According to today's state-of-art, ethical and legal issues of human design involve contemporary debate on cyborgs and robotics, rather than data protection through design policies.) On the other hand, as far as the modalities of design are concerned, the aim to prevent all sorts of harm-generating behaviour and, hence, conflicts of law at the international level, looks neither achievable nor desirable through the use of an allegedly infallible self-enforcing technology. While provisions on data protection include highly context-dependent normative concepts, which are hardly reducible to an automation process, the adoption of self-enforcing technologies would unilaterally end up determining people's behaviour on the basis of technology rather than by choices of the relevant political institutions.

Both these practical and ethical constraints do not imply that design policies should lower their goal to just changing the individual behaviour and decreasing the impact of harm-generating conducts. Embedding data protection safeguards in, say, hospitals' information systems, transports' video surveillance networks, or smart cards for biometric identifiers, is ultimately legitimized by the intention to strengthen people's rights and give a choice or widen the range of choices. Otherwise, combining compliance with regulatory frameworks and design policies would end up in paternalism modelling individual behaviour, or in chauvinism disdaining different national provisions of current legal systems. This stricter version of the principle of privacy by design finally addresses design choices that may result in conflicts among values and, vice versa, different interpretations of values that may impact on the features of design. Since most of the projects have to comply with often detailed and explicit guidance of legislators and privacy authorities, it is likely that the empirical evaluation and verification of design projects are going to play a crucial role in determining whether individual rights have been protected or not.

However, far from delivering any value-free judgement, such an experimental phase of assessment is intertwined with the political responsibilities grounding the guidance and provisions of law makers and privacy commissioners. At the end of the day, by insisting on the need to broaden the range of personal choices in digital environments, the stricter version of the principle makes it clear why matters of data protection do not only rely on technology.

REFERENCES

Abou-Tair, D., & Berlik, S. (2006). An ontology-based approach for managing and maintaining privacy in information systems. *Lecture Notes in Computer Science*, *4275*, 983–994. doi:10.1007/11914853_63

Ackerman, M. S., & Cranor, L. (1999). Privacy critics: UI components to safeguard users' privacy. In *Extended Abstracts of CHI* (pp. 258-259). New York: ACM Press.

Agre, P. E. (1997). Introduction. In P. E. Agre, & M. Rotenberg (Eds.), *Technology and privacy: The new landscape* (pp. 1–28). Cambridge, MA: The MIT Press.

Ardia, D. S. (2010). Free speech savior or shield for scoundrels: An empirical study of intermediary immunity under section 230 of the communications decency act. *Loyola of Los Angeles Law Review*, *43*(2), 373–505.

Bainbridge, D. (2008). *Introduction to information technology law*. London, UK: Pearson.

Borning, A., Friedman, B., & Kahn, P. (2004). Designing for human values in an urban simulation system: Value sensitive design and participatory design. In *Proceedings of Eighth Biennal Participatory Design Conference* (pp. 64-67). Toronto, Canada: ACM Press.

Breuker, J., Casanovas, P., Klein, M. C. A., & Francesconi, E. (Eds.). (2008). *Law, ontologies and the semantic web: Channelling the legal information flood*. Amsterdam, The Netherlands: IOS Press.

Brouwer, E. (2009). The EU passenger name record (PNR) system and human rights: Transferring passenger data or passenger freedom?. *CEPS Working Document, 320*.

CAPER. (2013). *Consolidated Review Report (EU Project 261712)*. CAPER.

Casanovas, P., Pagallo, U., Sartor, G., & Ajani, G. (2010). *AI approaches to the complexity of legal systems: Complex systems, the semantic web, ontologies, argumentation, and dialogue*. Dordrecht, The Netherlands: Springer.

Casellas, N., Nieto, J.-E., Meroño, A., Roig, A., Torralba, S., Reyes, M., & Casanovas, P. (2010). Ontological semantics for data privacy dompliance: The NEURONA project. In *Proceedings of AAAI Spring Symposium: Intelligent Information Privacy Management*. AAAI.

Cavoukian, A. (2010). Privacy by design: the definitive workshop. *Identity in the Information Society*, *3*(2), 247–251. doi:10.1007/s12394-010-0062-y

Deibert, R. J., Palfrey, J. G., Rohozinski, R., & Zittrain, J. (2008). *Access denied: The practice and policy of global internet filtering*. Cambridge, MA: The MIT Press.

Flanagan, M., Howe, D. C., & Nissenbaum, M. (2008). Embodying values in technology: Theory and practice. In J. van den Hoven, & J. Weckert (Eds.), *Information technology and moral philosophy* (pp. 322–353). New York: Cambridge University Press.

Floridi, L. (2006). Four challenges for a theory of informational privacy. *Ethics and Information Technology*, *8*(3), 109–119. doi:10.1007/s10676-006-9121-3

Floridi, L. (2014). *The fourth revolution – The impact of information and communication technologies on our Lives*. Oxford, UK: Oxford University Press.

Friedman, B. (1986). Value-sensitive design. *Interaction*, *3*(6), 17–23.

Friedman, B., Howe, D. C., & Felten, E. (2002). Informed consent in the Mozilla browser: Implementing value-sensitive design. In *Proceedings of 35th Annual Hawaii International Conference on System Sciences*. IEEE Computer Society.

Friedman, B., & Kahn, P. H. Jr. (2003). Human values, ethics, and design. In J. Jacko, & A. Sears (Eds.), *The human-computer interaction handbook* (pp. 1177–1201). Mahwah, NJ: Lawrence Erlbaum Associates.

Friedman, B., Kahn, P. H., Jr., & Borning, A. (2006). Value sensitive design and information systems. In Human-computer interaction in management information systems: Foundations (pp. 348-372). New York: Armonk.

Garfinkel, S., & Spafford, G. (1997). *Web security and commerce*. Sebastopol, CA: O'Reilly.

Ginsburg, J. (2003). From having copies to experiencing works: The development of an access right in US copyright law. *Journal of the Copyright Society of the USA*, *50*, 113–131.

Goldsmith, J. (1998). Against cyberanarchy. *The University of Chicago Law Review. University of Chicago. Law School*, *65*(4), 1199–1250. doi:10.2307/1600262

Heide, T. (2001). Copyright in the EU and the U.S.: What access right? *European Intellectual Property Review*, *23*(8), 469–477.

Holtzman, D. H. (2006). *Privacy lost: How technology is endangering your privacy*. New York: Jossey-Bass.

Jarfinkel, S. (2000). *Database nation: The death of privacy in the 21st century*. Sebastopol, CA: O'Reilly.

Johnson, D. (1985). *Computer ethics*. Englewood Cliffs, NJ: Prentice-Hall.

Katyal, N. (2002). Architecture as crime control. *The Yale Law Journal*, *111*(5), 1039–1139. doi:10.2307/797618

Katyal, N. (2003). Digital architecture as crime control. *The Yale Law Journal*, *112*(6), 101–129.

Lessig, L. (1999). *Code and other laws of cyberspace*. New York: Basic Books.

Lioukadis, G., Lioudakisa, G., Koutsoloukasa, E., Tselikasa, N., Kapellakia, S., & Prezerakosa, G. et al. (2007). A middleware architecture for privacy protection. *The International Journal of Computer and Telecommunications Networking*, *51*(16), 4679–4696.

Lloyd, I. (2008). *Information technology law*. Oxford, UK: Oxford University Press.

Maner, W. (1996). Unique ethical problems in information technology. *Science and Engineering Ethics*, *2*, 137–154. doi:10.1007/BF02583549

Mitcham, C. (1995). Ethics into design. In R. Buchanan, & V. Margolis (Eds.), *Discovering design* (pp. 173–179). Chicago: University of Chicago Press.

Mitre, H., González-Tablas, A., Ramos, B., & Ribagorda, A. (2006). A legal ontology to support privacy preservation in location-based services. *Lecture Notes in Computer Science*, *4278*, 1755–1764. doi:10.1007/11915072_82

Moor, J. (1985). What is computer ethics? *Metaphilosophy*, *16*(4), 266–275. doi:10.1111/j.1467-9973.1985.tb00173.x

Nissenbaum, H. (2004). Privacy as contextual integrity. *Washington Law Review (Seattle, Wash.)*, *79*(1), 119–158.

Pagallo, U. (2008). *La tutela della privacy negli USA e in Europa: Modelli giuridici a confronto*. Milano: Giuffrè.

Pagallo, U. (2010). As law goes by: topology, ontology, evolution. In P. Casanovas et al. (Eds.), *AI approaches to the complexity of legal systems* (pp. 12–26). Dordrecht, The Netherlands: Springer. doi:10.1007/978-3-642-16524-5_2

Pagallo, U. (2011). ISPs & rowdy web sites before the law: Should we change today's safe harbour clauses? *Philosophy and Technology*, *24*(4), 419–436. doi:10.1007/s13347-011-0031-x

Pagallo, U. (2012). On the principle of privacy by design and its limits: Technology, ethics, and the rule of Law. In S. Gutwirth, R. Leenes, P. De Hert, & Y. Poullet (Eds.), *European Data Protection: In Good Health?* (pp. 331–346). Dordrecht, The Netherlands: Springer. doi:10.1007/978-94-007-2903-2_16

Pagallo, U. (2013). Robots in the cloud with privacy: A new threat to data protection? *Computer Law & Security Report*, *29*(5), 501–508. doi:10.1016/j.clsr.2013.07.012

Pagallo, U., & Bassi, E. (2011). The future of EU working parties' the future of privacy and the principle of privacy by design. In M. Bottis (Ed.), *An Information Law for the 21st Century* (pp. 286–309). Athens, Greece: Nomiki Bibliothiki Group.

Pagallo, U., & Durante, M. (2009). Three roads to P2P systems and their impact on business practices and ethics. *Journal of Business Ethics*, *90*(4), 551–564. doi:10.1007/s10551-010-0606-y

Pagallo, U., & Durante, M. (2014). *Legal memories and the right to be forgotten*. In L. Floridi (Ed.), Protection of Information and the Right to Privacy - A New Equilibrium? (pp. 17-30). Dordrecht: Springer.

Palmirani, M., Pagallo, U., Sartor, G., & Casanovas, P. (2012). *AI approaches to the complexity of legal systems: Models and ethical challenges for legal systems, legal language and legal ontologies, argumentation and software agents*. Dordrecht, The Netherlands: Springer.

Post, D. G. (2002). Against against cyberanarchy. *Berkeley Technology Law Journal*, *17*(4), 1365–1383.

Reed, C. (2012). *Making laws for cyberspace*. Oxford, UK: Oxford University Press.

Shneiderman, N. (2000). Universal usability. *Communications of the ACM, 43*(3), 84–91. doi:10.1145/332833.332843

Simon, H. A. (1996). *The sciences of the artificial*. Cambridge, MA: The MIT Press.

Stein, A. R. (1998). The unexceptional problem of jurisdiction in cyberspace. *International Lawyer, 32*, 1167–1194.

Sykes, C. (1999). *The end of privacy: The attack on personal rights at home, at work, on-line, and in court*. New York: St. Martin's Griffin.

Tavani, H. T. (2007). Philosophical theories of privacy: Implications for an adequate online privacy policy. *Metaphilosophy, 38*(1), 1–22. doi:10.1111/j.1467-9973.2006.00474.x

Tavani, H. T., & Moor, J. H. (2001). Privacy protection, control of information, and privacy-enhancing technologies. *Computers & Society, 31*(1), 6–11. doi:10.1145/572277.572278

van Schewick, B. (2010). *Internet architecture and innovation*. Cambridge, MA: The MIT Press.

Whitbeck, C. (1996). Ethics as design: doing justice to moral problems. *The Hastings Center Report, 26*(3), 9–16. doi:10.2307/3527925 PMID:8736668

Yeung, K. (2007). Towards an understanding of regulation by design. In R. Brownsword, & K. Yeung (Eds.), *Regulating technologies: Legal futures, regulatory frames and technological fixes* (pp. 79–108). London, UK: Hart Publishing.

Zittrain, J. (2007). Perfect enforcement on tomorrow's internet. In R. Brownsword, & K. Yeung (Eds.), *Regulating technologies: legal futures, regulatory frames and technological fixes* (pp. 125–156). London, UK: Hart Publishing.

Zittrain, J. (2008). *The future of the internet and how to stop it*. New Haven, CT: Yale University Press.

ADDITIONAL READING

Aarts, E., & Encarnacao, L. K. (Eds.). (2006). *True visions: the emergence of ambient intelligence*. Berlin, Dordrecht: Springer. doi:10.1007/978-3-540-28974-6

De Cew, J. W. (1997). *In pursuit of privacy: law, ethics, and the rise of technology*. Ithaca, NY: Cornell University Press.

Etzioni, A. (2005). Limits of privacy. In A. I. Cohen, & C. H. Wellman (Eds.), *Contemporary debates in applied ethics* (pp. 253–262). Oxford: Blackwell.

Floridi, L. (2007). A look into the future impact of ICT on our lives. *The Information Society, 23*(1), 59–64. doi:10.1080/01972240601059094

Fried, Ch. (1990). Privacy: a rational context. In M. D. Ermann, M. B. Williams, & C. Gutierrez (Eds.), *Computers, ethics, and society* (pp. 50–63). New York: Oxford University Press.

Friedman, B. (Ed.). (1997). *Human values and the design of computer technology*. Cambridge, UK: Cambridge University Press.

Gavison, R. (1980). Privacy and the limits of the law. *The Yale Law Journal, 89*(3), 421–471. doi:10.2307/795891

Grodzinsky, F. S., & Tavani, H. T. (2005). P2P networks and the Verizon v. RIAA case: implications for personal privacy and intellectual property. *Ethics and Information Technology, 7*(4), 243–250. doi:10.1007/s10676-006-0012-4

Gutwirth, S., Leenes, R., & De Hert, P. (2014). *Reloading data protection. Multidisciplinary insights and contemporary challenges*. Dordrecht: Springer. doi:10.1007/978-94-007-7540-4

Hildebrandt, M. (2011). Legal protection by design: objections and refutations. *Legisprudence, 2*(5), 223–248. doi:10.5235/175214611797885693

Hongladarom, S., & Ess, C. (Eds.). (2006). *Information technology ethics: cultural perspectives*. Hershey, Pennsylvania: Idea Publishing. doi:10.4018/978-1-59904-310-4

Hughes, T. (2004). *Human-built world: how to think about technology and culture*. Chicago: University of Chicago Press.

Katyal, S. (2004). Privacy vs. piracy. *Yale Journal of Law and Technology*, *7*, 222–345.

Krug, S. (2005). *Don't make me think*. Indianapolis: New Riders.

Lacy, S. (2001). *Crypto: how the code rebels beat the government – saving privacy in the digital age*. New York: Viking.

Lessig, L. (2002). Privacy as property. *Social Research*, *69*(1), 247–269.

Murray, A. (2007). *The regulation of cyberspace: control in the online environment*. Abingdon, UK: Routledge Cavendish.

Nissenbaum, H. (1998). Protecting privacy in an information age: the problem of privacy in public. *Law and Philosophy*, *17*(5-6), 559–596.

Norman, D. A. (2007). *The design of future things*. New York: Basic Books.

Post, D. (2009). *In search of Jefferson's moose: notes on the state of cyberspace*. New York: Oxford University Press.

Prosser, W. (1960). Privacy. *California Law Review*, *48*(3), 383–423. doi:10.2307/3478805

Regan, P. M. (1995). *Legislating privacy: technology, social values, and public policy*. Chapel Hill, NC: University of North Carolina Press.

Roessler, B. (2005). *The value of privacy*. Cambridge, UK: Polity Press.

Rosen, J. (2001). *The unwanted gaze: the destruction of privacy in America*. New York: Knopf.

Schmidt, E., & Cohen, J. (2013). *The new digital age*. London: John Murray.

Slobogin, Ch. (2007). *Privacy at risk: the new government surveillance and the fourth amendment*. Chicago: The University of Chicago Press. doi:10.7208/chicago/9780226762944.001.0001

Solove, D. J. (2004). *The digital person: technology and privacy in the information age*. New York: The New York University Press.

Solove, D. J., Rotenberg, M., & Schwartz, P. M. (2006). *Privacy, information, and technology*. New York: Aspen.

Solum, L. B. (2009). Models of internet governance. In L. A. Bygrave, & J. Bing (Eds.), *Internet governance: infrastructure and institutions* (pp. 48–91). New York: Oxford University Press. doi:10.1093/acprof:oso/9780199561131.003.0003

van den Hoven, J., & Weckert, J. (Eds.). (2008). *Information technology and moral philosophy*. New York: Cambridge University Press.

Volkman, R. (2003). Privacy as life, liberty, property. *Ethics and Information Technology*, *5*(4), 199–210. doi:10.1023/B:ETIN.0000017739.09729.9f

Wright, D. (2012). The state of the art in privacy impact assessment. *Computer Law & Security Report*, *28*(1), 54–61. doi:10.1016/j.clsr.2011.11.007

Wright, D., & De Hert, P. (Eds.). (2012). *Privacy impact assessment*. Dordrecht: Springer. doi:10.1007/978-94-007-2543-0

Wu, T. (2006). *Who controls the internet*. New York: Oxford University Press.

KEY TERMS AND DEFINITIONS

Data Protection: The ideal condition regarding the processing of personal information, in order to assure the protection of the individual right to access, modify, delete, and refuse the processing of data at any given time. Individual rights to data protection entail obligations for the entities processing and controlling personal data, *e.g.*, the duty of processing personal data fairly and lawfully, by informing the data subjects, so that they can give their consent when required by the law.

Design: The traditional act of working out the form of something or someone, which has been broadened by the current capacities of computers to draw upon the tools of artificial intelligence and operations research. Design can aim to encourage the change of social behaviour, decreasing the impact of harm-generating conducts, or preventing harm-generating behaviour from occurring. Spaces and places, processes and products, down to biological organisms like plants, animals, and other human fellows, may be the objects of design.

Jurisdiction: In Ancient Roman law, the power to "say the law" (*dicere ius*); *i.e.*, to interpret and give law to a certain territory over which that power is exercised. In modern private and public international law, several criteria may be adopted to solve conflicts of law between national legal systems. In the absence of consensual international solutions, the state claims a right to control events within its territory so as to regulate the local effects of extraterritorial acts.

Privacy: The old "right to be let alone" that technology has updated by including a need to protect personal data of those who live, work, and interact in digital environments. While, in the U.S., a property standpoint still prevails, making consent the cornerstone in most of the current debate, in Europe privacy is mainly associated with the principle of human dignity and, therefore, considered an inalienable right of the person.

Privacy by Design: The idea of embedding data protection safeguards in ICT and other types of technologies, with the aim to process and control personal data in compliance with current regulatory frameworks. In accordance with today's state-of-art, the principle prohibits the redesigning of other human fellows in order to protect their personal data. The goal is rather the implementation of data protection safeguards in places and spaces, products and processes, so as to strengthen people's rights and widen the range of their choices.

Responsibility: The moral force binding people to their obligations and making them respond to their conscience and, eventually, to other fellows' expectations. From a legal viewpoint, we distinguish between legal irresponsibility, strict liability, and responsibility due to personal fault. While people are mostly liable for what they voluntarily agree upon through strict contractual obligations, there are also obligations imposed by the government to compensate for damage caused by wrongdoing or other damaging behaviour, so as to distribute risk among consociates.

Technology: The know-how of tools that *Homo sapiens* have developed over the last hundred thousand years, and that are entwined with our species' capacity to adapt to the challenges of natural environment by reducing its complexity. *Pace* techno-determinism, mutual interaction between values and technological development exists: value concepts influence possible developments of technology, while technology reshapes these values and their environmental framework. Significantly, the Aztecs knew the wheel but preferred not to employ it in the making of their pyramids.

Unexceptionalism: A popular opinion among legal scholars in the mid 1990s, according to which settled principles and traditional tools of international law could successfully grasp the new generation of cases emerging from digital technology (computer crimes, data protection safeguards, and provisions on digital copyright). The overall idea is that "activity in cyberspace is functionally identical to transnational activity mediated by other means, such as mail or telephone or smoke signal" (Goldsmith 1998).

Uniqueness or Exceptionalism Advocates: Scholars who reckon we are in the midst (or at the very beginning) of an information revolution, so that, contrary to unexceptionalism, new legal issues are actually arising with the generation of digital cases. While the failure to find satisfactory non-computer analogies confirms the exceptional character of such issues like identity thefts, spamming, or click-and-point contracts, the ubiquity of information on the internet explains why virtually all events and transactions have a transnational impact on current legal systems.

Chapter 10
Gambling with Laws and Ethics in Cyberspace

Lee Gillam
University of Surrey, UK

Anna Vartapetiance
University of Surrey, UK

ABSTRACT

The online world, referred to by some as cyberspace, offers a wide variety of activities: reading written content, interacting with others, engaging with multimedia and playing games of various kinds, and obtaining goods and services. However, many will ignore the written agreements and laws that govern these activities, and because of this, cyberspace can turn into a dangerous place. This chapter explores a small sample of legal and ethical issues that can arise in the interface between cyberspaces and real places for those not paying attention to such matters. The authors note how laws in the physical world remain applicable in the "virtual" world, requiring knowledge of jurisdiction, and discuss the potential for creating systems that protect the user from harm by embedding adherence to laws and providing support for ethics. The authors further explain how such embedding can work to address the complex legalities and potential need for intervention in addictive situations in online gambling.

1. INTRODUCTION

Cyberspace. A consensual hallucination experienced daily by billions of legitimate operators, in every nation, by children being taught mathematical concepts... A graphic representation of data abstracted from the banks of every computer in the human system...(Gibson, 1984, p.51)

The advent of "cyberspace" has led to traditional geographical boundaries being transcended. Cyberspace offers people the illusion that most things are available cheaply or for free, and all actions undertaken are acceptable everywhere. Sitting in front of a computer, a person accessing the internet is virtually relocated to a "generalized elsewhere" of distant places and "non-local" people (Jewkes, 2003). While the person inhabits this generalized

DOI: 10.4018/978-1-4666-6122-6.ch010

everywhere, they may be incorrectly extending the rules and social norms that are applicable in their own physical location across the geographical boundaries, or believing there is a relaxation of regulations and restrictions. They may also be erroneously enlarging their personal security perimeter, acting under a false impression that the limit of communication is with the computer screen itself, or is restricted to specific intended set of interested people. In this generalized elsewhere, people can be whoever, whatever, and wherever they wish, presenting themselves and re-inventing themselves as they desire. Unfortunately, this also offers the opportunity for those with fewer scruples to pretend to be people who already exist, based on information they have managed to obtain from unsuspecting users who are under such illusions and who become susceptible to such problems.

A key difficulty for cyberspace users is in this rapid but undistinguished crossing of jurisdictions, often but not always country boundaries, that include the legal, ethical and religious, amongst others. For tourists in the physical world, there are often certain clear indications of when geographical boundaries have been crossed, and other symbols may identify such a difference. In cyberspace, one can rapidly move across boundaries of geography without ever being aware of the fact. This can create significant difficulties for software builders and internet users alike in understanding what applies, where it applies, when it applies, and, most difficult of all, why.

Over time, geographical entities have introduced, updated, replaced and even discarded laws that enforce or supplement societal and cultural norms. As technologies have emerged, lawmakers have attempted to keep pace. Unfortunately, reinterpreting through legal cases and through the crafting of new legislation where old was insufficiently encompassing can be awkward and appear ill-informed. During such processes, typically elongated if anything remotely useful is to emerge, the technology has usually moved on: the present pace of technological innovation is vastly outstripping the ability of the majority to keep up with new products, let alone for lawmakers to keep up with problems created by new products. If laws are found wanting, those developing such technologies have to make reference to ethics and professional standards while the gaps are closed, and must hope for the best outcomes when courts decide whether their use of new technologies is acceptable or not. The jurisdictional framing of laws introduces yet another issue: the illusion of the generalized everywhere is not reflected in any kind of generalized law. Cyberspace has no set of unified laws governing all actions, enabling the fight against the crimes, or for promoting the wellbeing of society and prevention of harm. There may be some degree of commonality in law, for example when European Union member states implement certain directives, but these can happen over varying time spans, and even the transfer to a national implementation may be considered incomplete (Ashford, 2010).

Cyberspace offers up many benefits, but many more substantial risks. It may be possible to trust in well-known brands, but there are many others attempting to deceive through masquerading as these trused brands using, for example, phishing attacks. By compromising weakly secured systems, it is possible to construct botnets (Weber, 2007) that can co-ordinate attacks against yet other systems, act as spam generators to catch the unwary, deploy ransomware (Net-Security, 2010) or obtain and distribute personal data contained within such systems. By the time such systems are detected and blocked, yet further such botnets will have been spawned. Meanwhile, those who compose phishing emails or construct such systems are difficult to identify and bring to justice. Personal data obtained via such approaches can include credit card numbers, bank account details, potentially even DNA profiles (Vartapetiance & Gillam, 2010). Such data becomes valuable in being able adopt the identity of another, to obtain credit in their name, and consequently to impact on their credit records; the first that the innocent party knows of this is when they genuinely approach an organization to obtain credit, are refused, and have to uncover reasons why.

Such problems as outlined above raise the need for combined consideration of ethics, law and professionalism. Jurisdiction-based legislation seems ill-formed for problems that cross geographical boundaries, potentially with vast physical separation of these boundaries. Professional and ethical standards may transcend boundaries, but these also require systematic adoption by a majority in order to be effective. Conversely, acting legally may not guarantee that the actions are ethically or professionally acceptable.

In this chapter, we explore some of the legal difficulties that emerge in cyberspace and activities directly related to it. Our motivation is the development of computer systems that can embody ethical support and prevent unfortunate consequences due to a lack of, or incorrect, legal knowledge. Prevention of harm is key. In section 2, we explore examples of legal problems that relate strongly to cyberspace in order to emphasize the existence of certain differences due to geographical variations in laws. Section 3 makes consideration for Machine Ethics as involving the embodiment of legal, ethical and professional standards in computer programs for various purposes. In Section 4, we discuss how online gambling offers difficulties that span the legal, ethical and religious. We explore the argument that companies might not (knowingly) infringe laws, yet in attempting to increase their profits by allowing certain behaviours, can give rise to addiction, which is certainly unethical. In Section 5, we describe a system we have designed and implemented, called EthiCasino, that demonstrates how to promote legally and ethically acceptable online gambling. Large parts of sections 3, 4 and 5 are reproduced from our previous publication on "Machine Ethics for Gambling in the Metaverse: an "EthiCasino"", featured in Volume 2, Issue 3 of the Journal of Virtual Worlds Research (JVWR) and recontextualized in this chapter. They are included here by kind permission of the Editor in Chief of JVWR. We conclude by suggesting future directions for this work and identify areas which need more specific attention.

2. BACKGROUND

As discussed, people who believe in the supposed liberty afforded by cyberspace must take care in what they do. While there is nothing necessarily sinister in the technology itself, the actions that technologies facilitate may not always be harmless. Most cybercrimes are reasonably common offences and computer technologies have simply provided new ways to commit old crimes (Jewkes, 2003). However, laws have often been found wanting when considering what new technologies provide for. In real places, a person can only be in one place at one time; in cyberspaces, a person will be in a single physical location but can be simultaneously undertaking actions that link them with systems located in disparate locations, each of which has implications for laws, acceptable behaviors, and so on. In addition, the usual markers of identity are missing: cyberspace makes possible the creation of an identity so fluid and multiple that it strains the very limits of the notion (Jewkes, 2003). Those who can hide or fake an identity can act without fear of reprisal, surveillance, or legal intervention, and there is much supporting software for doing this. The powerful tools of investigation and identification, including surveillance cameras, fingerprints and DNA databases, can be as impotent in cyberspace as laws based on physical borders of countries being applied to people who are physically elsewhere. Additionally, for honest citizens, knowing the rules of one's own country provides insufficient coverage.

The 'Terms of Use' on websites bears testament to the need to understand the location of the servers being used, and this implies knowing the relevant laws of the country being virtually visited by making use of these servers. These terms of use can be many pages long if printed, and typically comprise legal-like phrasing that makes it something of a challenge for an everyday user to understand. This is brought into sharp focus when 'clickwrap' software is prevalent, and companies have opportunities for mischief here (Richmond, 2010; FoxNews, 2010). The physical correlate is apparent whenever people travel abroad, and some are occasionally surprised to find themselves incarcerated for violating laws or traditions that they claim no knowledge of (BBC News, 2010).

With some 192 registered countries, according to the United Nations, understanding all of these legal systems across the various languages would be a substantial undertaking. Yet even variations amongst countries apparently close in spirit can be stark. We demonstrate this by two relatively simple examples: the first relating to Copyright Law, the second to Computer Crime and additional legislation which can be brought to bear.

2.1 Copyright Law

Are you allowed to download the songs from your CD to your MP3 player in America? What if you do this same in the UK? What if you do this in America and travel to the UK?

The well-documented case of the file sharing website Napster (e.g. Baase, 2008) shows how technological advances can lead to new "old" crimes. The UK Copyright Act 1988 contains clauses relating to providing the apparatus for copyright infringement (secondary infringement), so would likely have been a starting point against a UK-based Napster. In its application, this law may well have needed to be reinterpreted as it appears the US law was. Since file sharing websites became a principal focus, those in cyberspace would be expected to be generally wary of the dangers, and typically huge fines, involved with large-scale distribution of music tracks. However, the very act of 'ripping' tracks from a CD into different formats in order to create shareable music files, may or may not be legal depending on your location. A visitor to the UK may be surprised to find that, at the time of writing, the same law also makes it illegal to make such copies. A Survey by Consumer Focus in 2010 showed that almost three in four (73%) of the UK population did not know that it was illegal to convert from CD to MP3. A majority would think that copying songs from the CD they own, having paid for it, is entirely legal: a CD can be played by an individual for their own listening pleasure on any suitable device and is readily portable, and it could be argued that this simply extends the range of suitable devices.

As noted in the Hargreaves report (2011, sections 5.27 et seq), and because this is the present legal position, adverts in the UK for the Brennan J7 - a device that copies CDs to HDD, so essentially performs the same function - have to warn that using it involves copyright infringement. The Hargreaves report recommended "a limited private copying exception which corresponds to what consumers are already doing" given that this is apparently "already priced into the purchase," to clarify the position. The report also quotes Martin Brennan, the inventor of the J7, as saying "I face the cost of reassuring customers that record companies will not sue them. It's daft because US companies Apple and Microsoft have been selling format shifting products in the UK for a decade.". So whilst not, strictly speaking, legal, the high cost for the music industry of being involved with very large numbers of civil cases to address each and every format shift would be prohibitive.

In America, the Recording Industry Association of America (RIAA) (n.d.) suggests that it's acceptable to copy music if the CD is legitimately owned and it's for personal use only:

- *It's okay to copy music onto an analog cassette, but not for commercial purposes.*
- *It's also okay to copy music onto special Audio CD-R's, mini-discs, and digital*

tapes (because royalties have been paid on them) – but, again, not for commercial purposes.

- *Beyond that, there's no legal "right" to copy the copyrighted music on a CD onto a CD-R. However, burning a copy of CD onto a CD-R, or transferring a copy onto your computer hard drive or your portable music player, won't usually raise concerns so long as:*
    - *The copy is made from an authorized original CD that you legitimately own*
    - *The copy is just for your personal use. It's not a personal use – in fact, it's illegal – to give away the copy or lend it to others for copying.*
- *The owners of copyrighted music have the right to use protection technology to allow or prevent copying.*
- *Remember, it's never okay to sell or make commercial use of a copy that you make.* Source: RIAA,2013

RIAA includes over 1.600 record companies, representing over 85% of the US market, so has some considerable influence. However, it is unclear how much of RIAA's advice is provided for by the Digital Millennium Copyright Act 1998, and how much is RIAA-specific and may invite interest from the remaining 15% who RIAA do not represent. However, that is probably of lesser concern that would be the case if present UK legislation were to become more robustly enforced: the unwary American tourists, visiting the UK with CDs and, on RIAA advice, converting these for use on their MP3 players and laptops could find themselves in trouble.

In this discussion, the problems of cyberspace became amplified through focus on Napster, and in that process the importance of an act that also breaches copyright law in the UK is forgotten about. We suggested, above, that while inhabiting the generalized everywhere a person may be incorrectly extending the rules and social norms that are applicable in their own physical location or believing there is a relaxation of regulations and restrictions. This example demonstrates that people may even take such apparent relaxation back to the real world from cyberspace, and this may have its own attendant problems.

2.2 Computer Crime

You live in the UK and hack into a system in America. Whose laws apply?

Hacking into computers is generally considered a crime unless, in the case of ethical hacking, explicit permission has been given to do so. When UK citizen Gary McKinnon, an "unemployed system administrator" was caught having hacked into 97 United States military and NASA computers without such permission, including US Army, US Navy, Department of Defense and the US Air Force (McKinnon v. U.S.), it was always likely that some kind of legal proceedings would follow. However, the nature of these proceedings has been highly controversial.

In 2002, he was interviewed by the UK National Hi-Tech Crime Unit (NHTCU)[1], and was indicted by a federal grand jury in the Eastern District of Virginia in November 2002 on seven counts of cybercrime (U.S. v. McKinnon). It was claimed that he caused damages costing over $700,000, and had deleted US Navy weapons logs on about 300 computers after September 11th attacks. However, he denied most of the charges and argued that he was only interested in UFOs, believing that such information should be publicly accessible. Furthermore, as there was little or no specific security in place on these systems, he thought what he did wouldn't be considered as hacking. McKinnon committed his hacking activities from his home computer in London. This makes the UK's Computer Misuse Act 1990, section 1, applicable as he had caused "a computer to perform any function with intent to secure access to any program or data held in any computer", and he was not authorized to do so.

On 7th October 2004, an extradition request was made so that McKinnon could be prosecuted in America (US v. McKinnon). It was convenient that the Extradition Act 2003 had provided for such an action without needing *prima facie* evidence; asymmetric provisions relating to this Act would not allow the same in the opposite direction. Nor was this the first instance of the application of this law: following the collapse of Enron in late 2001, the Natwest Three (US v. David Bermingham, et al.) were indicted in 2002, arrested in the UK in 2004, and extradited to the US in 2006 using this 2003 law. Arguably, this represents an *ex post facto* relaxation on the requirement for evidence; retrospective application of law is frowned upon in some places, and simply unacceptable in others. At the time of writing (December 2013), McKinnon no longer faces extradition, nor is likely to be prosecuted in the UK. A timeline of events in the McKinnon and Natwest Three cases is shown in Table 1.

Here, the focus is on the prioritization of real world laws in relation to problems that have occurred in cyberspace. Gary McKinnon is trying to imply relaxations of laws into cyberspace to support his own ends. The unwary must be careful that the targets of their crimes do not try to apply unexpected, and even *ex post facto*, laws in such situations to try to bring the perpetrators to justice. Hacking a system in the US from the UK is already evidence of risky behaviour, and extradition runs the risk of a relatively short imprisonment of a few years under UK law being inflated to some 70 years or so in a US prison.

3. MACHINE ETHICS

No sensible decision can be made any longer without taking into account not only the world as it is, but the world as it will be (Asimov, 1983, p.5).

Given the difficulty for everyday users in interpreting laws, there are two related possibilities: (i) to have incredibly well-specified, unambiguous, and highly readable laws so that they are readily understood by all users of cyberspaces; (ii) to codify laws into location-aware computer programs such that users are protected from attendant harms. Very long documents are very likely not going to be read, limiting the entire purpose of (i), so this brings us towards computerization of laws. Computerized laws need not be fully explicable, but they must allow for certain behaviours, warn about certain others, and deny yet others in certain locations, for certain users, and at certain times. The key principle being embodied here is the avoidance of harm to users, and this indicates ethics generally and, for us, Machine Ethics specifically.

Machine ethics is concerned with defining how machines should behave towards human users and other machines, with emphasis on avoiding harm and other negative consequences of autonomous machines, or unmonitored and unmanned computer programs. Researchers in machine ethics aim towards constructing machines whose decisions and actions will honour privacy, protect civil rights and individual liberty, and further the welfare of others (Allen, Wallach and Smit, 2005). However, the study of machine ethics is still undertaken by a relative minority, despite its apparent importance as autonomous machines become more prevalent. To produce ethical machines, it is necessary to understand how humans deal with ethics in decision making, and then try to construct appropriate behaviors within machines, or autonomous avatars, which with continuous availability and unemotional responses might start to replace human (ethical) advisors in a near future.

Steps towards ethical machines have been taken that focus on medical ethics, attempting to ensure human safety and social health. Such systems are intended towards understanding, and possibly reducing or avoiding, the potential for harm to an individual from, for example, un-

Table 1. Timeline of McKinnon and NatWest three cases

Gary McKinnon		Year	Natwest three	
Interviewed by police	19 March	**2002**	2002	Indicted on seven counts of wire fraud by Huston, Texas
Interviewed by NHTCU	8 August			
Indicted of seven counts of computer crime by a federal grand jury in the eastern	November			
Extradition Act 2003 (early 2004)				
Subject to bail conditions	June	**2005**	2002-2006	Series of courts in UK
US began extradition proceedings	Later in			
Extradition approved by UK	July	**2006**	2006	Extradition to US
Appeal to the High Court in London	February	**2007**	November	Plead guilty
House of Lords agree to hear the appeal	July			
Case presented in House of Lords	16 June	**2008**	22 February	Sentenced to 37 months in prison
Rejected	July			
Appeal to European Court of Human Rights against extradition – rejected	August		November	To spend the rest of their sentences in the UK
Won permission from High Court to apply for judicial review against extradition	23 January	**2009**	End of 2009	Released
Lost the appeal	31 July			
The Extradition had been blocked	16 October	**2012**	Free	
Dropping all the charges	14 December			
Free		**2013**		

NB: this table has been constructed based on well known news websites (e.g. BBC News and Guardian). Imprecise dates due to lack of readily available information.

necessary or incorrect medical intervention or in considering the kinds and nature of treatments being administered and any limiting factors that may consequentially emerge. In these systems, the final decision remains one of a human decision-maker, who becomes informed by such ethical considerations. The mainstream literature largely discusses using Case-Based Reasoning and machine learning techniques to implement systems that can mimic the responses of the researchers (Anderson, Anderson and Armen, 2005; McLaren and Ashley, 2000).

A machine-based ethical advisor has the following anticipated advantages, many of which are familiar arguments in the development of intelligent systems:

- Always available
- Unemotional
- Employ mixture of ethical theories
- Can explain reasoning
- Capacity for simulations
- Capacity for range of legal considerations
- No hypothetical limits on the number of situations assessed

A synthesized overview of many of the systems reported in the literature as ethical machines is shown in Table 2. Each of these systems has a specific ethical approach and related techniques that provide for solutions to ethical dilemmas, targeted at particular audiences and challenges for those audiences. The majority of these exist-

Table 2. Comparison of some of the existing applications

Name	Developed by	Ethical approach	Techniques	Suitable	Ethical area
Ethos	Searing, D.	Moral DM	Not AI Some ethical samples	Engineering Students	Practical- ethical problems
Dax Cowart	Multiple writers	Moral DM	Not AI	Students, Teachers	Biomedical ethics, Right to die
Metanet	Guarini, M.	Particularism Motive consequentialism	Pair case (SRN), Case base, Neural network (training), Three layers	Problems in flagging	Killing or allowing to die
	Robins, R. And Wallach, W.	Desire-intention	Multi-agent	Not implemented	
Truth-Teller	McLaren, B. M.	Casuistry	Pair case, Case-Based Reasoning,	Ethical advice	Pragmatic or hypothetical cases
HYPO	Ashley, K. D.	Legal- reasoning	Case base	Legal advice	Hypothetical cases
SIROCCO	McLaren, B. M	Casuistry	Pair case, Case-Based Reasoning, Simulating "moral imagination"	Ethical device	NSPE Code of Ethics
Jeremy	Anderson, M. Anderson, S. Armen, C.	Hedonistic act utilitarianism	"Moral arithmetic"		Rule generalization
W.D.	Anderson, M. Anderson, S. Armen, C	Prima facie duty, Casuistry	Inductive-logic programming, Learning algorithm, Reflective equilibrium		Rule generalization
MedEthEx	Anderson, M. Anderson, S. Armen, C.	W.D. Medical ethics, Casuistry, Prima facie duty	Inductive-logic programming, Machine learning, Reflective equilibrium	Health care workers	Biomedical ethics
EthEl	Anderson, M. Anderson, S.	Prima facie duty, Casuistry, W.D., Medical ethics	Inductive-logic programming, Learning algorithm, Reflective equilibrium	Eldercare	Biomedical ethics

ing systems focus on medical ethics, with limited coverage for other applications of ethics.

To develop a machine-based ethical advisor, we must make reference to such systems, but need not be constrained in our thinking by them. The problem with systems inherently based on cases and rules is that adaptation to other domains requires a new set of domain-specific rules or cases to be developed or captured; this is a well-known bottleneck in artificially intelligent systems and it may be more appropriate to attempt to avoid such constraints. Further, we need to determine the key ethical and legal points which are being embodied within such a system so that there is more than simply an informing approach: the ethical advisor must also be able to intervene when the need arises. Essentially, such a system must offer the best approach grounded in ethics and law, rather than simply offer up its reasoning. We explore this in the following sections in relation to online gambling.

4. ONLINE GAMBLING

Are you allowed to gamble online in America? What if you go to UK or any other country in the world?

In Section 2, we discussed two examples in which differences in international laws were key. We could also have explored these situations from the perspective of a number of ethical theories, although the ethical dimensions are less obviously relevant in contrast to the question posed above. In gambling, and in online gambling in particular, gamblers should be "responsible". Some can become addicted, lose large amounts of money and be used by others as a means to earn money to support their addiction or resort to crime. Meanwhile, those providing for online gambling are making substantial revenues: online casinos alone reached $4.7 billion by end of 2010 (iGambling Business, n.d.), and it has been estimated that the gambling business generally will hit $125 billion by 2015 (Young, 2006).

In cyberspace, gambling may or may not be a crime; it may or may not be taboo; knowing which it is and when is the first challenge. Prevention of harm then becomes important. In this section, we discuss gambling, its issues in general, and the reality for those seeking to gamble in cyberspace.

4.1 Gambling Ethics

Gambling is described as:

... betting or staking of something of value, with consciousness of risk and hope of gain, on the outcome of a game, a contest, or an uncertain event whose result may be determined by accident. Commercial establishments such as casinos ... may organize gambling when a portion of the money wagered by patrons can be easily acquired by participation as a favoured party in the game, by rental of space, or by withdrawing a portion of the betting pool (Gilmne, n.d.).

This simple definition leads to further observations about people's behaviour in relation to possibility of winning extra money or goods:

1. People may play for money not for fun.
2. The odds of losing are much higher than wining and outcome is uncertain.
3. People may chase the game to win back the money they lost.
4. People may lose the sense of their own "will" because they are playing a game in which "will" is not an element.
5. They do not feel they need to gain the knowledge about how to play before they begin because they consider it just a "GAME."

Such observations can lead towards corruption, addiction and organized crime. To reduce the risk of harm both to players and society, different countries have (different) regulations for the industry – in the UK, this is the Gambling Act 2005. Although each country approaches these issues differently, many have taken steps to raise public awareness about such problems. For example, the UK's law discusses limiting the number of casinos and forcing members of the industry to demonstrate their plans for contributions to research, for raising public awareness about the problems gambling can cause, and for helping to treat those affected (Russell, 2006). America has approached awareness issues by introducing The National Gambling Impact Study Commission Act 1996 (NGISCA; H.R.5474), which included a comprehensive legal and factual study of the social and economic impacts of gambling. Other steps for raising awareness have been taken by NGOs, promoting the idea that players should be aware of the time and money that they spend on gambling, and the consequences and risks involved.

According to "Ethical Corporation" there are three reasons the online gambling industry should take its responsibilities seriously (Saha, 2005):

- To clear up the industry's traditional image
- To attract potential customers who otherwise steer clear because of this image, and
- To comply with regulations

However, often when gambling websites are demonstrating that they are being "responsible", such a demonstration may simply be encapsulated in a document containing the kinds of rhetoric presented below:

- We are there to help whenever you realize that you need a control over the money that you spend
- We can decrease the amount of money you can put into your account if you ask.
- You can increase it again if you feel you are in control.
- If you think you need a break from gambling, you can use self-exclusion tool
- If you suspect that you may have a gambling problem, you may seek professional help from the following links
- Make sure gambling does not become a problem in your life and you do not lose control of your play.
- Make sure that the decision to gamble is your personal choice.

Such statements require individuals who may be experiencing addiction to be aware of the fact, and to be in sufficient control to do something about it. The "problem" is for the end user to deal with, and the organization has effectively absolved itself of responsibility. Gambling addiction is identified as one of the most destructive addictions which is not physically apparent - an "invisible addiction" (Comeau, 1997). Psychologists believe that online gamblers are even more prone to addiction because they can play without distraction or recognition.

Internet gambling, unlike many other types of gambling activity, is a solitary activity, which makes it even more dangerous: people can gamble uninterrupted and undetected for unlimited periods of time (Price, 2006).

Furthermore, the capability for adopting multiple false identities in cyberspace means that simply blocking user accounts is going to be ineffective. It is unlikely, then, that self-control could be exerted or easily enforced for online gambling.

Websites such as gambleaware.co.uk offer potential players and gamblers information about the odds of winning, the average return to players, "house edge", a gambling fact and fiction quiz, and more, to offer increased awareness. Gambleaware (n.d.) define a responsible gambler as a person who:

- Gambles for fun, not to make money or to escape problems.
- Knows that they are very unlikely to win in the long run.
- Does not try to 'chase' or win back losses.
- Gambles with money set aside for entertainment and never uses money intended for rent, bill or food.
- Does not borrow money to gamble.
- Does not let gambling affect their relationships with family and friends.

Defining measures to differentiate between responsible players and addicted gamblers would certainly help to prevent addiction, corruption and crime, but what do such measures look like?

4.2 Gambling Laws and Regulations

Though countries do have laws relating to gambling, people tend to ignore such laws in cyberspace. Huge growth in online gambling[2] often brings it in line with physical forms of gambling. Some countries have created new laws for online gambling, while others extended old laws to cover it. Interestingly, America has a restrictive law in the Unlawful Internet Gambling Enforcement Act 2006. This law has reportedly had a detrimental impact on its economy: it has been estimated that US could have created between 16,000 and 32,000 jobs and generated a total gross expenditure in the nation's economy of $94bn over the first five years and $57.5bn in domestic taxation (H2 Gambling Capital, 2010).

Such new and extended laws include:

- **US - The Unlawful Internet Gambling Enforcement Act 2006 (UIGEA, H.R.4411):** Prohibits financial institutions from approving transactions between U.S.-based customer accounts and offshore gambling merchants.
- **US - Internet Gambling Regulation and Enforcement Act 2007 (IGREA, H.R.2046):** "Providing a provision for licensing of internet gambling facilities by the Director of the Financial Crimes enforcement network"
- **US - Skill Game Protection Act 2007 (SGPA, H.R.2610):** "Legalize internet skilled games where players' skills are important in winning or losing games such as poker, bridge and chess"
- **US - Internet Gambling Regulation and Tax Enforcement Act 2007 (IGRTEA, HR 2607):** "Legalize internet gambling tax collection requirements"
- **US - Internet Gambling Regulation, Consumer Protection, and Enforcement Act 2009 (H.R. 2267):** (under discussion) provide licensing of Internet Gambling activities, consider consumer protection and tax enforcement. If passed, it will create a an exception to the Unlawful Internet Gambling Enforcement Act of 2006 (UIGEA) for poker
- **Australia - Interactive Gambling Act 2001 (IGA):** Provides protection for Australian players from the harmful effects of gambling
- **UK - Gambling Act 2005 (c. 19):** "it is not illegal for British residents to gamble online and it is not illegal for overseas operators to offer online gambling to British residents (though there are restrictions on advertising)"

Variation leads towards three principal divisions:

1. Countries where gamblers are free to play online, or there is no specific regulation, e.g. UK
2. Countries where gamblers are not allowed to play online, e.g. USA (GAO, 2002), and
3. Countries where gambling of any kind is prohibited, e.g. Islamic countries (Lewis, 2003)

A glimpse of the legality of online gambling in 100 countries is shown in Table 3.

Where gambling of any kind is legal, or not legislated against, it may also be age-restricted. Mostly, the gambling age – whether in cyberspaces or in real places - varies between 18 and 21; specific exceptions are Greece and some provinces in Portugal with minimum ages of 23 and 25 respectively, where online gambling is not currently allowed but a government might wish to allow it if it could improve the economic situations of either countries. So, whilst in one country, it may be perfectly possible for you to gamble online; go to another country and visit the same website, and you might not be allowed to or might be age restricted when doing so.

The cyberspace problems of online gambling are wider still. Suppose that a US company was operating an online gambling website from UK servers, allowing gamblers to play from all over the world with terms of use that say that if security breaches occur then the company is responsible and American courts are primary. A time later, a UK citizen on holiday on a Caribbean Island hacks the system and takes credit card numbers and information about gamblers from Saudi Arabia and the US. The hacker obtains large amounts of money from players and disappears.

Table 3. Online gambling regulations in different countries

Countries and territories where online gambling is legal							
1	Aland Islands	19	Dominican Republic	37	Lithuania	55	Seychelles
2	Alderney	20	Estonia	38	Luxembourg	56	Singapore
3	Antigua	21	Finland **	39	Macau	57	Slovenia
4	Argentina	22	France ***	40	Malta	58	Solomon Islands
5	Aruba	23	Germany	41	Mauritius	59	South Africa
6	Australia *	24	Gibraltar	42	Monaco	60	South Korea
7	Austria	25	Grenada	43	Myanmar	61	Spain
8	Bahamas	26	Hungary	44	Nepal	62	St. Kitts and Nevis
9	Belgium	27	Iceland	45	Netherlands Antilles	63	St. Vincent
10	Belize	28	India	46	Norfolk Island	64	Swaziland
11	Brazil	29	Ireland	47	North Korea	65	Sweden
12	Chile	30	Isle of Man	48	Norway	66	Switzerland
13	Colombia	31	Israel	49	Panama	67	Taiwan
14	Comoros	32	Italy	50	Philippines	68	Tanzania
15	Costa Rica	33	Jamaica	51	Poland	69	United Kingdom
16	Czech Republic	34	Jersey	52	Russia	70	US Virgin Islands
17	Denmark	35	Kalmykia	53	Sark	71	Vanuatu
18	Dominica	36	Latvia	54	Serbia	72	Venezuela
Countries where online gambling is illegal							
1	Afghanistan	8	Greece	15	New Zealand	22	Taiwan
2	Algeria	9	Hong Kong	16	Nigeria	23	Thailand
3	Bahrain	10	Indonesia	17	Pakistan	24	The Bahamas
4	Brunei	11	Iran	18	Portugal	25	The Netherlands
5	China	12	Japan	19	Saudi Arabia	26	Turkey
6	Cyprus	13	Jordan	20	South Korea	27	United States
7	Dubai	14	Libya	21	Sudan	28	Vietnam

* For Australia, different laws might be applied in different states.

** Must be a Finnish resident with a Finnish bank account to play.

***FRANCE'S National Assembly has voted in favour of a bill to legalize online gambling. Approval is still needed by the European Union and France's Conseil d'Etat (Supreme Court) and Conseil Constitutionnel (Constitutional Council). Previously, France did not allow such companies within its borders, but its citizens could gamble from other companies internationally.

Note: Table first constructed in 2010 and information may have changed in the intervening time. It would be important to keep re-appraising the situation in every jurisdiction to avoid emergent problems.

1. Can the company report a cybercrime? And if so, who to?
2. Can Saudi Arabian and US gamblers report the crime? Gambling of any kind is prohibited for Saudis and online gambling is forbidden for US players.
3. Can players claim their money back from the company in America, based on the terms of use?
4. For the hacker, which laws should be applicable? How does this compare to the Gary McKinnon case?

Companies who have not taken full account of the applicable laws will end up paying the consequences. One well-known example relates to casinos in the Second Life (SL) virtual world. All casinos were forced to close, effectively overnight, in SL by the FBI. Companies that had invested substantially in their virtual world presence suddenly lost revenues, and the virtual world itself lost users as a consequence. The country of origin of the gambling companies was irrelevant since the servers were operating in the US. Players from all over the world had been gambling in online casinos in SL, making it one of the strongest businesses in that virtual environment, and therefore having the most significant impact on the associated virtual world economy.

Inevitably, it will be suggested that universal legislation should be required for online gambling, but this will favour those with the most restrictive practices and therefore be readily unacceptable to large numbers of people. There may be the argument that users should take responsibility for knowing whether or not to play based on their country's laws. However, it might be considered unreasonable to ask users or potential users to read the regulations from all over the world in order to know whether or not they are affected. This, then, leads back to controls that must be put in place by the online gambling providers (Price, 2006). These responsibilities can be implemented by ensuring that servers are sensibly located, and controlling who enters on the basis of country of request and acceptable age for gamblers (both of which can change). There remains, then, the need to prevent harm, and here we also propose the embedding of such prevention in computer programs. We discuss such embedding in the next section.

5. MACHINE ETHICS FOR ONLINE GAMBLING

Machine ethics has not yet been applied by others for avoidance of harm in relation to online gambling. Alongside a number of other pursuits, and because gambling has potential for addiction, it could be claimed that a system for ethical gambling may be as effective for humans and social health as medical ethics. Machine ethics may not cure addiction, but it may be able to act to reduce the likelihood of addiction. Our consideration here is how Machine Ethics may support responsible gambling and lead towards an Ethical Corporation.

We base the design of EthiCasino on prior literature and systems in Machine Ethics as shown in Table 2. We have been inspired in particular by three of these systems, W.D., MedEthEx and EthEl, that have used Ross' *prima facie* duties (1930), extended by Garrett (2004). Ross introduced seven "prima facie duties" as guidelines for solving ethical dilemmas but these are not rules without exception. If an action does not satisfy a "duty", it is not necessarily violating a "rule"; however if a person is not practising these duties then he or she is failing in their duties. Garrett (2004) believed there to be aspects of human ethical life not covered by Ross, and extended this list with three further duties. MedEthEx uses a series of questions with a three responses, "Yes", "No" and "Don't know", to decide the outcome in relation to three of Ross' and Garrett's duties: *non-injury*, *beneficence* and *freedom* (autonomy). By weighting outcomes between -2 and +2, the application explains the likely impact on the patient ability to clarify the areas in which decisions will be made. EthEl takes two kinds of actions based on decisions made: (i) reminding users; (ii) notifying overseers. A system using Ross' and Garrett's duties for responsible gambling should consider the potential for the duties not being satisfied and act accordingly.

While MedEthEx and EthEl concentrate on three main duties of non- injury, beneficence and freedom, EthiCasino considers a wider range of duties; in particular, EthiCasino employs 6 of Ross' 7 duties and all 3 duties defined by Garret in different stages (Table 4). Using these Prima facie duties enables the system to learn from users' behaviour even if they might not match exactly the original definition of the duties.

Table 4. Duties of Ross and Garret in each stage

Stage	Name	Ross's duties involved
Stage 1: Legal considerations	Legal issues	Justice, Harm prevention, Non injury, Beneficence, Self-improvement
Stage 2: Knowledge of Risk	Ethical issues	Justice, Harm prevention, Non- injury
Stage 3: Boundaries for time and money	Boundaries	Justice, Harm prevention, Respect of freedom, Fidelity, Gratitude
Stage 4: Appropriate reminders: "nagware"	VIKIs reminders	Non-injury, Beneficence, Self improvement, Care
Stage 5: Boundary conditions	VIKIs alert	Justice, Harm prevention, Non-injury, Beneficence

EthiCasino takes certain actions that support the performance of these duties in order to assure users' safety and wellbeing by minimizing possibilities of problematic and addictive behaviour, providing ethically-acceptable support, and meeting the requirements of mimicking action of human ethical advisors. This aims at ensuring fair actions for both virtual gambler and virtual casino:

1. Gambler:
   a. Clarify the possible risks of gambling online
   b. Choose playing hours and amount of money they wish to gamble
   c. Remind the users of their playing hours and the amount money they are losing
2. Casino:
   a. Take decisions about whether or not to let specific persons play based on their answers
   b. Notify the company if a user is going over their own limitation
   c. Log the user off if they don't take action after being reminded by the system

We discuss the 5 main, often inter-dependent, stages involving legal and ethical considerations in the following steps:

Stage 1: Legal Considerations

Consideration of legal issues involves variations in acceptability of online gambling and associated age restrictions in 100 countries, as presented above. Here, system can attempt to capture the geographical location (DNS lookup) of the end user, and act accordingly, but because of the capacity for technological circumvention the gambler needs to self-certify. Self-certification is required, also, for confirming the age of the end user. Should the location of the end user change over time from the original registration, the legal situation may change accordingly and location information must be captured and verified for each session.

Stage 2: Knowledge of Risk

Decisions related to financial risks may be taken in a number of business environments, especially in relation to stock markets and world economies. Those involved in taking such decisions are usually considered well-informed and have a number of checks and balances against which to validate their decisions or off-set their risks and/or losses. The person's knowledge is the effective tool in making the final decision. Unfortunately, because of the purported "entertainment" aspect of gambling, it is less important for users to have such knowledge or to consider how to off-set risks and losses and more favourable to revenues if users are less well-informed.

To evaluate the risk behaviours of end users, we designed a questionnaire comprising 12 questions related to gambling fact and fiction and 8 related to risk and loss aversion. We offered L$10 to participants, equivalent to around 2½ hours camping, and obtained 61 responses to this questionnaire from Second Life users within a week. On average, 12.22 questions were correctly answered, with 7 and 17 as minimum and maximum. We *a priori* weighted questions based on our own perceptions of associated risk or negative impact on users in the absence of knowledge, leading to a division of questions into four categories:

1. **Low Risk:** users should be able to learn quickly or lack of knowledge will not have much negative impact. e.g. Q3: "Some people are luckier than others" (fact or fiction)
2. **Medium Risk:** users may believe in luck. e.g. Q6: "My lucky number will increase my chance of winning the lottery" (fact or fiction)
3. **Medium-High Risk:** questions relate to calculations and predictability of results e.g. Q14: "Assume you bet $1 on the toss of a coin the chances of heads or tails are 50/50. If you win and 'house edge' is 10% how much you will be paid? (10c, 50c, 90c, $1)"
4. **High Risk:** question regards perceptions of earning money and realistic facts of gambling. e.g. Q1: "Gambling is an easy way to make money" (fact or fiction)

User answers and weightings led to three distinct classes of users (Figure 1), important so that the system can help them to avoid negative impacts of incorrect decisions:

- **Group One:** Those who may only need additional information about the games (low and medium risk questions)
- **Group Two:** Those who need to be reminded about the facts (medium-high risk questions), and
- **Group Three:** Those who need full monitoring and potential intervention because they are less informed and might be more prone to addiction (high risk questions)

To evaluate these behaviour profiles, we analysed the correlations between the 20 questions for 50 users (Table 3), hoping that diversification would exist across the various responses. The resulting correlation matrix showed maximum correlation between 18 of the questions of less than 0.5 (-1/+1), suggesting that the questions themselves had a reasonable degree of independence. On this basis, the risk classification becomes the important factor since the individual questions themselves do not act as a reliable predictor for others in the same class.

Stage 3: Boundaries for Time and Money

For a user to stay in control, part of the main challenge of gambling, the system should allow them to opt for boundaries. Considering that each user background and experience is different, and that there is such variation across responses to 20 questions about gambling, it could be unethical to enforce boundaries without end user permissions. Users are asked to define their own boundaries both for the amount of time and the amount of money they plan to spend: these two elements are core in addiction and harm. The user's choice of boundaries is checked against their apparent riskiness. For users with profiles in Groups 1 and 2, the system will allow users to participate with limited interference; users in Group 3 will

Figure 1. Risk groups based on responses to questions on gambling

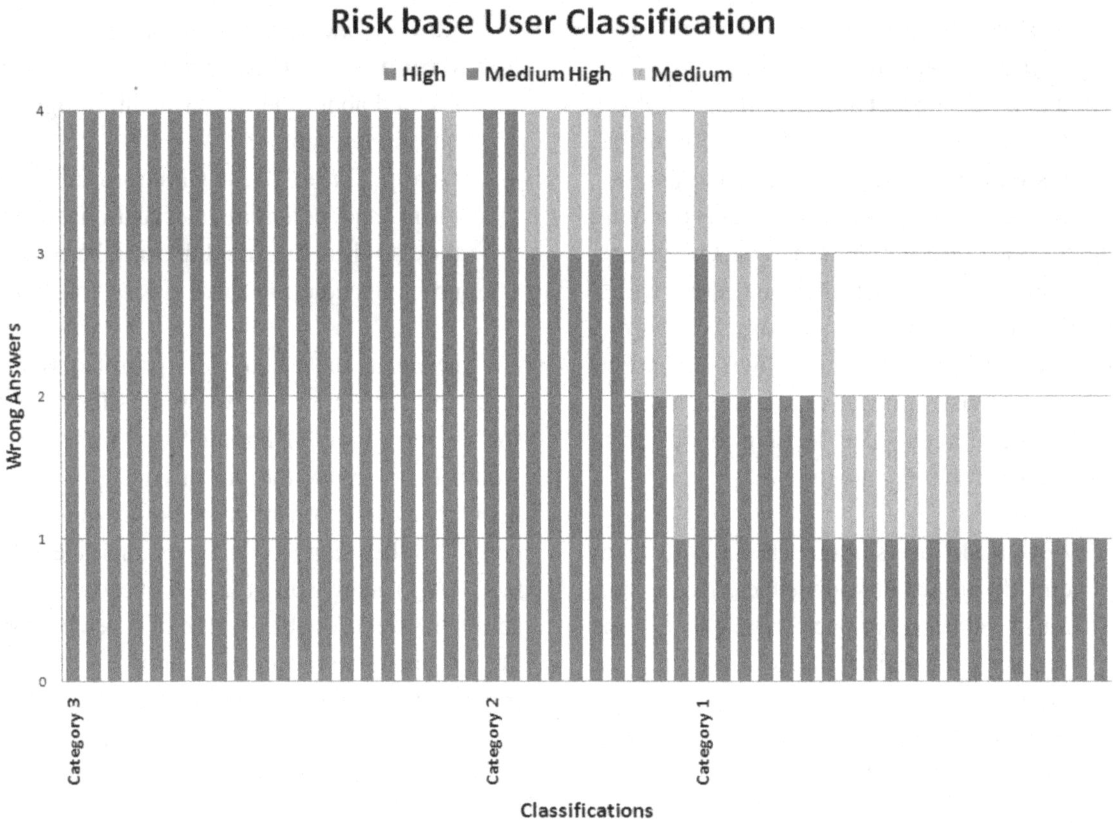

receive a moderated limit as the maximum boundary (Figure 2).

Stage 4: Appropriate reminders: "Nagware"

In EthiCasino, to minimize the potential for destructive behaviours, we adopt the idea of "nagware" [3] as used by a number of software providers to remind users of specific actions, e.g. that they should pay for the software they have been using. In EthiCasino, this nagware has been called **VIKI**[4] and undertakes specific responsibilities:

- **Artificial Ethical Conscience:** Suggestions allied to risk taking and user's circumstances, e.g. "high risk of losses, do you still what to bet?"
- **Educational:** Providing access to information about each game, risks and odds associated to it, e.g. "roulette, your odds are 35 to 1."
- **Nagging:** Regularly reminding users, depending on their risk profiles, about the time and money spent, as both diminish.

Users receive reminders depending on how they approach their own specified limits. Those identified as having riskier behaviours will receive fewer reminders compared to other users. Those who have spent their money more quickly may be tempted to spend more, sometimes chasing losses. Those who manage not to make losses within the

Figure 2. Maximum boundaries for each category

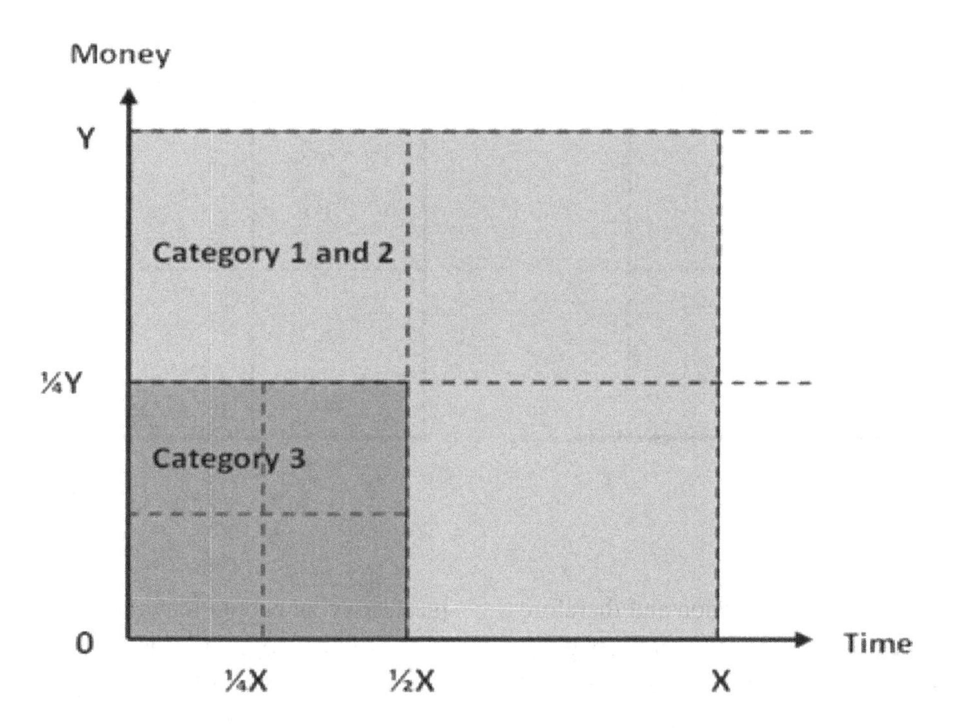

initial time period may be encouraged to continue and to make assumptions over the likelihood of larger future wins. Of course, user profiles may change over time depending on the increased or decreased risky behaviour of the end user (Fig. 3).

Stage 5: Boundary Conditions

After users receive their final reminder from VIKI, they will be prevented from further gambling. The purpose here is to ensure the user's own boundaries are enforced and to ensure the risky behaviours do not lead to harm; that is, EthiCasino acts to prevent behaviours that might lead to addiction. Those continuing beyond their own time and financial limits may also be going beyond their own limits of rational behaviour. A virtual doorman who ejects non-conforming end users is a possible future consideration.

6. CONCLUSION

In this chapter we have discussed the difficult interplay between cyberspaces and real places, the relationship of laws to both, and how the consideration of ethics can help to avoid some of the inherent problems. Cyberspace can be a dangerous place for ill-informed virtual tourists since they are rapidly crossing legal, typically geographic, boundaries without being aware of this fact. In so doing, the virtual tourist may be inappropriately projecting local laws and customs into cyberspace, or believing that such laws and customs are related. We provided examples of how specific laws related to cyberspace in different ways, and in particular how crime investigations become contentious. We then introduced the importance of Machine Ethics and applied this to Online Gambling. The motivation here is to prevent legal problems as would arise due to variations in international law, for example relating to acceptability and age for online gambling. Further, the importance of re-

Figure 3. Possible users' behaviour

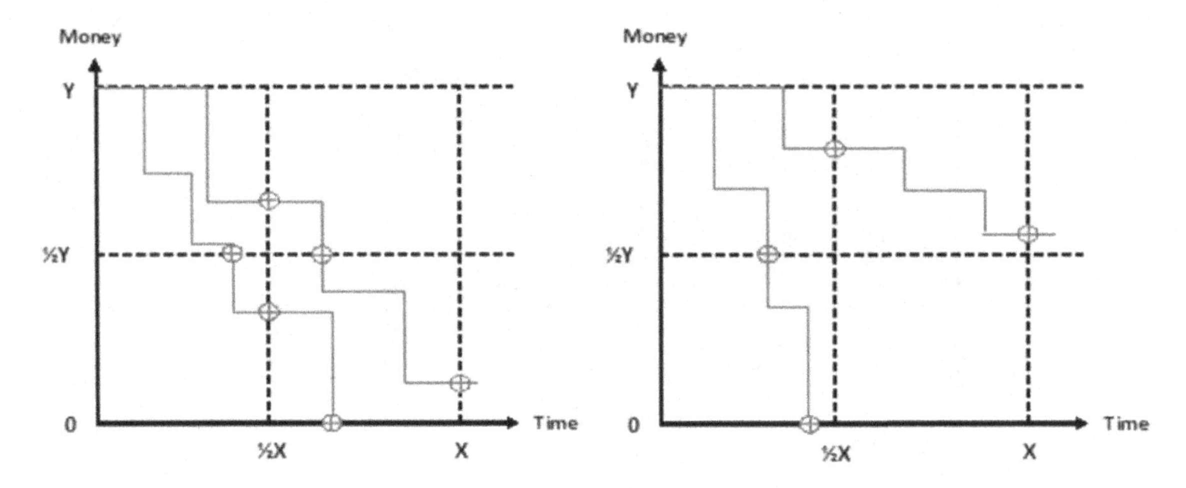

ducing the likelihood of addiction and therefore improving, at least, the autonomy of users gives rise to the system discussed acting as an ethical advisor which can intervene when necessary. Such considerations, if applied within virtual worlds, could reverse the substantial decline in turnover seen due to the hosting of online gambling systems being illegal in the US; more widely, this could offer up new opportunities for the advent of gambling elsewhere. We have demonstrated how to offer a framework via EthiCasino[5], a prototype created in the Second Life virtual world. EthiCasino implements specific ethical theories and learns about the risky behaviour and (lack of) knowledge of its users. While most other ethical systems considered in this paper are either conceptual or prototype conceptual models, never tested with actual users, or are otherwise unavailable or have been discarded by their creators, the ethical principles behind EthiCasino have undergone limited testing with users.

With the substantial revenues estimated for online gambling, a system such as EthiCasino may help to ensure that the ethical side of gambling remains to the fore by addressing issues relating to the impulse to gamble (Cutter and Smith, 2008). Reactive and non-intervening systems may be ineffective as people may not be aware of their lack of knowledge of odds, can adopt alternate identities to work around being restricted, and because problem gamblers may not realize that they have a problem. EthiCasino requires users to demonstrate their levels of gambling knowledge in order to assess their likely risk profile, and also requires them to provide limits before they become caught up in the actions. If their risk profile increases, or they exceed their self-imposed limits, intervention should occur; those who demonstrate low levels of gambling knowledge may be unable to increase their limits. We claim that EthiCasino demonstrates how users can be helped to avoid problems with addiction. The prototype framework of EthiCasino is relatively well-developed, and EthiCasino has been evaluated by a number of machine ethicists and experts in philosophy, computer science and business. A larger-scale evaluation is needed, but offering this in Second Life would entail allowing for online gambling, ruling out such an effort. The move to an alternative virtual world, the creation of a private virtual world, or the transference of the principles to web-based gambling system may allow for such an evaluation.

7. FUTURE RESEARCH DIRECTIONS

This chapter offers several avenues for future work, both with a narrow focus on ethical and legal online gambling and wider focus on ethical and legal cyberspace. Some are directly driven from the principles of EthiCasino, while others emerge from the concepts of machine ethics for cyberspace. Examples are:

1. **Ethical data mining:** An artificial agent can be designed to deal with data and rules which not only will increase the accuracy of the system but its ethicality (fairness) as well.
2. **Ethical gambling websites:** The framework of EthiCasino could be adopted for online gambling of all kinds.
3. **Ethical and legal advisors:** Formulated around specific ethical and legal problems in other domains, and grounded in applicable laws. For example, advisors for business ethics could act to prevent unethical actions (some of which may be similar to gambling) or promote adherence to codes of ethics and professional practice.

In all of these, ethics fill the gaps between laws, technologies, and what people know and/or are allowed to do and/or should be prevented from doing for their own good.

REFERENCES

H2 Gambling Capital. (2010). United States Internet Gambling: Job Creation – Executive Summary. *H2 Gambling Capital*. Retrieved from www.safeandsecureig.org/news/InternetGamblingStudy_04_15_2010.pdf

Allen, C., Wallach, W., & Smit, I. (2006). Why Machine Ethics? *IEEE Intelligent Systems*, *21*(4), 12–17. doi:10.1109/MIS.2006.83

Anderson, M., Anderson, S., & Armen, C. (2005). Towards Machine Ethics: Implementing Two Action-Based Ethical Theories. In *Proceeding of AAAI 2005 Fall Symposiom on Machine Ethics*. AAAI Press.

Ashford, W. (2010, June 9). Data Protection Act is out of kilter with EU law, warns privacy lawyer. *Computer Weekly*. Retrieved from http://www.computerweekly.com/Articles/2010/06/09/241513/Data-Protection-Act-is-out-of-kilter-with-EU-law-warns-privacy.htm

Asimov, I. (1983). Asimov on Science fiction. London: Granada.

Baase, S. (2008). *Gift of fire: Social, legal, and ethical issues for computing and the internet* (3rd ed.). Cambridge, MA: Pearson.

Boyd, C. (2008). Profile: Gary McKinnon. *BBC News*. Retrieved from http://news.bbc.co.uk/1/hi/technology/4715612.stm

Comeau, S. (1997). Getting high on gambling. *McGill Reporter*. Retrieved from http://reporter-archive.mcgill.ca/Rep/r3004/gambling.html

Consumer Focus. (2010). *Outdated copyright law confuses consumers*. Retrieved from http://www.consumerfocus.org.uk/news/outdated-copyright-law-confuses-consumers

Cutter, C., & Smith, M. (2008, April). Gambling addiction and problem gambling, signs, symptoms and treatment. *helpguide.org*. Retrieved from http://www.helpguide.org/mental/gambling_addiction.htm

FoxNews. (2010, April 15). 7,500 Online Shoppers Unknowingly Sold Their Souls. *Fox News*. Retrieved from http://www.foxnews.com/scitech/2010/04/15/online-shoppers-unknowingly-sold-souls/

Gambleaware. (n.d.). Responsible gambling. *Gambleaware*. Retrieved from http://gambleaware.co.uk/responsible-gambling

GAO. (2002). Internet Gambling: An Overview of the Issues. *United States General Accounting Office*. Retrieved from http://www.gao.gov/new.items/d0389.pdf

Garrett, J. (2004). *A Simple and Usable (Although Incomplete) Ethical Theory Based on the Ethics of W. D. Ross*. Western Kentucky University. Retrieved from http://www.wku.edu/~jan.garrett/ethics/rossethc.htm

Gibson, W. (1984). *Neuromancer*. New York: Ace Books.

Glimne, D. (n.d.). Gambling. *Encyclopaedia Britannica*. Retrieved from http://www.britannica.com

Hargreaves, I. (2011). *Digital Opportunity: A Review of Intellectual Property and Growth*. Retrieved from http://www.ipo.gov.uk/ipreview-finalreport.pdf

iGambling Business. (n.d.). Rosy future for online casino industry. *iGambling Business*. Retrieved from http://www.igamingbusiness.com/content/rosy-future-online-casino-industry

Jewkes, Y. (Ed.). (2003). *Dot.cons: Crime, deviance and identity on the internet*. Cullompton, UK: Willan.

Kennedy, M. (2012, December 14). Gary McKinnon will Face No Charges in UK. *The Guardian*. Retrieved from http://www.theguardian.com/world/2012/dec/14/gary-mckinnon-no-uk-charges

Lewis, E. (2003). *Gambling and Islam: Clashing and Co-existing*. Brigham Young University. Retrieved from http://www.math.byu.edu/~jarvis/gambling/student-papers/eric-lewis.pdf

McKinnon (Appellant) *v* Government of the United States of America (Respondents) and Another (2008) UKHL 59. House of Lords.

McLaren, B. M., & Ashley, K. D. (2000). Assessing Relevance with Extensionally Defined Principles and Cases. In *Proceedings of the 17th National Conference of Artificial Intelligence*. AAAI Press.

Net-Security. (2010). Botnets drive the rise of ransomware. *Net Security*. Retrieved from http://www.net-security.org/secworld.php?id=9095

News, B. B. C. (2010, April 4). Jailed Dubai kissing pair lose appeal over conviction. *BBC News*. Retrieved from http://news.bbc.co.uk/1/hi/uk/8602449.stm

Pasick, A. (2007). FBI checks gambling in Second Life virtual world. *The Reuters*. Retrieved from http://www.reuters.com/article/technologyNews/idUSHUN43981820070405

Price, J. (2006). Gambling - Internet Gambling. *The Ethics and Religious Library Commission*. Retrieved from http://erlc.com/article/gambling-internet-gambling

RIAA. (n.d.). Piracy: Online and On The Street. *Recording Industry Association of America (RIAA) homepage*. Retrieved from http://www.riaa.com/physicalpiracy.php

Richmond, S. (2010, April 17). Gamestation Collects Customers' Souls in April Fools Gag. *Telegraph*. Retrieved from http://blogs.telegraph.co.uk/technology/shanerichmond/100004946/gamestation-collects-customers-souls-in-april-fools-gag/

Ross, W. D. (1930). *The Right and the Good*. Oxford, UK: Oxford University Press.

Russell, J. (2006). Europe: Responsible gambling: A safer bet? *Ethical Corporation*. Retrieved from http://www.ethicalcorp.com/content.asp?ContentID=4291

Saha, P. (2005). Gambling with responsibilities. *Ethical Corporation*. Retrieved from http://www.ethicalcorp.com/content.asp?ContentID=3774

United States of America v. David Bermingham, et al., 18 U.S.C. §§ 1343 & 2. CR-02-00597.

United States of America v. Gary McKinnon. 18 U.S.C. §§ 1030(a)(5)(A) & 2 Indictment.

Vartapetiance, A., & Gillam, L. (2010). DNA dataveillance: Protecting the innocent? *Journal of Information. Communication and Ethics in Society*, *8*(3), 270–288. doi:10.1108/14779961011071079

Weber, T. (2007, January 25). Criminals 'may overwhelm the web'. *BBC News*. Retrieved from http://news.bbc.co.uk/1/hi/business/6298641.stm

Young, A. (2006). *Gambling or Gaming, Entertainment or Exploitation?* Ethical Investment Advisory Group of the Church of England. Retrieved from http://www.cofe.anglican.org/info/ethical/policystatements/gambling.pdf

ADDITIONAL READING

Anderson, M., & Anderson, S. (2006). Computing an Ethical Theory with Multiple Prima Facie Duties, Proccedings of the 10th international Conference on the Simulation and Synthesis of Living Systems, Indiana University.

Anderson, M., & Anderson, S. (2008). EthEl: Toward a Principled Ethical Eldercare Robot *Robotic Helpers: User Interaction, Interfaces and Companions in Assistive and Therapy Robotics Workshop at the Third ACM/IEEE Human-Robot Interaction Conference*. ACM/IEEE. Amsterdam, NL, 33-39.

Arkin, R. C. (2008). Governing Lethal Behaviour: Embedding Ethics in a Hybrid Deliberative/Reactive Robot Architecture - Part I: Motivation and Philosophy, *Robotic Helpers: User Interaction, Interfaces and Companions in Assistive and Therapy Robotics Workshop at the Third ACM/IEEE Human-Robot Interaction Conference*. ACM/IEEE. Amsterdam, NL.

Bentham, J. (1948). *Introduction to the Principles of Morals and Legislation,* Oxford, UK, (origial work published at 1781). Retrieved from http://socserv.mcmaster.ca/econ/ugcm/3ll3/bentham/morals.pdf

Bringsjord, S., Arkoudas, K., & Bello, P. (2006). Toward a General Logicist Methodology for Engineering Ethically Correct Robots. *IEEE Intelligent Systems*, *21*(4), 38–44. doi:10.1109/MIS.2006.82

Carlson, D. (2007). Internet gambling law a success, but faces scrutiny. *The Ethics and Religious Library Commision*. Retrieved from http://erlc.com/article/internet-gambling-law-a-success-but-faces-scrutiny

EETL. (n.d.). Engineering Ethics Transcription Exercise. *University of Pittsburgh*. Retrieved from http://www.pitt.edu/~bmclaren/ethics/

Guarini, M. (2006). Particularism and Classification and Reclassification of Moral Cases. *IEEE Intelligent Systems*, *21*(4), 22–28. doi:10.1109/MIS.2006.76

Kilcullen, R. (1996). Rawls, A Theory of Justice. *Macquarie University*. Retrieved from http://www.humanities.mq.edu.au/Ockham/y6411.html

McCarty, L. T. (1997). Some Arguments About Legal Arguments, *Proceeding of the Sixth Internationa Conference of Artificial Intelligence and Law,* ACM, New York, 215-224.

McLaren, B. M. (2003). Extensionally Defining Principles and Cases in Ethics: an AI Model. *Artificial Intelligence Journal*, *150*(November), 145–181. doi:10.1016/S0004-3702(03)00135-8

McLaren, B. M. (2006). Computational Models of Ethical Reasoning: Challenges, Initial Steps, and Future Directions. *IEEE Intelligent Systems*, *21*(4), 29–37. doi:10.1109/MIS.2006.67

Moor, J. H. (2006). The Nature, Importance and Difficulty of Machine Ethics. *IEEE Intelligent Systems*, *21*(4), 18–21. doi:10.1109/MIS.2006.80

Power, T. M. (2006). Prospects for a Kantian Machine. *IEEE Intelligent Systems*, *21*(4), 46–51. doi:10.1109/MIS.2006.77

Robbins, R. W., & Wallace, W. A. (2006). Decision Support for Ethical Problem Solving: A Multi-agent Approach. *Decision Support Systems*, *43*(4), 1571–1587. doi:10.1016/j.dss.2006.03.003

Sharkey, N. E. (2007). Automated Killers and the Computing Profession. *IEEE Computer Society Press*, *40*(11), 122–127.

SIROCCO. (n.d.). System for Intelligent Retrieval of Operationalized Cases and COdes (SIROCCO). *SIROCCO home page*. Retrieved from http://sirocco.lrdc.pitt.edu/sirocco/index.html

UVA Video Library. (2002). Dax's Story: A Severely Burned Man's Thirty-Year Odyssey. *University of Virginia*. Retrieved from http://www.researchchannel.org/prog/displayevent.aspx?rID=3619

KEY TERMS AND DEFINITIONS

Cybercrime: Or crime mediated by computer.

Cyberspace: Introduced by William Gibson in Neuromancer (1984) as "*A consensual hallucination experienced daily by billions of legitimate operators*". Now used interchangeable with World Wide Web as it meets criteria mentioned by Gibson.

EthiCasino: Our prototype system for demonstrating ethical and legal approaches to online gambling.

Machine Ethics: A research area concerned with defining how machines should behave towards human users and other machines, with emphasis on avoiding harm and other negative consequences of autonomous machines, or unmonitored and unmanned computer programs.

Online Gambling: Using online websites, virtual worlds, and other such software to gamble, a opposed to physically entering a casino or betting shop.

Responsible Gambling: Gambling whilst being fully aware of the facts of gambling, and staying in control of both money and time.

Second Life: A multi-user virtual environment developed by Linden Labs.

ENDNOTES

1. NHTCU has been moved under SOCA e-crime unit.
2. Internet gambling revenue for offshore companies was estimated to be $5.4 billion in 2009 from players in the United States and $25.8 billion from players worldwide, according to H2 Gambling Capital.
3. The idea to describe this as "nagware" was introduced by Prof. Allen, Indiana University (private correspondence, 16/6/2008).
4. Virtual interactive Kinetic Intelligence (VIKI) is a fictional computer introduced by Isaac Asimov. She serves as a central computer for robots to provide them with a form of "consciousness" recognizable to humans.
5. A prototype was built in Second Life purely for purposes of evaluation.

Chapter 11
A Study of Cyber Crime and Perpetration of Cyber Crime in India

Saurabh Mittal
Asia Pacific Institute of Management, India

Ashu Singh
Asia Pacific Institute of Management, India

ABSTRACT

The world is facing a new era of criminal activities in cyber space, which are being committed across the world, irrespective of geographical boundaries. These cybercrime acts may be financially driven acts, related to computer content, or against the confidentiality, integrity, and accessibility of computer systems. The relative risk and threat differs between governments and businesses. The level of criminal organization represents a defining feature of the human association element behind criminal conduct. India accounts for close to $8bn of the total $110bn cost of global cyber crime. The Information Technology (IT) Act, 2000, specifies the acts that are punishable. Cyber crime has also affected the social media. A crime prevention plan with clear priorities and targets needs to be established, and government should include permanent guidelines in its programmes and structure for controlling crime and ensuring that clear responsibilities and goals exist within government for the organization of crime prevention. This chapter seeks to find out the motives and suspects of cyber crime perpetration and suggests measures for crime prevention.

DOI: 10.4018/978-1-4666-6122-6.ch011

INTRODUCTION

Cyber Crimes are a new class of crimes rapidly increasing due to extensive use of Internet and I.T. enabled services. Computer crime is one of the fastest-growing types of illegal activity, both in the U.S. and internationally. While the Internet links people together like never before, it also provides endless opportunity to criminals seeking to exploit the vulnerabilities of others. There are several different types of computer crime, many of which overlap. Below are a few of the most commonly reported cyber crimes:

- **Phishing:** Phishing is the practice of sending fraudulent emails in an attempt to trick the recipient, usually for the purpose of obtaining money. The elderly are particularly vulnerable to these types of cyber crime.
- **Hacking:** Hacking is similar to digital trespassing. Hackers infiltrate online networks to illegally download confidential information, manipulate functions and in some cases steal identities that can be used to fraudulently purchase goods online.
- **Stalking and/or Harassment:** Not all types of cyber crime involve money. Some cyber criminals use the Internet as a cover for other illegal behaviors like stalking, harassment and in lesser cases, bullying.

The Information Technology (IT) Act, 2000, specifies the acts which are punishable. Since the primary objective of this Act is to create an enabling environment for commercial use of I.T., certain specific omissions and commissions of criminals while using computers have not been included. Several offences having bearing on cyber arena are also registered under the appropriate sections of the IPC with the legal recognition of Electronic Records and the amendments made in several sections of the IPC vide IT Act, 2000.

NATURE OF CYBERCRIME ACTS

Cybercrime acts may be financially-driven acts, related to computer content, or against the confidentiality, integrity and accessibility of computer systems. The relative risk and threat may vary between Governments and businesses. Individual cybercrime victimization is significantly higher than for 'conventional' crime forms especially in countries with lower levels of development, highlighting a need to strengthen prevention efforts in these countries. Private sector enterprises in Europe report victimization rates of between 2 and 16 percent for acts such as data breach due to intrusion or phishing. Criminal tools of choice for these crimes, such as botnets, have global reach. More than one million unique IP addresses globally functioned as command and control servers for botnets in 2011 (United Nations Office on Drugs and Crime, 2013).

- Internet content targeted for removal by governments includes child pornography and hate speech, but also defamation and government criticism, raising human rights law concerns in some cases.
- Some estimates place the total global proportion of internet traffic estimated to infringe copyright at almost 24 per cent.

CYBERCRIME PERPETRATORS

Cybercrime perpetrators no longer require complex skills or techniques, due to the advent and ready availability of malware toolkits. Upwards of 80 per cent of cybercrime acts are estimated to originate in some form of organized activity, with cybercrime black markets established on a cycle of malware creation, computer infection, botnet management, harvesting of personal and financial data, data sale, and 'cashing out' of financial information. Cybercrime often requires a high degree of organization to implement, and may lend itself to small criminal groups, loose ad hoc networks, or organized crime on a larger scale (United Nations Office on Drugs and Crime, 2013). The typology of offenders and active criminal groups mostly reflect patterns in the conventional world. In the developing country context in particular, sub-cultures of young men engaged in computer-related financial fraud have emerged, many of whom begin involvement in cybercrime in their late teenage years. The demographic nature of offenders mirrors conventional crime in that young males are the majority, although the age profile is increasingly showing older (male) individuals, particularly concerning child pornography offences. While some perpetrators may have completed advanced education, especially in the computer science field, many known offenders do not have specialized education. There is a lack of systematic research about the nature of criminal organizations active in cyberspace; and more research is needed regarding the links between online and offline child pornography offenders. The full depiction of a 'cybercrime perpetrator' may contain many elements. Age, sex, socio-economic background, nationality, and motivation are likely amongst the core characteristics. In addition, the level of criminal organization – or the degree to which individuals act in concert with others – represents a defining feature of the human association element behind criminal conduct. Understanding cybercrime as a 'socio-technological' phenomenon, based on an appreciation of the characteristics of persons who commit such crimes, represents a broader approach to prevention than that focused solely on technical cyber security concepts.

Cyber Crime in the World

The list of the most dangerous countries for computer attacks is based on the figures for 2012. "China and the U.S. may dominate the headlines when it comes to hacker attacks, but countries in the developing world are the most vulnerable to online assaults... When targeting consumers, cyber criminals are likely to go where there are fewer defenses. Developing markets provide such an opportunity, with millions of new Internet users every year and fewer resources to devote to security," the agency reported. However, the U.S. was also included in the list - being put in 19th place with 45% of people who faced cyber attacks last year. About 78 percent of the U.S.'s population, or an estimated 245 million people, were online last year (Ukrinform, 2012).

The most threatening situation in terms of computer attacks is in Russia, where 59% of Internet users faced online assaults last year. Only about 48% of Russian citizens (68 million people) were online last year (Ukrinform, 2012).

Cyber Crime in the US

The Internet Fraud Complaint Center — a partnership between the FBI and NW3C (funded by BJA) — was established May 8, 2000, to address the ever-increasing incidence of online fraud. Just three years later, in response to the exponential increase in cybercrime of all types, the Center changed its name to the Internet Crime Complaint Center (IC3). Today, the IC3 accepts more complaints in a single month than it received in its first six months. With more than two million complaints received since its inception, the IC3 serves as the nation's portal for reporting Internet crime and suspicious activity (The Internet Crime Complaint Center (IC3), 2012).

Frequently Reported Internet Crimes in the US

- **Auto Fraud:** In fraudulent vehicle sales, criminals attempt to sell vehicles they do not own. An attractive deal is created by advertising vehicles for sale on various online platforms at prices below market value.

- **FBI Impersonation E-mail Scam:** Officials have been used in spam attacks in an attempt to defraud consumers. Government agencies do not send unsolicited e-mails.
- **Intimidation/ Extortion Scams:** Intimidation and extortion scams have evolved over the years to include scams like Telephone Calls, Payday loan, Process server, the Grandparent Scam, Hit-man scam.
- **Scareware/ Ransomware:** Extorting money from consumers by intimidating them with false claims pretending to be the federal government watching their Internet use and other intimidation tactics have evolved over the years to include scams like Pop-up Scareware Scheme, Citadel Malware, IC3 Ransomware.
- **Real Estate Fraud:** Rental Scams - Criminals search websites that list homes for sale and take information from legitimate ads and post it with their own e-mail addresses.
- **Timeshare Marketing Scams:** Timeshare owners across the country are being scammed out of millions of dollars by unscrupulous companies that promise to sell or rent the properties. In the typical scam, timeshare owners receive unexpected or uninvited telephone calls or e-mails from criminals posing as sales representatives for a timeshare resale company.
- **Loan Modification Scams:** A loan modification scam often starts when a bogus loan company contacts a distraught homeowner and offers a loan modification plan via phone call, e-mail or mailing. A homeowner may reach out to these companies after seeing an ad online or in the newspaper
- **Romance Scams:** Perpetrators use the promise of love and romance to entice and manipulate online victims. A perpetrator scouts the Internet for victims, often finding them in chat rooms, on dating sites and even within social media networks. These individuals seduce victims with small gifts, poetry, claims of common interest or the promise of constant companionship. Once the scammers gain the trust of their victims, they request money, ask victims to receive packages and reship them overseas or seek other favors. These cyber criminals capitalize on the vulnerabilities of their victims. This crime not only affects the victims financially, there are emotional and mental implications as well. The IC3 received 4,467 complaints and the victims' losses totaled more than $55 million (The Internet Crime Complaint Center (IC3), 2012).

Complainant Demographics

During the recent years, the number of male and female complainants is equalizing. The age groups between 20-60 years file the maximum number of complaints. India stands at 5th position with 0.5871% with respect to victim complainants among the top five countries of the world with U.S. being the first with 91.1907%. (The Internet Crime Complaint Center (IC3), 2012)

Major findings on the cyber attacks on the Social network (Norton, 2013):

- 1/10 Social network users have fallen victim to a scam or fake link on social network platforms.
- 4/10 Social network users have fallen victim to cybercrime on social networking platforms.
- 3/4 Believe cybercriminals are setting their sights on social networks.
- 1/6 Social network users report that someone has hacked into their profile and pretended to be them.

In early February 2013, Twitter made an official announcement that malicious users had managed to steal user data (including hash passwords) from 250,000 members of the social network. Two weeks later, Facebook issued a statement that the computers of several of the company's staff members had been infected by exploits during visits to a mobile developer site. Facebook described this as a sophisticated, targeted attack, noting that the goal was to penetrate Facebook's corporate network. Luckily, company representatives stated that Facebook was able to avoid any leakage of user information. Just a few days later, Apple announced that several company employees had fallen victim to the very same attack while visiting a website for mobile developers. Apple's data showed that no data leakages had taken place. In early March 2013, Evernote reported that their internal network had been hacked and that malicious users had attempted to gain access to their data.

Facebook, like every significant internet service, is frequently targeted by those who want to disrupt or access our data and infrastructure. As such, we invest heavily in preventing, detecting, and responding to threats that target our infrastructure, and we never stop working to protect the people who use our service. The vast majority of the time, we are successful in preventing harm before it happens, and our security team works to quickly and effectively investigate and stop abuse. In Jan 2013, Facebook Security discovered that their systems had been targeted in a sophisticated attack. This attack occurred when a handful of employees visited a mobile developer website that was compromised. The compromised website hosted an exploit which then allowed malware to be installed on these employee laptops. The laptops were fully-patched and running up-to-date anti-virus software. (Norton, 2013).

Cybercrime presents particular crime prevention challenges. These include the increasing ubiquity and affordability of online devices leading to large numbers of potential victims; the comparative willingness of persons to assume 'risky' online behaviour; the possibility for anonymity and obfuscation techniques on the part of perpetrators; the transnational nature of many cybercrime acts; and the fast pace of criminal innovation. Each of these challenges has implications for the *organization, methods* and *approaches* adopted for cybercrime prevention. Organizational structures, for example, will need to reflect the need for international and regional cooperation in cybercrime prevention. *Methods* will need to ensure a constantly updated picture of cyber threats, and *approaches* will need to involve a range of stakeholders – in particular the private sector organizations that own and operate internet infrastructure and services.

Cyber Crimes in India

Incidence of Cyber Crimes (IT Act + IPC Sections) has increased by 57.1% in 2012 as compared to 2011 (from 2,213 in 2011 to 3,477 in 2012). Cyber Fraud accounted for 46.9% (282 out of 601) and Cyber Forgery accounted for 43.1% (259 out of total 601) were the main cases under IPC category for Cyber Crimes. 61.0% of the offenders under IT Act were in the age group 18-30 years (928 out of 1,522) and 45.2% of the offenders under IPC Sections were also in the age group 18-30 years (248 out of 549) (National Crime Records Bureau, 2012).

Cyber Crimes: Cases of Various Categories under It Act, 2000

2,876 cases were registered under IT Act during the year 2012 as compared to 1,791 cases during the previous year (2011), thereby reporting an increase of 60.6% in 2012 over 2011. 50.1% (1,440 cases) of the total 2,876 cases registered under IT Act 2000 were related to Loss/damage to computer resource/utility reported under hacking with computer systems. 612 persons were arrested for committing such offences during 2012. There were 589 cases of obscene publications/transmission in electronic form during the year2012 wherein 497 persons were arrested. Out of the total (1,875) hacking cases, the cases relating to loss / damage

Table 1. Cyber Crimes/cases registered and persons arrested under IT Act during 2009-2012

SL. NO.	Crime heads	Cases registered				% Variation in 2012 over 2011	Persons arrested				% Variation in 2012 over 2011
		2009	2010	2011	2012		2009	2010	2011	2012	
1	Tampering computer source documents	21	64	94	161	71.3	6	79	66	104	57.6
2	Hacking with computer system										
	i) Loss/damage to computer resource/utility	115	346	826	1,440	74.3	63	233	487	612	25.7
	ii)Hacking	118	164	157	435	177.1	44	61	65	137	110.8
3	Obscene publication/transmission in electronic form	139	328	496	589	18.8	141	361	443	497	12.2
4	Failure										
	i) Of compliance/orders of certifying authority	3	2	6	6	0.0	6	5	4	4	0.0
	ii) To assist in decrypting the information intercepted by govt. agency	0	0	3	3	0.0	0	0	0	3	-
5	Un-authorised access/attempt to access to protected computer system	7	3	5	3	-40.0	16	6	15	1	-93.3
6	Obtaining licence or digital signature certificate by misrepresentation/suppression of fact	1	9	6	6	0.0	1	4	0	5	-
7	Publishing false digital signature certificate	1	2	3	1	-66.7	0	2	1	0	-100.0
8	Fraud digital signature certificate	4	3	12	10	-16.7	6	4	8	3	-62.5
9	Breach of confidentiality/privacy	10	15	26	46	76.9	5	27	27	22	-18.5
10	Other	1	30	157	176	12.1	0	17	68	134	97.1
	Total	**420**	**966**	**1,791**	**2,876**	**60.0**	**228**	**779**	**1,184**	**1,522**	**28.5**

(National Crime Records Bureau, 2012)

of computer resource/utility under Sec 66(1) of the IT Act were 76.8% (1,440 cases) whereas the cases related to hacking under Section 66(2) of IT Act were 23.2% (435 cases). The age-wise profile of persons arrested in Cyber Crime cases under IT Act, 2000 showed that 61.0% of the offenders were in the age group 18 – 30 years (928 out of 1,522) and 28.6% of the offenders were in the age group 30 - 45 years (436 out of 1522). Crime head-wise and age group wise profile of the offenders arrested under IT Act, 2000 reveals that 40.2% (612 out of 1,522) of the offenders arrested for 'Loss/damage to computer resource/utility under hacking with computer systems' of which 62.3% (381 out of 612 were in the age-group 18 –30 years. 55.9% (278 out of 497) of the total persons arrested for 'obscene publication/transmission in electronic form' were in the age-group of 18 - 30 years. (National Crime Records Bureau, 2012)

INCIDENCES OF CYBER CRIMES REGISTERED UNDER IPC

Information on the cases registered under various sections of IPC which were considered as cyber crimes at all-India level is presented in Table 1.

CYBER CRIMES: CASES OF VARIOUS CATEGORIES UNDER IPC SECTION

A total of 601 cases were registered under IPC Sections during the year 2012 as compared to

Table 2. Persons Arrested Under Cyber Crimes (IT Act + IPC Sections) By Age Group During 2012 (All-India)

Sl. No	Crime Head	Below 18 Years	Between 18 – 30 Years	Between 30 –45 Years	Between 45 – 60 Years	Above 60 Years	Total (All Age Groups)
(1)	(2)	(3)	(4)	(5)	(6)	(7)	(8)
	A. Offences under IT Act						
1	Tampering computer source documents	0	64	36	3	1	**104**
2	Hacking with Computer Systems						
	i) Loss/damage to computer resource/utility	21	381	175	33	2	**612**
	ii) Hacking	2	98	35	2	0	**137**
3	Obscene publication/transmission in electronic form	37	278	147	35	0	**497**
4	Failure						
	i) Of compliance/orders of certifying Authority	0	1	1	2	0	**4**
	ii) To assist in decrypting the information intercepted by Govt. Agency	0	3	0	0	0	**3**
5	Un-authorised access/attempt to access of protected Computer system	0	1	0	0	0	**1**
6	Obtaining License or Digital Signature Certificate by misrepresentation/suppression of fact	0	4	0	1	0	**5**
7	Publishing false digital Signature Certificate	0	0	0	0	0	**0**
8	Fraud Digital Signature Certificate	0	0	3	0	0	**3**
9	Breach of confidentiality/privacy	1	17	3	1	0	**22**
10	Other	4	81	36	13	0	**134**
11	**Total (A)**	**65**	**928**	**436**	**90**	**3**	**1522**
	B. Offences under IPC						
1	Offences by/Against Public Servant	0	1	2	1	0	**4**
2	False electronic evidence	0	1	1	0	0	**2**
3	Destruction of electronic evidence	0	11	3	2	0	**16**
4	Forgery	1	130	109	19	4	**263**
5	Criminal Breach of Trust/Fraud	1	80	86	46	2	**215**
6	Counterfeiting						
	i) Property/mark	0	7	3	3	0	**13**
	ii) Tampering	0	15	11	0	0	**26**
	iii)Currency/Stamps	0	3	6	1	0	**10**
7	**Total(B)**	**2**	**248**	**221**	**72**	**6**	**549**
	Grant Total (A+B)	**67**	**1176**	**657**	**162**	**9**	**2071**

(National Crime Records Bureau, 2012)

Figure 1. Cyber crimes / cases registered and persons arrested under IT Act during 2008-2012
(National Crime Records Bureau, 2012)

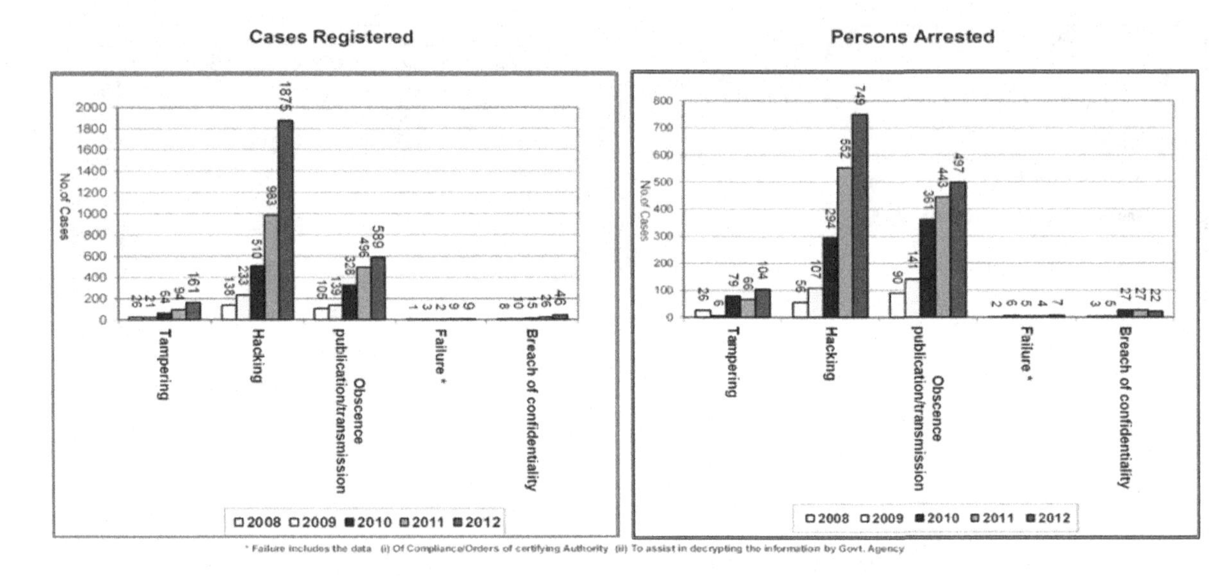

422 such cases during 2011, thereby reporting an increase of 42.4%. Majority of the crimes out of total 601 cases registered under IPC fall under 2 categories viz. criminal breach of trust or fraud (282) and forgery (259). Although such offences fall under the traditional IPC crimes, these cases had the cyber overtones wherein computer, Internet or its enabled services were present in the crime and hence they were categorised as Cyber crimes under IPC. Cyber forgery (259 cases) accounted for 0.27% out of the 94,203 cases reported under cheating. Cyber frauds (118) accounted for 0.66% of the total criminal breach of trust cases under IPC (17,201). A total of 549 persons were arrested in the country for cyber crimes under IPC during 2012. 47.9% (263) of these offenders were arrested for offences under 'cyber forgery', 39.2% (215) for 'criminal breach of trust/fraud' and 2.9% (16) for 'distribution of electronic evidence'.

CYBERCRIME SUSPECTS IDENTIFIED BY POLICE

In one country in Southern Asia, published national police statistics contain details of recorded cybercrime offences and suspects. Suspects are classified in reported statistics through a number of categories, according to relationship with the victim and other characteristics. While a high proportion of suspects remain unclassified, national police statistics show that:

- Over 10 per cent of recorded cybercrime suspects are known to the victim as neighbours, friends, or relatives;
- 'Disgruntled employees' and 'crackers' each constitute around 5 per cent of recorded cybercrime perpetrators;
- A significant number of cybercrime suspects are enrolled in higher education and other learning programmes.

Table 3. Cyber crimes/cases registered and persons arrested under IPC during 2009-2012

SL. NO.	Crime heads	Cases registered				% Variation in 2012 over 2011	Persons arrested				% Variation in 2012 over 2011
		2009	2010	2011	2012		2009	2010	2011	2012	
1	Offences by/against public servant	0	2	7	2	-71.4	0	3	3	4	33.3
2	False electronic evidence	0	3	1	4	300.0	0	4	1	2	100.0
3	Destruction of electronic evidence	3	1	9	9	0.0	0	0	10	16	60.0
4	Forgery	158	188	259	259	0.0	161	257	277	263	-5.1
5	Criminal breach of trust/fraud	90	146	118	282	139.0	79	100	129	215	66.7
6	Counterfeiting										
	i) Property Mark	1	1	6	21	250.0	3	2	8	13	62.5
	ii) Tampering	3	8	5	19	280.0	0	12	7	26	271.4
	iii)Currency/stamps	21	7	17	5	-70.6	20	16	11	10	-9.1
7	**Total**	**276**	**356**	**422**	**601**	**42.4**	**263**	**394**	**446**	**549**	**23.1**

(National Crime Records Bureau, 2012)

HISTORICAL DEVELOPMENTS

Cyber Crime in India

India accounts for close to $8bn of the total $110bn cost of global cyber crime. By 2017, the global cyber security market is expected to cross $120bn (India Infoline News Service, 2013). Bangalore accounted for the majority of the cases with around 83 per cent cases being booked here. That India remains one of the most violent nations to live in has been further reinforced by NCRB. Cyber crimes (IT Act and IPC) in India have increased by 57.11 per cent in 2012. Cyber forgery with 47.9 per cent (263 out of total 549) and cyber fraud with 39.1 per cent (215 out of 549) were the main cases under IPC category for cyber crimes in 2012 (National Crime Records Bureau, 2012).

Maharashtra remains the centre of cyber crime with a maximum cases reported in 2012. Banglore is the IT capital and most networked city in India. But these IT-fication leads to being obvious

Table 4. Cases registered under cyber crimes categorized by motives during 2012 (India)

Revenge / Settling scores	Greed/ Money	Extortion	Cause Disrepute	Prank/ Satisfaction of Gaining Control	Fraud / Illegal Gain	Eve Teasing / Harassment	Others	Total
87	624	48	117	45	668	587	1301	3477

Table 5. Cases registered under cyber crimes categorized by suspects- 2012 (India)

Foreign National / Group	Disgrunted Employee/ Employee	Cracker /Student /Professional learners	Business Competitor	Neighbours / Friends & Relatives	Others	Total
104	117	160	103	413	2580	3477

(National Crime Records Bureau, 2012)

Table 6. Incidence Of Cases Registered And Number Of Persons Arrested Under Cyber Crimes (IT Act + IPC Sections) During 2012 (All-India)

Sl. No	Crime Head	Cases Registered	Persons Arrested
(1)	(2)	(3)	(4)
	A. Offences under IT Act		
1	Tampering computer source documents	161	104
2	Hacking with Computer Systems		
	i) Loss/damage to computer resource/utility	1440	612
	ii) Hacking	435	137
3	Obscene publication/transmission in electronic form	589	497
4	Failure		
	i) Of compliance/orders of certifying Authority	6	4
	ii) To assist in decrypting the information intercepted by Govt. Agency	3	3
5	Un-authorised access/attempt to access of protected Computer system	3	1
6	Obtaining License or Digital Signature Certificate by misrepresentation/suppression of fact	6	5
7	Publishing false digital Signature Certificate	1	0
8	Fraud Digital Signature Certificate	10	3
9	Breach of confidentiality/privacy	46	22
10	Other	176	134
12	**Total (A)**	**2876**	**1522**
	B. Offences under IPC		
1	Offences by/Against Public Servant	2	4
2	False electronic evidence	4	2
3	Destruction of electronic evidence	9	16
4	Forgery	259	263
5	Criminal Breach of Trust/Fraud	282	215
6	Counterfeiting		
	i) Property/mark	21	13
	ii) Tampering	19	26
	iii)Currency/Stamps	5	10
7	**Total(B)**	**601**	**549**
	GrandTotal (A+B)	**3477**	**2071**

(National Crime Records Bureau, 2012)

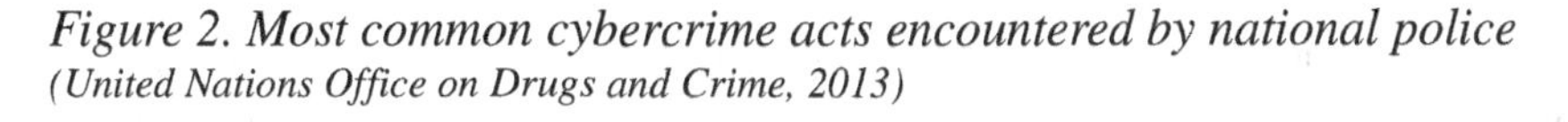

Figure 2. Most common cybercrime acts encountered by national police
(United Nations Office on Drugs and Crime, 2013)

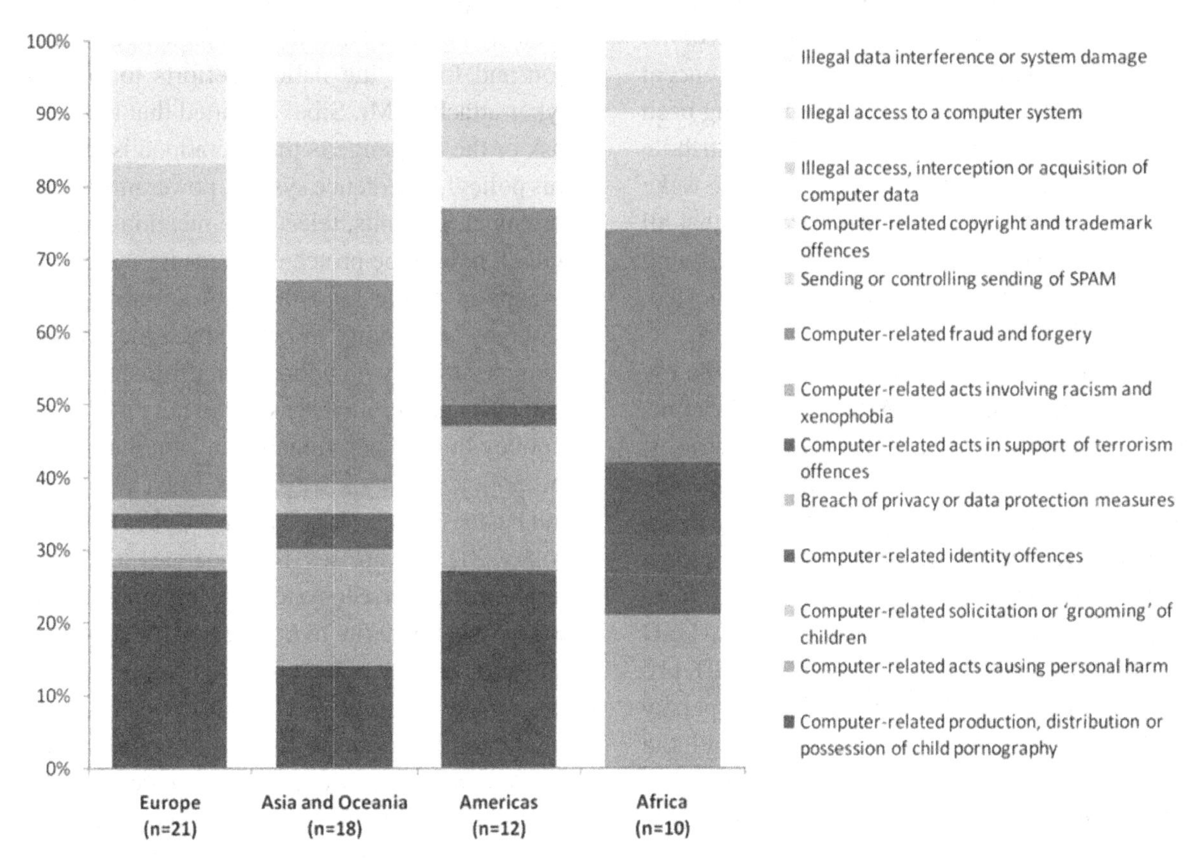

top position in national charts when it comes to cyber crime. Bangalore city itself accounts for 24.4 per cent of cyber crimes booked under the IT Act among 53 'megacities' across India and 11.4% of the total cyber crime in country. Year-on-year increase of 192% was registered, with 342 cases booked in 2012, up from 117 in 2011. Visakhapatnam registered 153 cases in year 2012. Interestingly, the numbers are much more modest when it comes to cyber crimes booked under provisions of the IPC — only seven compared to 72 in Mumbai. While 'greed/money' and 'fraud/illegal gain' account for the majority of motives behind cybercrime a majority of offences recorded come under the hacking category. Majority of recorded crimes under the IT Act or the Indian Penal Code fall under two heads: 'loss/damage to computer resource' and 'hacking' (National Crime Records Bureau, 2012).

In a detailed written response to the Lok Sabha, India's Minister of State for Communications & IT Mr. Milind Deora revealed that a total of 78 government websites have been hacked up to June 2013. Mr. Deora was communicating information reported to and tracked by the Indian Computer Response Team (CERT-In), who also found that in the past two years (2011 and 2012) a total of 308 and 371 websites had been comprised, respectively. The Minister informed the Lower House that as many as 16,035 instances of scanning, spam, malware infection, DDoS attacks and system break-in affecting the government, defence and public sector undertakings were reported till June2013. He added that the nation also witnessed as many as 13,301 instances of security breaches in 2011, and 22,060 in 2012. Deora added that it is known that miscreants were compromising computer systems located in different parts of the world and that they use hidden servers to conceal the true identity of the system that is being used to further the attacks. "It is difficult to attribute the origin of cyber attacks," he said. In the wake of such a situation, the Minister stressed that all new government websites must be audited with respect to cyber security before they are hosted. (Economic Times, 2013)

42 million cyber crimes happen in India every year (Norton, 2013). About 52% of victims have suffered attacks such as malware, viruses, hacking, scams, fraud and theft. Over 80 people are victimized under various cyber crimes every minute across India. 7% of the estimated global price tag of cyber crimes every year comes from India. The estimated price tag in India is USD 8 billion as against the global bill of USD 110 billion (Norton, 2013). Security firm Kaspersky ranks India ninth on its list of countries with the highest percentage of computer attacks. India has also been a victim of cyber intrusions by neighbouring states (Kaspersky, 2012). As a recent media report by a leading channel reveals, China has repeatedly mounted cyber attacks on Indian websites through its cyber warfare army. India recently released its National Cyber Security Policy 2013 (NCSP 2013). However, according to EC-Council, a professional certification body, India is unlikely to meet the target of creating a workforce of 5,00,000 cyber security professionals in next five years (as laid out by the NCSP 2013) due to lack of infrastructure and investment (Ground Zero Summit 2013, Asia's Largest Information Security Conference, Launched, 2013).

THE ROAD AHEAD

Mr. Kapil Sibal (Cabinet Minister- Communication & IT, Govt. of India) says, "2013 National Cyber Security Policy will be a 'real task' to operate. The latest edition of the nation's cyber security policy is here and the National Cyber Security Policy 2013 aims to safeguard information and fortify the nation's efforts to prevent cyber-attacks". Mr. Sibal admitted that the "real task or the challenge is the operationalisation of this policy. Air defence system, power infrastructure, nuclear plants, telecommunications system will all have to be protected to ensure there is no disruption of the kind that will destabilise the economy...instability in cyber space means economic instability no nation can afford economic instability, therefore it is essential not just to have a policy but to operationalise it." On the whole, the policy will look to protect India's physical and business assets. The National Cyber Security Policy 2013 comprises 14 objectives including creating a cyber ecosystem in the country, offering fiscal benefits to businesses for taking up standard security practices and processes, building effective public private partnerships, and collaborative engagements through technical and operational cooperation. There is no denying the importance of having a strong cyber security in place, especially since there is always a possibility of an attack from any quarter. Sibal said it will not be possible to point fingers at any particular country. "In the ultimate analysis, we have to develop global standards because there is no way that we can have a policy within the context of India which is not connected with the rest of the world because information knows no territorial boundaries," he said. Over the next five years, the government plans to create a workforce of 5,00,000 professionals skilled in cyber security "through capacity building, skill development and training." (Anuradha Shetty, 2013)

Even as India continues its march towards being technologically adept, it continues to grapple with instances of rising cyber crime. National Cyber Security Policy 2013 - aims to safeguard information and fortify the nation's efforts to prevent cyber-attacks. On the whole, the policy will look to protect India's physical and business assets. The National Cyber Security Policy 2013 comprises 14 objectives including creating a cyber ecosystem in the country, offering fiscal benefits to businesses for taking up standard security practices and processes, building effective public private partnerships, and collaborative engagements through technical and operational co-operation. (Anuradha Shetty, 2013)

FINDINGS

1. Fragmentation at the international level, and diversity of national cybercrime laws, may correlate with the existence of multiple instruments with different thematic and geographic scope. While instruments legitimately reflect socio-cultural and regional differences, divergences in the extent of procedural powers and international cooperation provisions may lead to the emergence of country cooperation 'clusters' that are not always well suited to the global nature of cybercrime;
2. Reliance on traditional means of formal international cooperation in cybercrime matters is not currently able to offer the timely response needed for obtaining volatile electronic evidence. As an increasing number of crimes involve geo-distributed electronic evidence, this will become an issue not only for cybercrime, but all crimes in general;
3. In a world of cloud computing and data centers, the role of evidence 'location' needs to be re-conceptualized, including with a view to obtaining consensus on issues concerning direct access to extraterritorial data by law enforcement authorities;
4. Analysis of available national legal frameworks indicates insufficient harmonization of 'core' cybercrime offences, investigative powers, and admissibility of electronic evidence. International human rights law represents an important external reference point for criminalization and procedural provisions;
5. Law enforcement authorities, prosecutors, and judiciary in developing countries, require long term, sustainable, comprehensive technical support and assistance for the investigation and combating of cybercrime;
6. Cybercrime prevention activities in all countries require strengthening, through a holistic approach involving further awareness raising, public-private partnerships, and the integration of cybercrime strategies with a broader cyber-security perspective.

Options to strengthen existing and to propose new national and international legal or other responses to cybercrime include:

- Development of international model provisions
- Development of a multilateral instrument on international cooperation regarding electronic evidence in criminal matters
- Development of a comprehensive multilateral instrument on cybercrime
- Delivery of enhanced technical assistance for the prevention and combating of cybercrime in developing countries

NATIONAL APPROACHES TO CYBERCRIME PREVENTION: ROLE OF VARIOUS INSTITUTIONS

A crime prevention plan with clear priorities and targets needs to be established and government should include permanent guidelines in its programmes and structure for controlling crime, and ensure that clear responsibilities and goals exist within government for the organization of crime prevention. In addition to the traditional laws, legislation must also consider new concepts and object related to computer data. Criminalization, procedural powers, jurisdiction, international cooperation, and internet service provider responsibility and liability are crucial to prevent and combat cybercrime. Investigative measures, jurisdiction, electronic evidence and international cooperation can provide the much needed support in this direction.

The Private sector is more aware of the cybercrime risk assessment and uses cyber-security technology but many small and medium-sized companies incorrectly perceive they will not be a target and do not take sufficient steps to protect their systems. Some companies have taken proactive steps to counter cybercrime acts, including through the use of legal action. Internet service providers and hosting providers can play a key role in cybercrime prevention. They may retain logs that can be used to investigate criminal activity; help customers to identify compromised computers; block some kinds of illegal content such as spam; and in general support a secure communications environment for their customers. Academic institutions represent an important partner in cybercrime prevention through knowledge development and sharing; legislation and policy development; the development of technology and technical standards; the delivery of technical assistance; and cooperation with law enforcement authorities.

For effective cyber crime prevention practices, proper legislation, effective leadership, development of criminal justice and law enforcement capacity, education and awareness, the development of a strong knowledge base, and cooperation across government, communities, the private sector in the national and international spheres is required.

CONCLUSION

Cybercrime acts may be financially-driven acts, related to computer content, or against the confidentiality, integrity and accessibility of computer systems. The relative risk and threat differs between Governments and businesses. Age, sex, socio-economic background, nationality, and motivation are likely amongst the core characteristics of cybercrime perpetrators. The level of criminal organization represents a defining feature of the human association element behind criminal conduct. A crime prevention plan with clear priorities and targets needs to be established and government should include permanent guidelines in its programmes and structure for controlling crime, and ensure that clear responsibilities and goals exist within government for the organization of crime prevention. In addition to the traditional laws, legislation must also consider new concepts and object related to computer data. Criminalization, procedural powers, jurisdiction, international cooperation, and internet service provider responsibility and liability are crucial to prevent and combat cybercrime. Investigative measures, jurisdiction, electronic evidence and international cooperation can provide the much needed support in this direction. The Private sector is more aware of the cybercrime risk assessment and uses cyber-security technology but many small and medium-sized companies incorrectly perceive they will not be a target and do not take sufficient steps to protect their systems. Some companies have taken proactive steps to counter cybercrime acts, including through the use of legal action. Internet service providers and hosting providers can play a key role in cybercrime prevention. They may retain logs that can be used to investigate criminal activity; help customers to identify compromised computers; block some kinds of illegal content such as spam; and in general support a secure communications environment for their customers. Academic institutions represent an important partner in cybercrime prevention through knowledge development and sharing; legislation and policy development; the development of technology and technical standards; the delivery of technical assistance; and cooperation with law enforcement authorities. For effective cyber crime prevention practices, proper legislation, effective leadership, development of criminal justice and law enforcement capacity, education and awareness, the development of a strong knowledge base, and cooperation across government, communities, the private sector in the national and international spheres is required.

REFERENCES

Economic Times. (2013, August 7). *78 govt websites hacked till June this year: Milind Deora*. Retrieved from http://articles.economictimes.indiatimes.com/2013-08-07/news/41167675_1_cyber-attacks-national-cyber-security-policy-websites

Ground Zero Summit 2013, Asia's Largest Information Security Conference, Launched. (2013, August 9). Retrieved August 18, 2013, from http://www.prnewswire.co.in/: http://www.prnewswire.co.in/news-releases/ground-zero-summit-2013-asias-largest-information-security-conference-launched-218973741.html

India Infoline News Service. (2013, August 10). *10% of all cyber crimes globally happen in India: Study*. Retrieved December 12, 2013, from http://www.indiainfoline.com/Markets/News/10-percent-of-all-cyber-crimes-globally-happen-in-India-Study/5754085645

Internet Crime Complaint Center (IC3). (2012). *Internet Crime Report*. Retrieved from https://www.ic3.gov/media/annualreport/2012_IC3Report.pdf

Kaspersky. (2012, October 10). *Kaspersky Security Bulletin 2012*. Retrieved from http://www.securelist.com/en/analysis/204792255/Kaspersky_Security_Bulletin_2012_The_overall_statistics_for_2012

National Crime Records Bureau. (2012). *Crime in India 2012 Compendium*. New Delhi: NCRB, Government of India. Retrieved from http://ncrb.gov.in/CD-CII2012/Compendium2012.pdf

Norton. (2013). *Norton Symantec report on Cyber Crime, 2013*. Retrieved from http://now-static.norton.com/now/en/pu/images/Promotions/2012/cybercrimeReport/2012_Norton_Cybercrime_Report_Master_FINAL_050912.pdf

Shetty, A. (2013, July 2). *Sibal says 2013 National Cyber Security Policy will be a 'real task' to operate*. Retrieved August 18, 2013, from http://tech.firstpost.com/news-analysis/sibal-says-2013-national-cyber-security-policy-will-be-a-real-task-to-operate-101759.html

Ukrinform. (2012, June). *Ukraine among 20 countries with highest percentage of computer attacks*. Retrieved August 12, 2013, from http://www.ukrinform.ua/eng/news/ukraine_among_20_countries_with_highest_percentage_of_computer_attacks_305071

United Nations Office on Drugs and Crime. (2013). *Comprehensive Study on Cybercrime Draft*. United Nations Office on Drugs and Crime, Vienna. Retrieved from http://www.unodc.org/documents/organized-crime/UNODC_CCPCJ_EG.4_2013/CYBERCRIME_STUDY_210213.pdf

KEY TERMS AND DEFINITIONS

Counterfeiting: Ability of digital devices to render nearly perfect copies of material artifacts such as official documents, educational, currency notes, letterheads etc. For Example high-quality colour copiers and printers has brought counterfeiting to the masses. Ink-jet printers now account for a growing percentage of the counterfeit currency confiscated by the governments.

Cyber Crime: Criminal activity done using computers and the Internet. This includes anything from downloading illegal content, stealing money, creating and distributing viruses on other computers or posting confidential business information on the Internet.

Cyber Stalking: Cyber stalking is a crime in which the attacker harasses a victim using electronic communication, such as e-mail or instant messaging, or messages posted to a Web site or a discussion group. A cyberstalker relies upon the anonymity afforded by the Internet to allow them to stalk their victim without being detected. Cyberstalker generally targets a specific person/organisation/community with often threatening and insulting messages.

Cyber Terrorism: Cyberterrorism is any premeditated, politically motivated attack against information, computer systems, computer programs, and data which results in violence against non-combatant targets by sub-national groups. Also the use of information technology to organize and execute attacks against private/public networks, computer systems and telecommunications infrastructures, or for exchanging information or making threats electronically.

Digital Signature: A mathematical scheme for demonstrating the authenticity of a digital message or document. A valid digital signature gives a recipient reason to believe that the message was created by a known sender, such that the sender cannot deny having sent the message (authentication and non-repudiation) and that the message was not altered in transit (integrity). Digital signatures are commonly used for software distribution, financial transactions, and in other cases where it is important to detect forgery or tampering.

Hackers: In the computers world, a hacker is someone who seeks and exploits weaknesses in a computer system or computer network. Hackers may be motivated by a multitude of reasons, such as profit, protest, or challenge.

NCRB: National Crime Records Bureau is an attached office of Ministry of Home Affairs (MHA), Government of India, NCRB is mandated to empower the Indian Police with Information Technology for modernization of Indian Police. NCRB has the proud distinction of installing 762 server - based computer systems at every District Crime Records Bureau and State Crime Records Bureau, across the country, 'Crime Criminal Information system (CCIS)' project, with a view to maintain a National - level Database of Crimes, Criminals and Property related to crime.

Section 3
New Waves in Cyberspace

Chapter 12
Ethics, Media, and Reasoning:
Systems and Applications

Mahmoud Eid
University of Ottawa, Canada

ABSTRACT

Due to the rapidly changing norms and constant developments in technology, media and communication educators and practitioners are expected to (re)evaluate the functioning of ethics and reasoning in this field. This chapter discusses the relationship between ethics, reasoning, and the media, and the integral role of ethical reasoning education for communication and media professionals. Ethical systems and theories are discussed to inform the debate on the importance of ethics and reasoning education. Globalization and the growing interconnectivity of global media systems are presented, providing insight on how different media systems function around the world. The large impact that the media have on society necessitates the possession of rational and ethical skills; thus, the connection between reasoning and ethics is explained.

INTRODUCTION

Modern societies rely heavily on governance and infrastructure to dictate how people and groups function within this social system. These societies are primarily clusters of individuals, groups, and institutions working and living together and sharing several agreed-upon norms, cultural practices, and social values. Individuals continue to create new tools and processes for living, functioning, and developing that impact how people behave and interact. New technologies foster an environment of mass information sharing and gathering, which is changing how people exchange materials and knowledge. Communication and media industries are greatly impacted by technological advancements; they are granted new opportunities and face different challenges. However, as communication and media industries continue to dominate public and private spaces, modern societies are faced with new issues pertaining to the conduct of these highly pervasive entities. In this, the question of ethics and sound decision-making becomes paramount.

DOI: 10.4018/978-1-4666-6122-6.ch012

The purpose of this chapter is to discuss issues pertaining to ethics in this new environment and explain the importance of educating media professionals on ethical conduct. It discusses ethical theories and systems relevant to the teaching and practice of reasoning and ethics in the media. Being that media systems are becoming globalized, new ethical questions are raised, forcing frequent monitoring and evaluation of ethical issues. Codes of ethics are considered important tools in this task, as they provide principles and guidelines to individuals and organizations facing ethical qualms. Through awareness fostered by education, media professionals can be equipped with the knowledge and tools needed for ethical decision-making.

ETHICS THEORIES AND SYSTEMS

Understanding the scope of ethics requires an examination of existing definitions pertaining to this topic. However, defining ethics is not an easy task, as various methods for explaining and outlining this term exist. Despite varying definitions, the genealogy of the term "ethics" is considered an accurate point of departure in this endeavor. According to Larry Leslie (2000), the word "ethics" hails from the Greek word "ethos", which means "character". These origins lead us to understand that one's character, and the daily-life values attached to it, ultimately commands one's ethical behavior. Thus, Leslie provides this concise definition: "Ethics are moral principles for living and making decisions" (2000, p. 16). Morality is closely linked to ethics. It is derived from the Latin "moralis," meaning "customs and manners" (Leslie, 2000). While popular usage of ethics and morality can be synonymous, most philosophers deem them dissimilar. Ethics refers to the individual's thinking and conduct in matters of right and wrong, while morality pertains to a society's set of beliefs and customs concerning proper conduct (Markel, 2001). Successful *ethical* decision-making benefits from a set of *moral* beliefs that strives to serve both the individual and the society. Ethical behavior is defined as an accepted or preferred and agreed-upon practice (Hanson, 2002).

Philosophy is the scholarly discipline under which ethics falls, containing many approaches, subcategories, and perspectives (Eid, 2008). The dominant paradigm in ethics has traditionally been ethical rationalism: "through reason the human species is distinctive and through rationality moral canons are legitimate" (Christians, 2005, p. 3). Therefore, Christians (2005) explains, the conventions of particular societies are independent of timeless moral truths that are rooted in human nature; however, the concept of a common morality and the idea of "the good life" might not translate across all cultures. Despite the potential for varying views between cultures pertaining to what is ethical and what is not, ethical principles are ultimately based on what is deemed proper conduct, or "doing the right thing". For instance, when we face a decision that requires us to enact an outcome that may or may not hurt another person, we turn to ethical principles to decide on the best option.

Ultimately, defining ethics results in the investigation of its various incarcerations (Eid, 2008). Within the context of this chapter, it is deemed important to identify whether or not ethical theories guiding the field of communication exist. According to Donald Wright (1996), there are; he divides them into two types: classical ethical theory and moral reasoning theories. Classical ethical theory understands ethical obligation in two different ways: teleological and deontological. "*Teleological ethics* underscores the consequences of an act or decision, whereas *deontological ethics* emphasizes the nature of an act or decision" (Wright, 1996, p. 525). Moral reasoning theories are based on the shared values of freedom, justice, and wisdom. Wright (1996) explains four essential requirements that must be met prior to making ethical judgment. First, society must come to an agreement on moral conduct standards; second, these guidelines should be based on experience and reason; third, justice should be sought by a system of ethics; and finally, and ethical system should be rooted primarily in freedom of choice.

Although academics continue to debate over which approach is best suited for ethical thinking, the perspectives on dialogical ethics in relation to communication ethics is highly respected and understood. According to William Neher and Paul Sandin (2007), dialogical ethics is perhaps the most important system for communication ethics because it fosters relationship building—our human instinct—through interpersonal communication; it encourages the development of love and trust, and allows us to expand our knowledge of others and ourselves. Ethical reasoning exists within a framework. Various major systems of ethical reasoning have existed throughout history, such as virtue ethics, universalistic systems, egalitarian and utilitarian ethical theories, dialogical ethics, and postmodern and feminist perspectives. Neher's and Sandin's discussion on dialogical ethics demonstrates that dialogical ethics is an excellent way to understand ethics from an experiential, everyday point of view: "our personal ethics are formed by, and demonstrated in, the ways in which we *dialogue* with others" (2007, p. 87). This method requires us to seek honest, open communication, forcing us to abstain from the defence mechanisms and superficialities that we commonly use. Thus, this perspective emphasizes the importance of dialogue—especially the ways in which it can allow for sound communication between two moral agents. However, there are many large obstacles impeding our ability to communicate using this method, such as the concept of the "game", which allows us to evaluate the advantages of the different ethical systems. As described by Neher and Sandin, "the notion of *Playing the Game* introduces the possibility that lying and deception are acceptable if not necessary when advancing one's own interests" (2007, p. 86). To move forward, in some instances, deception is viewed as the only option available. At first glance, it is easy to dismiss this "game" as a practice for unethical people; however, it is revealed that this concept is present in many aspects of our daily lives.

Ethical theories and systems are intrinsically linked to many aspects of our lives. They can be a part of very individualistic thought processes that impact only our personal selves, or participate in the decision-making processes of governments and global institutions with pervasive impacts. Although ethics are present throughout our existence, how we understand and learn about the creation and functioning of the principles and guiding ideologies is essential to understand how people enact ethical thinking. Therefore, the following section explains the role of ethical reasoning and education in relation to media industries.

ETHICS REASONING AND EDUCATION

Educating media practitioners about ethics is becoming an increasingly important, as media industries have become a large element of modern society, developing a growing provocative influence on the public (e.g., Christians & Lambeth, 1996; Day, 2006). While the teaching of media ethics may not necessarily result in an ethical conduct, it is believed that ethics is an ongoing process integral to the human condition (Day, 2006). Educating aspiring media professionals is a unique task; although it is based in theory, the skills gained for the knowledge provided is highly applicable for their everyday functions in the media industry (Reis, 2000). Instead of being a subject-based topic, Reis (2000) asserts that teaching ethics is a type of outcomes-based education (OBE). The OBE model strives for easily identifiable outcomes and active learning that focuses on a specific set of knowledge: "OBE emphasizes problem solving, applied learning, active inquiry, and the use of multiple subject matter in the pursuit of knowledge and skills" (Reis, 2000, p. 195). This model of ethics education explains that students are provided with specific theories and knowledge that translate into the creation of a specific set of skills.

The teaching of ethics in educational institutions outside of the scholarly field of philosophy has not always been common (Christians & Lambeth, 1996). According to Christians (2008), the role of media ethics in academia emerged around 1980, which was sparked by texts such as the MacBride Report (MacBride, 1980) that reviewed communication and human rights, international media policies, cultural diversity, and professional journalism. This publication, along with others and with many events, motivated the parallel growth demonstrated between media ethics and professional ethics. As research continued to focus on the role of ethics in media and communication, higher education institutions followed this trend, integrating ethics courses into their media and communication programs (Lee & Padgett, 2000).

Thus, although media ethics courses in media and communication programs were not always present, they now are commonplace. The growth in the value placed on media ethics and the corresponding increase in media ethics courses at post-secondary institutions has been dramatic (Plaisance, 2007). This is considered a very important development in the training systems that are used to educate aspiring media practitioners, as knowledge on ethics is critical (Hanson, 2002). According to their survey of recent alumni from two different academic institutions, Kendra Gale and Kristie Bunton (2005) found that media ethics instruction corresponds with ethical leadership and ethical awareness. They found that graduates "who took media ethics courses were significantly more likely than those who did not to consider ethical issues in their profession important" (2005, p. 272). Further, graduates were more likely to place a higher value on ethics, demonstrate an ability to identify ethical ideas, and were more likely to view personal and professional ethics as indistinguishable.

To educate future media practitioners about ethics and ethical theory is to provide students with the necessary skills for autonomous ethical thinking. Media practitioners, especially journalists, must possess virtues such as honesty and telling the truth. This is because if journalism is to contribute anything beyond entertainment in a society, it must be rooted in truth telling (Phillips, 2013). While truth is an elusive concept, making it challenging to understand how this idea functions within society, Louis Alvin Day (2006) argues that it can be understood as adherence to uncompromising standards such as accuracy, a commitment to promoting understanding, and fair and balanced content. These standards foster public trust and respect. For Day, the idea of democracy is also a key element to this discussion, especially due to the highly influential nature of the media in modern societies. In a democracy, journalists play a pivotal role due to their ability to communicate events and issues to the public, free from external control. They have the ability to inform the public on elements of society that are meaningful and shape ideas. With this great power comes the responsibility to ensure proper conduct and practice. Essentially, journalists must act as moral agents. However, unfortunately many unjust acts among journalists and practitioners have come to the attention of the public in recent years. By choosing to engage in unethical conduct, a journalist fails not only their own ethical standards, but also the society they serve. Several media outlets have devoted great efforts and attention to abide by ethical standards enhancing their conduct, some still continue to conduct themselves in a destructive manner that begs ethical attention.

Media practitioners have the ability to seriously impact how audiences perceive experiences and events. Gary Hanson (2002) discusses that news directors and editors make daily ethical decisions that determine the newsroom's editorial style—and by virtue, these practitioners must be trained to make informed, ethical decisions. He warns that incongruence between the nature of the journalism profession and education exits, resulting in tension between these two entities. Newsroom recruiters and supervisors usually express a low regard for the expertise of journalism educators.

This disengagement is a serious obstacle faced by the media industry and media educators alike; for, if that which is taught is not applicable or respected in the workplace then adjustments are necessary. While concession is key in the resolution of discrepancies, it is suggested that the role of ethics should not be compromised, despite any form of resistance from the media industry. Enacting ethical behavior as a media practitioner is necessary for the functioning of a socially responsible media outlet, making this disconnect between teaching and practice particularly concerning.

In many ways, the media can be seen as a filter through which reality percolates; media practitioners make decisions in the process through which reality is refined for dissemination. This filter is an elemental tool that can allow for the packaging of ideas either in a clear, honest, and concise manner or in a distorted, manipulative, and stereotypical manner. In this, the media act as a "window" to the world or a "mirror" of reality—granting people access to images and events (de Mooij, 2014). The ways in which the media distil ideas is directly linked to ethics, as media practitioners are consistently faced with situations that require decision-making on, for example, how to share an idea, what information to publish, and how to assign meaning in their messages. An understanding of ethics and the ways in which abidance to ethical principles impact audiences is essential for media practitioners, especially as our world becomes increasingly globalized.

Thus, ethics education is paramount in the training of media professionals because it arms them with the ability to execute critical decision-making that abides by sound value systems. This training can be easily rationalized by simply surveying the massive impact that the media have on community, national, and global levels. Therefore, the provision of training that combines moral reasoning with ethical decision-making is fundamental. Moral reasoning is an integral element of sound decision-making and is "a *systematic* approach to making ethical decisions" (Day, 2006, p. 54). Thus, moral reasoning can be understood as a form of intellectual activity that can take the form of logical argument and persuasion. Ethical decisions must be carefully made, be justifiable through a reasoned and rational analysis of the situation, and take into account the rights and interests of others. Ethical conduct and rational thinking are two necessary weighs to maintain a balanced and effective communication for media professionals facing highly impactful decision-making on a regular basis (Eid, 2008). The process of ethical and rational decision-making is needed in all media systems regardless of their differences.

GLOBAL MEDIA SYSTEMS

The process of globalization is attracting interest among scholars because of the cultural convergence that is occurring due to a host of societal advancements. Globalization is perhaps the most profound characteristic of the 21st century that requires scholarly inquiry due to the growing implications of this process (McKenzie, 2006). As patterns in our exchange and communication continue to evolve, it becomes clear that physical boundaries and geography no longer define a community or nations' "natural limits" (Morely & Robins, 2013). Robert McKenzie describes globalization as "a kind of worldwide climate in which people, industries, governments, and countries across the world are being propelled into closer political, economic, and cultural unions" (2006, p. 14). As a consequence of new technologies, significant transformations are now occurring in information and communication exchange (Morely & Robins, 2013).

Globalization can also be understood in relation to the role of corporate profit making, which is understood as a main motivation behind this movement. McKenzie (2006) describes the four factors that are considered the main contributors to globalization (international travel, communication technologies, global media conglomerates, and audience curiosity), which define the driving forces of this movement. Although all of these factors are considered integral, McKenzie's exploration of global media conglomerates is of specific importance. While this is a common trend in the realm of contemporary media, global media conglomerates can be difficult to conceptualize as distinct entities because they are incredibly complex organizations. These massive conglomerates pose challenges to the conception of nationality as a primary defining influence on self-identity. That is, because global media conglomerates are decentralized and largely amorphous entities stretching across borders, audiences can be targeted regardless of their geographical location. McKenzie suggests that cross-border assembling of audiences has led to the conjecture that the role of national boundaries in helping to form individual identity and social cohesiveness is greatly being diminished.

In order to extract data that allow for a comparative analysis of media trends, McKenzie (2006) analyzes how newspaper, television, radio, and Internet practices differ among eight different countries (China, France, Ghana, Lebanon, Mexico, Sweden, United Kingdom, and United States) In this, he compares media trends in the eight countries by exploring and describing the meaning of culture for the context of his study, followed by detailed descriptions of each country, pertaining to language, society, government, and geography, In this, he is able to articulate the many differences that exist between media systems, such as regulation, financing, accessibility, content, reporting, importation and exportation, and audiences. As media systems are becoming global entities due to globalization, the issue of regulation is considered particularly pertinent—especially in relation to communication ethics. Regulation pertains to how media operations are controlled and carried-out, and although geographical boundaries are seemingly transparent due to communication technology, national boundaries that can possess unique regulatory systems are still intact. Additionally, although global communication borders have opened up, our cultural norms and practices pertaining to values and what is right and wrong have yet to be completely assimilated on a global scale. Advertising, for instance, plays a large role in the development and dissemination of messages in many societal capacities, but still faces issues pertaining to regulation. Moreover, advertising has a large persuasive and influential presence in many countries.

While North America, particularly the United States, is known for its great involvement in this industry, McKenzie (2006) explores an important element of advertising that spans across all countries. Although not found in all media around the world, advertising and sponsorship tends to act as a major funding source for media operations; in this, covert forms of regulation can occur. Specifically, advertisers (or sponsors) can make requests to media organizations with which they do business to refrain from delivering a certain kind of content, with the threat of withdrawing funds. Media organizations can also choose independently to refrain from producing content that might offend advertising or sponsorship clients. While this may be a common trend within the industry, many implications of this practice threaten the value and freedom of media outlets due to the influence of advertising on their material. McKenzie asserts that advertising tends to be used in countries influenced by a libertarian philosophy of media operations in which there is a market-based economy. While the involvement of advertising has been critiqued due to its ability to censor content in a country, McKenzie also suggests that this practice can be beneficial due to its ability to limit government involvement in production of media content. This can allow for a commercial marketplace in which individuals can freely select media content from the choices that are available, without being required to pay for media. Supporters of advertising as a method of financing media also argue that it provides producers with a financing model that allows media organizations to be more innovative and responsive to the tastes of audiences.

In addition to these implications and the actual comparisons of each country that are very beneficial to understand the interrelated elements of a media system, McKenzie (2006) extracts information on processes and influences of content. An interesting and insightful metaphor that McKenzie uses to describe the nature of understanding media systems is through conceptualizing them as trees. In this, it is possible to visualize what a media system looks like, particularly the "roots" or "underground" elements. Furthermore, in conceptualizing a media system as a tree, it is possible to relate imagery that aids in understanding how different organic aspects impact one another; that is, how one country can impact another. The three main points for comparison are exocentrism, ethnocentrism, and worldcentrism. These terms are directly related to information flow and the exchange of media content between countries. According to McKenzie, from a rhetorical perspective, the extent to which a country imports and exports media content has profound implications for the country's media system and for many of the people who live in the country. Worldcentric and exocentric countries that import a lot of media content tend to have more open and "organic" media systems—an organic system being one that thrives on an exchange of ideas with other media systems.

COMMUNICATION AND MEDIA ETHICS: THEORY AND PRACTICE

Being that media and communication systems are becoming increasingly intertwined, the theory and practice of ethics has never been more important. Across time and space, information is constantly in transit, messages are being disseminated, and people are connecting. Media systems are colliding and narratives from around the globe are more available to audiences than ever before. The Internet allows traditional news media outlets, for example, to search and share stories about events and people from faraway places, breaking down geographical barriers that have once impeded the spread of global news. Additionally, knowledge can be spread by non-traditional media forms, such as Facebook and Twitter, which can sometimes act as sources for traditional news outlets looking to stay on top of breaking news—as getting "news first" has become increasingly worrisome given the real-time speed of online publishing (Spence & Quinn, 2008).

As the media landscape continues to change, discussions on how ethics can fit into the present and the future of this field are pertinent. New questions are raised pertaining to issues of regulation and governance of the media, as they are no longer confined to physical boundaries. Globalizing media ethics is considered a great challenge at present; historically, media ethics have been largely Western and monocultural (Christians, 2005). McKenzie (2006) asserts that due to globalization, media scholars must adapt their perspective and scope of thought to incorporate the influence and implications of the interaction of cultures and media systems in today's world. Currently, media systems are deeply rooted in our lives, not only as a means for personal communication, but also as an educational tool when aiming to learn about the lives and practices of others.

To provide context for the practice of ethics, Neher and Sandin (2007) discuss the importance of codes of ethics. The use of codes of ethics in large organizations has become commonplace, as they aid in the creation of a community that has shared goals and values. However, the abuse of power exhibited in recent times by many organizations that have great responsibility linked to their great propensity for financial gain and their persuasive power demonstrates how some organizations fail to follow codes of ethics. The commercialization of the media can pose ethical challenges for media practitioners (e.g., Herrscher, 2002). Day (2006) discusses issues pertaining to money and power, looking into how many news media have adopted commercial personas, pushing products and services through overt and hidden advertising and marketing schemes. While some argue that this is simply a symptom of modern capitalist society, others argue that this practice exhibits a disappointing dilution of the true value of journalism. Therefore, in allowing for the commercial domination of media, the lines between true journalism and corporate initiatives become blurred. This is also evident in the disturbing growing trend of product placement in the news media. Day provides examples to show subtle and hidden presentation of products within the news, urging journalists and media practitioners to avoid involving themselves with this process, by outlining the negative outcomes that can result from being enticed by the alluring compensation provided by companies.

Educating media practitioners about ethical conduct and providing them with information about their important role in society can increase their ability to apply strong ethical reasoning skills. Gordon (1999) argues that media codes of ethics are useful and necessary both to the media and to society. Their purpose is deemed important, as they set standards and provide guidelines against which conduct can be evaluated and measured—this is an important service for both society and the media. However, it can be challenging to monitor the conduct of all media practitioners or find workers to properly abide by these codes. While communication and media codes of ethics can vary in how they approach their guidelines, Cooper (1990) outlines three strong and logical contenders as media ethics universals, which are: 1) the quest for truth; 2) desire for responsibility, and 3) a compulsion for free expression. While seemingly utopian, these guidelines are considered important principles to take into account, especially as the media become increasingly globalized.

CONCLUSION

Ethics are necessary in society; they allow us to understand what is right and wrong and how to make sound decisions. Ethics are fundamental to provide people with a rational sounding board for responsible conduct in society. As the media industry continues to change and grow, media professionals can face uncharted terrain that may challenge their abilities to make ethical and rational decisions. The role of ethics in communication and media industries is particularly important due to the pervasive role of these entities in modern society. Journalist, for instance, have the power to motivate people to understand situations based on how they decide to depict a story or idea. This important responsibility enacts ethical behavior by media professionals through ethical and rational decision-making. As the media's grip on society continues to strengthen, the ethical conduct of these industries must be measured and monitored against ethical theories and systems.

Despite variances among media systems around the globe, they are becoming increasingly interconnected. Technological advancements are propelling the globalization movement forward, bringing together people and media systems from different nations and cultures. Through ethical principles such as telling the truth, objectivity, balance, maintain context, accuracy, and so on (Eid, 2008), it can be possible to foster the development of a global media system. Although education is not the only avenue through which media practitioners can learn about ethics and reasoning, it is deemed an integral step towards having communication and media professionals to practice ethical and rational decision-making. Thus, gaps must be bridged between the academic and the media industry to build a strong relationship in which both parties mutually consider and appreciate the value of teaching and enacting ethics. Finally, ethical reasoning must not be lost in the decision-making process. Combining both reasoning and ethics in the decision-making process can enhance the effectiveness of communication and media performance.

REFERENCES

Christians, C. G. (2005). Ethical theory in communications research. *Journalism Studies*, *6*(1), 3–14. doi:10.1080/1461670052000328168

Christians, C. G. (2008). Media ethics in education. *Journalism & Communication Monographs*, *9*(4), 180–221.

Christians, C. G., & Lambeth, E. B. (1996). The status of ethics instruction in communication departments. *Communication Education*, *45*(3), 236–243. doi:10.1080/03634529609379052

Cooper, T. (1990). Comparative international media ethics. *Journal of Mass Media Ethics*, *5*(1), 3–14. doi:10.1207/s15327728jmme0501_1

Day, L. A. (2006). *Ethics in media communications: Cases and controversies*. Belmont, CA: Thomson Wadsworth.

de Mooij, M. (2014). Mass media, journalism, society, and culture. In M. de Mooij (Ed.), *Human and mediated communication around the world* (pp. 309–353). Cham, Switzerland: Springer International Publishing. doi:10.1007/978-3-319-01249-0_10

Eid, M. (2008). *Interweavement: International media ethics and rational decision-making*. Boston, MA: Pearson.

Gale, K., & Bunton, K. (2005). Assessing the impact of ethics instruction on advertising and public relations graduates. *Journalism & Mass Communication Educator*, *60*(3), 272–285. doi:10.1177/107769580506000306

Gordon, A. D. (1999). Media codes of ethics are useful and necessary to the mass media and to society. In A. D. Gordon, & J. M. Kittross (Eds.), *Controversies in media ethics* (pp. 61–68). New York: Longman.

Hanson, G. (2002). Learning journalism ethics: The classroom versus the real world. *Journal of Mass Media Ethics*, *17*(3), 235–247. doi:10.1207/S15327728JMME1703_05

Herrscher, R. (2002). A universal code of journalism ethics: Problems, limitations, and proposals. *Journal of Mass Media Ethics*, *17*(4), 277–289. doi:10.1207/S15327728JMME1704_03

Lee, B., & Padgett, G. (2000). Evaluating the effectiveness of a mass media ethics course. *Journalism and Mass Communication Educator*, *55*(2), 27–39. doi:10.1177/107769580005500204

Leslie, L. Z. (2000). *Mass communication ethics: Decision making in postmodern culture*. Boston, MA: Houghton Mifflin Company.

MacBride, S. (1980). *Many voices, one world: Towards a new more just and more efficient world information and communication order*. Paris: UNESCO.

Markel, M. (2001). *Ethics in technical communication: A critique and synthesis*. Westport, CT: Ablex Publishing.

McKenzie, R. (2006). *Comparing media from around the world*. Boston, MA: Pearson Education.

Morely, D.,,& Robins, K. (2013). *Space of identity: Global media, electronic landscapes and cultural boundaries*. London: Routledge.

Neher, W. W., & Sandin, P. J. (2007). *Communicating ethically: Character, duties, consequences, and relationships*. Boston, MA: Pearson Education.

Phillips, A. (2013). Transparency and the news ethics of journalism. In B. Franklin (Ed.), *The future of journalism* (pp. 289–298). London: Routledge.

Plaisance, P. L. (2007). An assessment of media ethics education: Course content and the values and ethical ideologies of media ethics students. *Journalism & Mass Communication Educator*, *61*(4), 378–396. doi:10.1177/107769580606100404

Reis, R. (2000). Teaching media ethics in a multicultural setting. *Journal of Mass Media Ethics*, *15*(3), 194–205. doi:10.1207/S15327728JMME1503-5

Spence, E. H., & Quinn, A. (2008). Information ethics as a guide for new media. *Journal of Mass Media Ethics*, *23*(4), 264–279. doi:10.1080/08900520802490889

Wright, D. K. (1996). Communication ethics. In M. B. Salwen, & D. W. Stacks (Eds.), *An integrated approach to communication theory and research* (pp. 519–535). Lawrence Erlbaum Associates Publishers.

KEY TERMS AND DEFINITIONS

Code of Ethics: A compilation of principles and ethical standards that guide the conduct of individuals within a community or organization.

Communication and Media Ethics: A strand of ethics dedicated to the field of communication and media.

Decision-making: The process through which people make choices.

Education: The teaching of ideas and concepts.

Ethics: A branch of philosophy whose principles help guide individuals (persons, organizations, etc.) toward good conduct.

Ethics Education: The teaching of information pertaining to the field of ethics.

Globalization: The process of international integration fostered by growing interconnectedness.

Journalism: The discipline of representing, reporting on, and narrating discourses about events.

Reasoning: A method of logical and rational thought processing.

Society: A community with agreed-upon values and norms.

Chapter 13
Antinomies of Values under Conditions of Information Age

Liudmila V. Baeva
Astrakhan State University, Russia

ABSTRACT

The information age with its unprecedented acceleration changes a person's world and creates new virtual environments, forms of communication, and creative work. The character of these changes is antinomic to a large degree: information technologies give rise to super-powers, entrust a person with the power in time and space, but they create new challenges to individuality, freedom, and intelligence. Virtualization of communication, education, leisure, art following the evolution of high technology production, and consumption contribute to the substitution of real relations and amenities with virtual versions and simulacra. This chapter is devoted to studying of value antinomies of the modern age: information and knowledge, virtuality and reality, feelings and game, friendship and contacts, etc. Since values are the projections of the future in the present, this chapter helps to elicit to some extent the main trend of the social-cultural dynamics of the modern high-tech society.

INTRODUCTION

Development of the society in the conditions of mass penetration of information technologies into culture, lifeworld of a modern person reveals unique manifestations. New possibilities which the digital space provides multiply the potential of a person's activity, create new forms of culture, transform phenomena and processes which had been forming for centuries. Informatization of such important spheres of the society existence as education, science, communication, art, leisure, along with the introduction of information technologies into production of goods and services, management, commerce and finances inevitably results in changes in the sphere of values of a person and society. Being the dominants of a person's consciousness and behavior guiding his activity to the image of the proper, values modify the person himself and his environment. Studying the values of the youth of our time allows not only to diagnose the current situation in culture, but also to detect the trends of socio-cultural dynamics in future.

DOI: 10.4018/978-1-4666-6122-6.ch013

Transformation of values related to the information stage of the society development is also in focus of attention of the leading scientists studying psychological, cognitive, moral aspects of the value dynamics, pluralistic view of the values of modern person is being developed. This view is specific for a person lives actually in three parallel worlds: in the world of nature and physical objects, the world of his own consciousness and the world of virtual life (synthesis of the world of ideas and high technologies). The role and importance of the virtual world is developing so rapidly that people cannot imagine their life without the electronic space. The research aims at defining the most significant transformations in basic spheres of person's life which are caused by the influence of this new form of existence. Characteristics of these changes will allow to determine the risks of the ultimate risk for a person and his value basis and thus for the development (or deformation) of culture.

For that purpose the following tasks should be solved: to specify the understanding of values as a polysemantic notion: to define the changes of the value priorities in cognitive, communicative, existential, interpersonal spheres. Multiplicity of the modern axiological picture will require application of the principle of antinomy in the analysis of values, this will help to avoid one-sided conclusions and will enable to forecast the alternative scenarios of development.

The research methods of our research are determined by the existential approach to understanding of values, application of axiological analysis of key anthropologic phenomena and processes in the society, papers of such theorists of the "information society" as D.Bell, M.Castells, etc.

The analysis of the values of the information age with be carried out by means of defining their antinomy. Antinomy as a principle underlies the human culture and mentality of a person who is choosing between various forms of worldview and behavior. The notion of "antinomy" was assigned philosophical meaning in the "Critique of Pure Reason" by I.Kant, where the author accentuates four main antinomies which arise from the contradictions in the subject's understanding when he tries to consider the world of phenomena and "things-in-themselves" as an integral unity. I. Kant's idea on antinomy was developed by G.Hegel who confirms the contradiction by the inherent objective characteristics of the Absolute idea. But unlike I. Kant Hegel supposed that the opposites are settled positively in their synthesis, their dialectical withdrawal. Further on the method of antinomy was efficiently applied in ontology and exact sciences (for example analysis of the theory of sets, antinomy of quantum mechanics, Bohr's complementarity principle etc). We suggest that the method of antinomy analysis should be applied to studying values, for at present their studying inevitably faces the confrontation of the oppositely directed tendencies within the whole.

The research findings of our research can be used in the work with the youth, in carrying out of psychological trainings for the users of social networks, having problems of real communication, and Internet-dependent teenagers, as well as in conducting the courses in culture and ethics of the information age.

BACKGROUND

Leading academic and collegiate centers in different countries of the world have been recently dealing in studying of the influence of informatization on different socio-cultural processes. Thus the issues of the development of electronic culture are the subject matter for the scientists in the University of Milan (A. Ronchi, 2009); McLuhan Institute (Virtual Maastricht McLuhan Institute (VMMI), the Netherlands (K.H. Veltman, 2004); studying ethical and anthropological issues of the information space are the subject matter for the researchers in the International Centre for Innovation in Education (ICIE) Karlsruhe, Germany (R.Capurro, 2006); London School of Economics, department of Media and Communication (Great Britain (L. Haddon, 2004); Centre for Computing and Social Responsibility (De Montfort University, Great Britain (S. Rogerson, 1998); Center for the Study of the Information Society of the University of Haifa, Israel (D.R. Raban, 2009); ethical, political and legal aspects of informatization are the subject matter for R. Baarda, L.Rocci (2012); Kallenberg, B. J. (2001), Chang, C.L. (2011) and others. The author's research deals with the development of modern technoethics trend of current interest as an study area of ethical, spiritual and ideological society development aspects in the conditions of high technologies, that has been covered in the scientific papers of such researchers as Rocci Luppicini (2012), Peter B. Heller (2012), GuiHong Cao (2013), etc in recent years. These researches define technoethics as a cross-disciplinary field that tries to determine an appropriate viewpoint, attitude or philosophy in the application of technology to real-life situations.

VALUES AS AN EXISTENTIAL CHOICE

We will assume the understanding of a value according to existential approach, allowing to take into account both the factor of the freedom of individual choice and the diversity of the world of values itself.

The theories by M. Heidegger (1927; 2000), V. Frankl (1963; 1997) and N. Abbagnano (1970; 1990) influenced the formation of the conception of existential axiology in different ways. The research of the structure and the essence of values by R. Hartman (1973), R. Frondizi (1971), S.O. Hansson (2001), symbolic and logical expression of a value by G. Vernon (1973), the correlation of meaning and significance of a value structure, search for subject and object of values by K. Baier (1967), the ethical content of values by R. Brumbaugh (1972), vital and existential analysis of values–A. Maslow (1966), natural values by Ph. Foot (2001), conceptions of values as cognitive representations of needs (Schwartz & Bilsky, 1987), values as desirable behaviours (Verplanken & Holland, 2002), values as desirable trans-situational goals (Schwartz, 1994), the crisis of classical values of the West and the search of new imperatives by E. Levinas (1979), the analysis of the priorities of postindustrial and information epochs Y. Masuda (1983), A. Giddens (1990), M. Castells (1997), B. Friedman (1997), B.J. Kallenberg (2001), G. Cockton (2004), etc. played an important role for our research.

The existential research of values starts with the issues of a subjective source of value representations and aims at the understanding of their influence on surrounding natural and social being. In existential aspect a value appears as a kind of a vector, representing the trend of those changes in social being, which do not belong to the category of objective, spontaneous and natural phenomena. In the cases, which we deal with directed self-development and the attempts to contribute a rational and ethic principle into the natural being, the appeal to the values–meaningfully-significant purposes–is inevitably. The modelling of the reality in the consciousness, its practical changing and the complement in terms of an adequate thing, appears as the sphere of true freedom of individuals to be able to go beyond the scope of their natural and social programme.

Values are an ideal phenomenon with a specific feature to belong to subjective perception and consciousness. In the most general case a value can be defined as a complex of the willing, emotional and intellectual feelings, directed from a subject to objective reality and representing, more relevant aspirations and ambitions with goals and significance. In this regard the same value is deeply individual, felt by each subject, and can exist objectively only as a result of the development of public conscience or collective unconscious.

To our opinion the searches for the universal meaning of values should be related to the understanding of the incompleteness of human beings and their ability to personal identity and self-development. We believe that more or less all kinds of values are associated with the solution of the key problems of the existence–mortality, solitude, absurdity of mechanical vital activity, and striving to the immortalization or consolidation of their presence in the objective life, filling it with meaning and sense, not given initially stands for as the united in the appraisement of the objective reality and the formation of person's values. The desire for the immortality or the achievement of new quality of life in any form lay in the different values, showing, what subjects consider the missing and the necessity for the strengthening their objective reality and its changing for the improvement. In this regard values are individual options of adjusting the substantial differences between the mortality of a body and the possession of consciousness.

Thus, values of material and vital welfare (life, health, a family, children, safety, comfort, wealth, etc.) express the aspiration for physical perfection and the prolongation of corporal objective reality (both in personal and descendants' life). Social and moral values (equality, respect, justice, peace, friendship, communication, etc.) express the striving for the strengthening the relations of individuals with the society, other persons, which individuals multiplies their objective reality in and through. Mystical and spiritual values prove the desire of spiritual immortality, if the subjectivity is even rejected in such a case. The vitally important values (love, creativity, freedom, spirituality, knowledge, etc.) reflect the aspiration to fill life with unique sense, allowing us realizing that a person can get new quality and mark the objective reality with creative work and thinking. In this connection values can be defined as subjective search for the overcoming of limitations of natural and social programme, obtaining the source, strengthening, multiplying, prolonging and improving individual objective reality, moving it on a new level of quality.

Values, being the base of person's world outlook, are able to influence on the external world, have their own objective reality and become a response to the life challenge and the absurdity of the existence without appraisal, estranging from a subject like pieces of spiritual creative work. Thus, we meet the definition of a value as a dominant of the consciousness and the existent, directed at the achievement of the ideal objective reality, creatively influencing on the person's inner development and visual environment through the filling them by meaningfulness and sense. Joining a person and the world with the ties of meaningfulness, values change both sides of this relationship in the line of proper things.

In such a way the existential vision of values is expressed through the understanding of its essence as manifestation of the subjectivity and its role as the possibility of the reality "decoding", filling the objective reality with sense and goals which have creative and transforming nature.

Unlike knowledge, values contain limited quantity of information on the inner objective reality of a subject and appear as its highest manifestation in the world. If knowledge is the manifestation of the being for a subject, values are the manifestation of a subject for the objective reality.

Relying on the researches of V. Frankl (1997) and A. Maslow (1966; 2002) we can note, that there are two main types of values. They are values, forming under the influence of the objective being, "from being" (values of being) and controlling the reality, forming "over being" (values of a subject), "being-values". The values of the first type express the significance of qualities of actual being, as well as the objective needs of a subject (values of life, health and possession of material welfare). The values of the second type denote the ability of a creative person to be free regarding the objective conditions of the existence and the possibility of the directed influence on individual and external being (freedom, knowledge, spirituality, creative work, love and nonviolence). On the basis of the above mentioned we can conclude (and this is different from Frankl's tradition) that values are preferences of one or another way of self-development, related to subject's aspiration towards the development. In this connection value formation or evaluation represents a phenomenon of the subjectivation of the external being for an individual, when a new meaning and sense are introduced to it, serving as the basis for its active and practical changing. On the one hand the subjectivation can be evaluated as introduction of a new spiritual and psychic component into the unconscious world, which can result in the formation of the noosphere in a positive way. On the other hand the subjectivation transforms objects of the external reality evaluated according to the specific needs and peculiarities of an individual. Thirdly, the subjectivation changes the one who represents its active side. Persons, who highly evaluate the objects of the external reality, associates the existence with the possession of them, what results in their deeper rooting into the natural, social and objective being. The subjectivation is creative in each of these aspects, as it deals with the being expansion, the creation of a new spiritual reality, which can become later the source of practical creative work. Evaluation represents more than application or appropriation of objects, it turns out to be their vivification – bringing sense and meaning into their existence. It refers only to the "noetic" world (Frankl, 1997), the world interpreted by subjects, but at the same time it becomes the being itself, as subjects have meaning and sense in themselves. The world of values of an individual transforms the objective being into the subjective reality, becoming a kind of removal of both of them. On the basis of an intuitive or a conscious idea of the proper reality variant, an individual adds the missing elements, modifying the whole universe. Values like a mirror reflect marks of subjects' life activity, but studying of their dynamics is important not only for understanding of the individual being, but also for the comprehension of social self-development, as the direction of the world history course can depend on the acts of individuals today, who use not only their force and abilities, but all the facilities of the modern science and technologies to achieve their meaningful purposes.

DYNAMICS OF VALUES IN CONDITIONS OF INFORMATIZATION

Modern high technologies afforded new opportunities to human beings: The borders of space and time expanded; conventionalities of language barrier were overcome; the boundless world of digital information appeared, etc. How did these achievements influence on their creators–human beings and their world of values?

First of all we should note that the value pattern of the modern society is distinguished by pluralism that characterizes not only social, aesthetic, cultural values, but moral, existential, religious as well – which used to have been standing for many epochs. To avoid categorical assessments we are going to apply the antinomy principle to studying of the values, that will allow to define different, even contradictory tendencies and images of the proper.

In terms of anthropological analysis information technologies are a human-developed system of processing, storage, transferring of information, accompanying with the assignment (delegation) of certain techniques of "intelligent" ("smart") information management to a computer machine.

A human becomes free to analyze and generate data, create new ideas, predict, carry out creative activity in a broad sense, transferring functions of calculation and information processing to a computer machine. In fact the introduction of computer technologies however causes complex consequences. To analyze the nature of influence of new technologies on a human and his values it is important to study the most significant sides of person genesis: freedom, communication, cognition, sensibility (emotionality) and first of all attitude to the environmental reality itself.

REALITY OR VIRTUALITY?

Reality is a core ontological category, which is considered as value of a person, basis for vital, social and moral values in the context of an axiological approach. Nowadays an attitude to reality undergoes essential changes under the influence of the factor of information and virtualization. Modern young people spend the greatest part of their free time, communicating virtually (in social networks, game space, doing computer creative work). For them the modern life is first of all the life in the Network. It means the presence of a subject in the virtual Internet-space as a source and a recipient of information, a person, expressing activity and feelings, an author of ideas and judgements. It is some kind of "the second life", an analogue and extension of a real person at the same time. This state can be compared with an Indian concept "illusiveness" of the external world which a person lives in. The modern virtual world, created by the age of high technologies, is associated in many ways with such characteristics: For "user", wandering in the Network, and a computer game player the virtual world is the illusion, being created by information technologies, though it is visible, perceptible, attractive, having a set of a great number of ways to delude, attract and relax a person. The Buddhists call it "tricks", by means of which a person is involved into the routine and a samsara cycle. A gamer understands that the world, attracting him, is unreal, as attitudes and relations, formed by it, are unreal. Step by step gamers begin interpreting the virtual world as more and more significant and real, as they have real feelings of this world environment. The virtual reality seems to cover individuals with one more coating, hiding the true basis of the universe from them. Illusions become more and more significant for individuals, and they are losing interest to the world of real people and feelings, replacing them with brighter artificial images. Coming into the world of the virtual samsara (turns of transformations) and dying in it, human beings strive not to rid of its chains, but to get higher levels.

The real life thus is less bright and amazing, as it involves a lot of problems, that's why it pales into insignificance. Real relationship with other people and own status, which is compensated with a striking avatar in the Net, also become less important. An avatar was originally a graphical representation of a gamer or the Net user that they chose as they wished. This image was a desirable form of self-presentation for others and a picture of that person, who a real person would like to seem to be. We mention, that self-presentation in the world of a virtual samsara becomes an extremely important engagement, taking a considerable amount of time and requiring exuberant imagination. Researchers even point out the development of special ethics of representation and behavior in the Network (Brey, 1999). The self-presentation becomes important in the context of the ontology as there is a principle in the world of modern communication: If you are not in the Network (i.e., There is no information on persons or their blog, profile in a social network etc.), you do not exist. The features of the self-presentation depend on the purposes of a person. First of all, the self-presentation is focused on the search of partnership in particular sphere where individuals search emotional, professional or intellectual communication. To achieve their purpose individuals choose appropriate nickname, style, image and content of the self-presentation. An avatar and a nickname in this reality is that person image which other users communicate with (Raban, 2009). These are moreover not users themselves, but their images of the self-representation. It strengthens a game element, the virtualization of relations between players of games or users of social networks.

So in the mentality of a modern person appear new antinomies caused by a new created reality – the virtual world. Antinomy related to the perception of the reality has anthological character: Thesis: reality has the status of the true existence, and virtuality – the one of illusive. Antithesis: reality is just a requirement for maintaining the virtual dimension that is of the primary importance in the context of the information age. A person is searching for an alternative to the existing reality and finds it in this new age not in the world of dreams or mystic experience, but in a controlled artificially created information environment. Virtual life is no longer an auxiliary for a person and his real life, it becomes dominating, intruding into the system of values of a person, influencing his freedoms and unfreedoms.

INFORMATION FREEDOM

Freedom remains to be the most important value of a person. The UNESCO Code of Ethics for the Information Society (2011) was developed for the absolute guarantee of the rights and the freedom of a person in the information society. Has a human of the Information Age become freer in such a way?

On the one hand, new technologies make possible high-speed and high-precision processing of the information, discharging a human from the necessity to do routine activity. Thus, it seems, a person has more free time, but in fact a modern human spends this free time for new activities, involving (a) maintenance of information systems (development and installation of new software programs, fight against viruses, protection of data and failure recovery), (b) implementation of volume of work increased in conditions of information boom, (c) uploading and downloading of data and their reprocessing, (d) search of necessary information and check of its validity, (e) processing of increased flow of documents, high tech innovations, etc.

Nowadays a modern specialist should have more competencies (language, intercultural, information and other competences), than last years and centuries, and much of free time is spent for the development of this knowledge and these skills. In its turn, it requires new methods of training, extension of professional skills, especially for brain workers, personnel managers of management of the personnel and workers, dealing with information processing (Hemingway & Gough, 2000). While developing of information facilities a person more often spends time to achieve a high level of high technologies. Human beings are developing together with them, in which case they are becoming unfreer in their activity.

What are the perspectives of freedom development in the Information Age? Information first of all has provided us with significant freedom for obtaining of information which should not be underestimated. That, what was earlier limited and was represented only for professionals, became available to the wide public. Politicians, businessmen and community leaders have to regard, that their acts have not already been confidential and evoke a broad public response at once. In such a case the quality of information and messages, presented by mass media, has sharply become lower. Information is full of plagiarism, unverified and unconfirmed facts. There are almost no prohibitions and taboos for the modern society of information freedom. Alistair Duff (2008) emphasized the situation of "normative crisis", that is characteristic for the information society, where traditional norms have not already been effective, but new principles have not formed yet. It veils one of its main dangerous tendencies. Risks and instability, produced by achieved freedom, are growing and result in instability, chronic crisis and bifurcation of the system. A new factor, that can break fragile peace and agreements, can appear at any moment of such system development. The maximum of freedom involves a high degree of responsibility and moral, though the last ones, as we have found out, are not valuable priorities of the new world.

Electronic credit cards, identification documents both are becoming opportunities for more freedom in the activity and produce new nonliberties for a person. As Simon Rogerson (1998) pointed out, human beings, not having such choice, cannot feel themselves comfortable in the modern world as nowadays their personalities exist in two levels: physical and electronic ones.

Freedom for new generation became the main achievement and welfare, but during its realization, the issue of sense and purpose of its use arose more acutely. Freedom that has been understood as an end in itself for a long time became a value and a facility during its implementation to achieve something else. Now freedom represents a value as a constant opportunity of a choice, where it is important not so much direction and result of a choice, as the state of choosing. Freedom in choosing is realized during consumption of goods, in communication with other people, as an opportunity of moving in space and a choice of forms of leisure, education, etc. The availability of freedom does not still mean its high function by itself. Freedom can turn out to be not a form of self-development or moral development, but only a choice of one commodity from many ones. The freedom of the information society is an opportunity to achieve a specified purpose; however, technologies do not set these purposes.

The second antinomy has existential aspect: a person has acquired unprecedented freedom in time and space but a dependence of a new character appears – he is dependent on involvement into virtual communication which threatens his individuality still more. In a person's system of values freedom has been always playing a significant role, but convenience, comfort, practical usefulness, speed of acquiring information and services becomes of more importance for a modern "electronic subject".

E-COMMUNICATION

The most considerable changes are related to the sphere of communication, which was changed by the Internet and social networks. Communication is an important sphere of a human life and performs numerous functions: adaptive, educational, searching for meaning of life, ethic etc. Communication is one of human's essential values, especially if it is filled with profound meaning and feelings. Modern information networks have tied people by means of numerous threads and chains, having overcome the borders of space, language and social conventionalities. The Internet has become, first of all, an instrument for human communication. For the most users it is its main value and opportunities.

The analysis of the development of social networks shows, how fast the way from the creation of a new Network for communication to the formation of the entire community is despite of the diversity of interactions (Zhou, Ding & Finin, 2011). Appearing communities in social networks are a specific new kind of communities of people, using mostly a virtual form of communication, striving to social activity and self-presentation in wider audience of the society and keeping some information about themselves confidential or replacing it with information about imagined persons.

According to A. Ronchi, Professor at Milan's Polytechnic University, now we should consider the Internet as a bulwark even in respect of social communication (Ronchi, 2009).

In the classical age human beings were connected with quite a small group of people, their "near and dear ones". Today individuals can make "friends" with hundreds and thousands of people unfamiliar to them in real life and learn about their interests. What has this given to them and what has it deprived them of? Let us consider antinomies related to the sphere of communication. On the one hand, it has given a lot: (a) an opportunity to express themselves among a larger group of people, (b) incredible choice of opportunities for getting acquainted, (c) keeping in touch with dear people who are far away, (d) constant possibility to maintain contact with other persons, (e) escape from problems caused by psychological or physical complexes during communication, etc.

But what has the E-communication deprived a modern human of? The losses are also numerous: (a) obligations and responsibility of real life communication, (b) "real" friends and family (because of lack of time to spend with them), (c) profoundness of communication (that is compensated for by the abundance of communication), (d) profoundness of understanding other persons and feeling with them, (e) feeling of being important for Another one, (f) long-standing friendly relations and feelings, (g) fidelity in communication, etc.

It is obvious that quantitative possibilities of communication have replaced their "quality" and profoundness. Human beings are unable to love passionately many other people and devote all their heart to thousands of virtual friends. Wide choice and escape from the real difficulties in communication substitute profound interpersonal relations. Will this change a human being? For sure, loss of real human communication and its substitution with the virtual one make individuals more oriented to their "Egos", distracted from sufferings and feelings of other people. Virtualization of communication evokes the sensation of its being unreal, weakening the call of duty, responsibility for themselves and other persons. Loss of the profoundness of communication, its reduction to an exchange of phrases with no particular value, deprives human beings of the environment where they get values for finding the essence. When enduring real problems human beings do not obtain support of virtual friends, that aggravates depression, suicide among the youth, existential vacuum, when human beings absolutely cannot understand the purpose of their existence in the world, and this inevitably decreases the value of life itself. The possibilities brought forth by new technologies are virtual in their essence, e.g., they are illusive and devoid of true value.

KNOWLEDGE OR INFORMATION?

Another type of antinomy is related to the cognitive sphere and states the following: Truth and Knowledge are the primary values in the cognitive sphere and information is just a tool; however in the modern society it is Information, not Knowledge that plays the primary role and can be defined as "Knowledge minus person". Now we are going to consider which changes are related to these antinomic approaches.

The role of knowledge extremely increases in the Information Age, associated with intelligent technologies in economy. The information society differs qualitatively from the previous epoch; its basis is first of all not a material factor, but an ideal one. For the first time for all history of the mankind knowledge are getting the highest status and estimation that allows to hope for the positive solution of a variety of problems, caused by a new epoch. The values (from spiritual and cultural to material and economic ones) are re-estimated, people are striving from quantitative indexes of the economic development to the improvement of the life quality, sustainable development and environmental safety, the production of mass standard goods is being shifted to the production of goods and services for personal consumption, to creation of scientific and technical complexes by means of reduction of an intermediate product rate in fuel and raw material branches. At the same time information as impersonal volume of data of objects, which can be used by a human to achieve certain goals, along with knowledge is being exalted.

In the cognitive sphere it is important to distinguish concepts "data", "information", "knowledge", "understanding", "wisdom". R.L. Ackoff (1989) believed, "data" are symbols, primary material, not meaning anything, "information" is data which can mean something in the certain context or interrelation, "knowledge" is an answer to a question "how?", it is the systematized information which can be used, "understanding" is an answer to a question "why?", and "wisdom" is the developed understanding with elements of foresight.

According to Russell Ackoff (1989), a systems theorist and professor of organizational change, the content of the human mind can be classified into five categories: (a) *data,* symbols, (b) *information,* data that are processed to be useful and provides answers to "who", "what", "where", and "when" questions, (c) *knowledge*, application of data and information, answers "how" questions, (d) *understanding*, appreciation of "why", (e) *wisdom,* evaluated understanding.

As information plays a dominating role in the modern society, the value of rational, the importance of rational level and level of understanding is decreasing, ethical essence is replacing, giving way to esthetic things. The handling of "information" is the activity, concerned with understanding, evaluation, revealing of sense in a less degree, it does not practically have moral and evaluating elements. This activity is a product of machine operation (a computer) or a person as a machine. As a consequence of that the considerable part of routine computing activity is transferred to the sphere of computer competences. However, these processes are not limited to it. The computer is becoming an analogue of a person even in creative types of activity: It composes music, writes novels, plays chess, etc. In their turn individuals are becoming similar to a computer: "downloads", "stores" and "transfers" information, arranges the connections to power units and the worldwide Network to use them, interacts with other systems for updating options, etc. Personal responsibility is decreasing as a subject of information is becoming more and more massive; creative uniqueness, understanding thinking and "wisdom" are replacing with standardization and data replication. Change of accents appears more essential in terms of world view: The True and Wisdom give the place to Knowledge, and then it is replaced by Information, human beings "does not learn" the world, but they use it; the relations with other people turn out to be similar to these ones. The sphere of cognition is developing from elite one into mass one, and moral regulators distinguishing the elite of knowledge media in the last centuries, are eliminated at the same time.

INTERNET LOVE: FEELINGS OR GAME?

The most important values of a person are those, that are related with sensual and emotional spheres act, they are first of all love and friendship. Feelings of sympathy, empathy, care, respect, attachment to another person form the basis of human communication. This sphere also undergoes certain changes, first of all, concerning the virtualization of emotions. "Feelings" transmitted via SMSs or Internet messages cannot differ from the ones perceived during the personal contact, but it do. Intervening of high technologies into the most intimate sphere of human life–the sphere of sentiments, love–results in quite ambiguous consequences. On the one hand, the modern age of high technologies has produced new means of communication, bringing people closer to each other, provided an opportunity to be with close and dear people for almost all the time. On the other hand, anonymous communication changes the character of interpersonal relations which become still more intended to satisfaction of consumption requirements and sexual needs.

Such form of communication is peculiar due to the virtuality of the object and subject who present themselves as people they would like to seem, but not the ones they are in real life. Each of the partners creates an own image, supplementing it with the desired features (often those, which are promoted by the mass media), and becomes almost an ideal screen hero/heroine (Kerdellant & Crezillon, 2009). Such relations can be quite long-standing (virtual acquaintance sometimes transforms into family life), but more often they are a form of temporary expression of some affection. These relations are easy to sever (one may simply finish the chat without even explaining such a decision), they are filled with lots of bum information (that undermine love, which presupposes sincerity).

Modern researchers note that the number of romantic relations via networks has been increased, and virtual relations (called "the Second Life") sometimes can improve the mental state of persons, help to solve their real problems (Gilbert, Murphy & Ávalos, 2011).

Virtual love more likely can be defined as the illusion or game, occurring together with sharp desire of spiritual relations with an imagined partner which writes what a person wants reading. The virtualization of love simplifies it essentially, Internet-affair does not involve joint every day routine, responsibility for each other; it is some kind of a simulacrum of a new age. We also can agree with the words of the Italian philosopher and existentialist N.Abbagnano, today in the struggle between sex and moral love has suffered a defeat, theoretical and practical pansexualism seems to dominate on a stage, not only allowing the widest adulterous relationship, but also praising all forms of sexuality as manifestation of limitless freedom (Abbagnano, 1990). The Internet also favors this process. The world with huge flows of information and the all-time intensive communication in conditions of freedom is led to new understanding of love. Now it is not the only thing for all the life, the feeling granted by destiny, joining people together, but only one of many kinds of pleasure which a person can enjoy. The virtualization multiplied these opportunities, transformed qualitative feelings (associated with depth and duration of feelings) into quantitative features (where to have a number of partners became a criterion of success). Such changes can contribute to the further pragmatization of gender relationship and reduce it to pleasure and "consumption" that risks to result in the destruction of the family institute (including both parents) as prevailing form of people's life. Love as a key anthropological ability, an immanent feature of a person, which forms a basis for the moral development (as a restriction of one's own egoism and allotting the supreme value to the Other's life) still exists in the ideal world of literature and cinema, but in the context of the modern virtual freedom it is being replaced with its antinomy: love as an anachronism which requires too much efforts and is easier in the conditions of the high tempo-rhythm of our age to be substituted by the "love game" in the virtual world.

It should be noted, that the specified features are still characteristic only for a part of the society (for youth, mainly living in megacities) which plays the most creative role and forms the direction of the socium development in many ways.

SOLUTIONS AND RECOMMENDATIONS

What are the main risks connected with transformation values under condition of development of information technology? We suppose that first of all the next ones among them should be mentioned:

The most dangerous tendencies probably lead to the loss of traditional culture and vital world of man which have existed for thousands of years and which are now disappearing under the impact of globalization and informatization of culture. The annihilation of traditional values, models and norms of behaviour formed at the level of ethnic groups, states and social communities leads to the change of lifestyle of man moulded by new eclectic, informationally open cultural conditions which causes a greater openness and instability of the individual himself. Culture itself, the legacy created by humanity which aren't nowadays absolute values but are ejected and regularly renewed, are threatened with neglect.

Serious risks are also connected with the transfer of values in virtual worlds, social networks, which complicates the real communication and the capability to solve vital problems, develops the destructive way of thinking and self-destructive behaviour.

The dangerous contents of the websites with suicide clubs and so on are the factor that contributes strongly to the self-destructive behaviour of the youth. Internet addiction disorder, loss of real vital emotions caused by perception of oneself and his milieu as virtual world characters can be other reasons of such behaviour. Interpersonal communication is being replaced by its virtual imitation, such important spheres of relations as love and friendship being ejected also.

The spiritual sphere appears to be nowadays the zone of risk. It suffers crisis under conditions of the consumer society and hedonism. Fundamental science, education, art in their traditional forms lost in many respects their leading position in the postmodern culture. The commercialization and pragmatization of these spheres called into being by the post-industrial age and the post-industrial society eliminate their humanistic, educational contents, their developmental and ennobling influence over man. In spite of a high-speed growth of scientific technologies and an active expansion of scientific influence over civilization it may be mentioned that theoretical science as well as humanities which are less connected with quick applied results are at the periphery of the entertainment and consumer society development. As the consequence of the humanistic knowledge crisis the liberalization of morality, spiritual imperatives comes; upbringing and education are replaced by forming applied skills and abilities; art doesn't elevate man but entertains him cultivating the lowest needs to an even greater degree than the highest ones.

In the age of information some risks are caused by the development of the value of freedom, which under new conditions acquired such form as information freedom. The term "information freedom" refers to the ability of man to be the source, bearer or transmitter of information not being responsible for its contents and the consequences of its translation in mass media. Every web user can be a source of unverified and what is more falsified information which can considerably influence not only over social processes but also political stability.

Openness of information and the dependence of its contents from each subject multiply risks for modern culture and engender new types of unfreedom such as manipulation of public opinion, misinformation of large social groups, dark public relations, interference with citizens privacy, etc.

FUTURE RESEARCH DIRECTIONS

The most promising for the development of this research topic are the essence of e-culture, new forms of communication, virtualization of person's lifestyle. We also plan to study existential risks and freedoms offered by individual virtual technologies.

CONCLUSION

The system of values is subjected to considerable deformations resulted by the influence of a new developing type of existence – information reality, which the modern age has produced. Classical view of the world and the person himself is replaced by a new view of reality where the information technologies complete and visualize fantasies and dreams of a person about his abilities. The limits of the reality expand, a new level of existence where a person finds his continuation and development arises. The classical antinomy of the worldview – spiritual or material – is replaced with a modern antinomy – real or virtual. In this case material and spiritual is more related to "reality" and the new form of existence – virtual – is associated with their electronic versions, substitutes, simulators. Interpersonal communication, fellowship, friendship, love, knowledge are being gradually replaced with the electronic forms which deprive them of individual personal content, replacing it with artificial standards, programmed structures. Instead of acquiring knowledge we search for information, instead of making friends we "add contacts", love is substituted with the internet correspondence with people we hardly know. On the one hand a person acquires new freedoms, soaring to a level of a superman, on the other hand he only spends his freedom virtually – on "existence" in the Internet, games, social networks, and acquire maximum of real emotions and feelings from them. Contradictory and antinomic character of such worldview arise new abilities, achievements and a certain risk for a person. Modern society becomes a conventional form of existence of independent people distanced from each other and interacting in a new reality, where family is substituted with virtual society, education – by obtaining information, self-actualization – by various pleasures, selfperfectioning – by new levels of game etc. This transforms the essence of communication, socialization, worldview, the subject of these transformations himself and the results of them we will have to estimate in future.

Having emphasized the essential tendencies and risks of the values dynamics let us sum up. Society and man are in a very active phase of their development, they are in the situation of a high degree of uncertainty and bifurcation, when nobody knows what factor can be the most dangerous. New ways of development connected with high technologies can contribute to qualitative changes in man's character and in the essence of his relations with Others; they can contribute to the change of the social interaction character.

REFERENCES

Abbagnano, N. (1990). *Ricordi di un filosofo: A cura di Marcello Staglieno*. Milano: Rizzoli.

Ackoff, R. L. (1989). From Data to Wisdom. *Journal of Applies Systems Analysis*, *16*(4), 3–9.

Baarda, R., & Rocci, L. (2012). The Use and Abuse of Digital Democracy. *International Journal of Technoethics*, *3*, 50–69.

Baeva, L. V. (2012). Existential Axiology. *Culture: International Journal of Philosophy of Culture and Axiology*, *l*(9), 73–85.

Baeva, L. V. (2012). Anthropogenesis and Dynamics of Values under Conditions of Information Technology. *International Journal of Technoethics*, *3*(3), 33–50.

Baier, K. (1967). Concept of Value. *The Journal of Value Inquiry*, *1*(1), 1–11.

Brey, P. (1999). The Ethics of Representation and Action in Virtual Reality. *Ethics and Information Technology*, *1*(1), 5–14.

Brumbaugh, R. S. (1972). Changes of Value Order and Choices in Time. In *Value and Valuation: Axiological Studies in Honor of Robert S. Hartman*. The University of Tennessee Press.

Capurro, R. (2006). Towards an Ontological Foundation of Information Ethics. *Ethics and Information Technology*, *8*(4), 175–186.

Castells, M. (1997). *The End of the Millennium, the Information Age: Economy, Society and Culture* (Vol. 13). Cambridge: MA, Oxford, UK: Blackwell.

Chang, C. L. (2011). The Effect of an Information Ethics Course on the Information Ethics Values of Students–A Chinese guanxi culture perspective. *Computers in Human Behavior*, *27*, 2028–2038.

Cockton, G. (2004). From quality in Use to Value in the World. In *Proceedings of the CHI'04 Extended Abstracts on Human factors in Computing Systems* (CHI'04). New York: ACM Press.

Code of Ethics for the Information Society proposed by the Intergovernmental Council of the Information for All Programme (IFAP). (2011). *General Conference for consideration UNESCO, 36th session 2011*. Retrieved from http://unesdoc.unesco.org/images/0021/002126/212696e.pdf

Duff, A. (2008). The Normative Crisis of the Information Society. *Cyberpsychology. Journal of Psychosocial Research on Cyberspace*, *2*(1).

Foot, P. (2001). *Natural Goodness*. Oxford, UK: Blackwell.

Frankl, V. E. (1963). *Man's Search for Meaning: An Introduction to Logotherapy* (I. Lasch, Trans.). New York: Washington Square Press.

Frankl, V. E. (1997). *Man's Search for Ultimate Meaning*. New York: Peruses Book Publishing.

Friedman, B. (1997). *Human Values and the Design of Computer Technology*. CSLI Publications.

Frondizi, R. (1971). *What is Value? An Introduction to Axiology by Risiery Frondizi*. La Salle.

Gabbay, D., Thagard, P., & Woods, J. (2008). *Philosophy of Information*. North-Holland.

Giddens, A. (1990). *The Consequences of Modernity*. Cambridge, MA: Polity.

Gilbert, R. L., Murphy, N. A., & Ávalos, C. M. (2011). Communication Patterns and Satisfaction Levels in Three-Dimensional Versus Real-Life Intimate Relationships. *Cyberpsychology, Behavior, and Social Networking*, *14*(10), 585–589. PMID:21381970

Haddon, L. (2004). *Information and Communication Technologies in Everyday Life: A Concise Introduction and Research Guide*. Berg.

Hansson, S. O. (2001). *The Structure of Value and Norms*. New York: Cambridge University Press.

Hartman, R. S. (1973). Formal Axiology and the Measurement of Values. In *Value Theory in Philosophy and Social Science*. New York: Gordon and Breach.

Heidegger, M. (1927). *Sein und Zeit*. Frankfurt am Main: Vittorio Klostermann Verlag.

Heidegger, M. (2000). *Being and Time* (J. Macquarrie, & E. Robinson, Trans.). London: Blackwell Publishing Ltd.

Hemingway, C. J., & Gough, T. G. (2000). The Value of Information Systems Teaching and Research in the Knowledge Society. *Informing Science*, *4*(3), 167–184.

Kallenberg, B. J. (2001). *Ethics and Grammar: Changing the Postmodern Subject*. Notre Dame Press.

Kerdellant, C., & Crezillon, G. (2006). *Children of the Processor: How Internet and Video Games Form the Adults of Tomorrow* (A. Luschanova, Trans.). U-Faktoriya.

Levinas, E. (1979). The Contemporary Criticism of the Idea of Value and the Prospects for Humanism. In *Value and Values in Evolution*. New York: Gordon and Breach.

Maslow, A. (1966). *The Psychology of Science: A Reconnaissance*. New York: Harper & Row.

Maslow, A. (2002). *The Psychology of Science: A Reconnaissance*. Chapel Hill, NC: Maurice Bassett.

Masuda, Y. (1983). *The Information Society as Post-Industrial Society*. Washington, DC: Academic Press.

Ott, M., & Pozzi, F. (2011). Towards a New Era for Cultural Heritage Education: Discussing the Role of ICT. *Computers in Human Behavior*, *27*, 1365–1371.

Raban, D. R. (2009). Self-Presentation and the Value of Information in Q&A Websites. *Journal of the American Society for Information Science and Technology*, *60*(12), 2465–2473.

Rogerson, S. (1998). *Social Values in the Information Society*. Retrieved from http://dehn.slu.edu/courses/fall06/493/rogerson.pdf

Ronchi, A. M. (2009). *E-Culture*. New York: Springer-Verlag, LLC.

Schwartz, S. H. (1994). Are There Universal Aspects in the Structure and Contents of Human Values? *The Journal of Social Issues*, *50*(4), 19–46.

Schwartz, S. H., & Bilsky, W. (1987). Toward a Psychological Structure of Human Values. *Journal of Personality and Social Psychology*, *53*, 550–562.

Turner, R., & Eden, A. H. (2008). The Philosophy of Computer Science. *Journal of Applied Logic*, *6*, 459–626.

Veltman, K. H. (2004). Towards a Semantic Web for Culture. *Journal of Digital Information*, *4*(4).

Vernon, G. M. (1973). Values, Value Definitions, and Symbolic Interaction. In *Value Theory in Philosophy and Social Science*. New York: Gordon and Breach.

Verplanken, B., & Holland, R. W. (2002). Motivated Decision Making: Effects of Activation and Self-Centrality of Values on Choices and Behavior. *Journal of Personality and Social Psychology*, *82*(3), 434–447. PMID:11902626

Zhou, L., Ding, L., & Finin, T. (2011). How is the Semantic Web evolving? A Dynamic Social Network Perspective. *Computers in Human Behavior*, *27*, 1294–1302.

KEY TERMS AND DEFINITIONS

Antinomies: Contradictions between the two judgments, each of which is equally valid within a knowledge system (scientific theory).

E-Communication: Electronically communication, information communication is a form of social communication, involving the use of information technology, characterized by remotely possible anonymity of the subjects.

E-Culture: Electronically culture represents cumulative results of creative activity and communication of people under the conditions of the information technology implementation, characterized with creating of free information space, a virtual form of expression, distant technology, and content liberality.

Existential Axiology: Philosophical theory of values as the individual responses to the existential problems of life (death, loneliness, absurdity, etc.)

Information Technologies: Terms of anthropological analysis, are a human-developed system of processing, storage, transferring of information, accompanying with the assignment (delegation) of certain techniques of "intelligent" ("smart") information management to a computer machine.

Values: Dominants of the consciousness and the existent, directed at the achievement of the ideal objective reality, creatively influencing on the person's inner development and visual environment through the filling them by meaningfulness and sense.

Virtual Reality: Artificial reality, the reality of electronic, computerized model of reality, created by technical means the world (objects and subjects) transmitted to humans through his feelings. Virtual reality simulates exposure as well as the effect of the reaction.

Virtualization: A process of moving objects and processes of material reality to virtual (synthesized by the human mind, and information technology).

Chapter 14
Shaping Digital Democracy in the United States:
My.barackobama.com and Participatory Democracy

Rachel Baarda
University of Ottawa, Canada

Rocci Luppicini
University of Ottawa, Canada

ABSTRACT

Ethical challenges that technology poses to the different spheres of society are a core focus within the field of technoethics. Over the last few years, scholars have begun to explore the ethical implications of new digital technologies and social media, particularly in the realms of society and politics. A qualitative case study was conducted on Barack Obama's campaign social networking site, my.barackobama.com, in order to investigate the ways in which the website uses or misuses digital technology to create a healthy participatory democracy. For an analysis of ethical and non-ethical ways to promote participatory democracy online, the study included theoretical perspectives such as the role of the public sphere in a participatory democracy and the effects of political marketing on the public sphere. The case study included a content analysis of the website and interviews with members of groups on the site. The study's results are explored in this chapter.

INTRODUCTION

In November 2008, *Barack Obama's* election as president of the United States was heralded as the beginning of a new era of citizen participation in American politics (Carpenter, 2010; Marks, 2008; Tapscott, 2009). Writers attributed Obama's victory to his use of the Internet, which he used to engage and mobilize supporters and raise funds (Scherer, 2009; Talbot, 2008). Theorists also argued that Obama's use of digital media allowed citizens to participate in politics as never before.

DOI: 10.4018/978-1-4666-6122-6.ch014

In 2008, Marc Ambinder wrote in *The Atlantic*: "Obama clearly intends to use the Web, if he is elected president, to transform governance just as he has transformed campaigning…What Obama seems to promise is, at its outer limits, a participatory democracy in which the opportunities for participation have been radically expanded." Andrew Raseiej, co-president of techpresident.com, declared shortly after Obama's victory: "there's a newly engaged and empowered citizenry that is ready, able, and willing to partner with the Obama administration on rebooting American democracy in a 21st-century model of participation" (Marks, 2008).

Based on the widespread claims about a new age of citizen participation under President Obama, this study explores the use and abuse of digital media in promoting participatory democracy. More specifically, the study examines the ways in which Obama's social networking site, my.barackobama.com, influences participatory democracy. The study draws on interviews with members of my.barackobama.com, as well as the text of the website itself.

BACKGROUND

The Rise of E-Democracy

What is *participatory democracy* and what good for? Participatory democracy has been defined as a democratic system which "involves extensive and active engagement of citizens in the self-governing process; it means government not just for but by and of the people" (Barber, 1995, p. 921). Common themes in participatory theory include the belief in a widespread loss of faith in representative democracy; the idea that increased democratic participation can educate and transform humankind; and that people should participate more directly not only in government, but also in the workplace and other spheres of society (Barber, 1984; Kaufman, 1960; Miller, 1987; Pateman, 1970). Participatory theorists also emphasize community over individualism, and they highlight the importance of public debate and deliberation (Barber, 1984; Mansbridge, 1983; Miller, 1987).

In the 1990s, with the advent of the Internet, some theorists began to argue that electronic technology could create participatory democracy through digital communications media (Budge, 1996; Dahl, 1989; Hague & Loader, 1999). "Digital media" describes media types, like film and audio, which "traditionally were accessed in analog format, but…are being converted to a digital format" (Dillon & Leonard, 1998, p. 72). Digital media includes both offline and transmitted media (Feldman, 1997). Digital media transmitted over the Internet is expected to enhance democracy.

The use of digital technology to revitalize democracy is called e-democracy, also known as *digital democracy*, teledemocracy, and cyberdemocracy (Hague & Loader, 1999). The UK Hansard Society, which promotes increased political participation, offers one definition of e-democracy: "The concept of e-democracy is associated with efforts to broaden political participation by enabling citizens to connect with one another and with their representatives via new information and communication technologies" (Hansard Society, as cited in Chadwick, 2006, p. 84). Dahl (1989) argues that digital technology can prompt a "third democratic transformation" which will restore the characteristics of the Athenian participatory democracy.

Digital democracy theories have five characteristics that, in some ways, parallel the characteristics of participatory democracy theories. Firstly, digital democracy can supposedly reverse declining trust in political processes and institutions and voter apathy (Hague & Loader, 1999). Theorists of digital democracy present the Information Age as an opportunity to "rethink and, if necessary, radically overhaul or replace those [liberal democratic] institutions, actors and practice" (Hague & Loader, 1999) that are causing the crisis of participation. Secondly, digital media can supposedly create a more educated citizenry (Budd & Harris, 2009; Hague and Loader, 1999). Thirdly, digital technologies are expected to allow citizens to participate more directly in politics and other areas of society (Fuchs, 2007; Malina, 1999). Fourth, digital technologies can create virtual communities and an online public sphere (Dahlberg, 2001; Rheingold, 1993); and finally, digital technologies are expected to encourage debate and deliberation (Hague & Loader, 1999; Bryan et al., 1998).

The United States has been a forerunner in the political use of digital technology. Barack Obama was the first presidential candidate whose victory has been attributed to his use of the Internet (Carpenter, 2010). Obama not only used familiar Internet campaign strategies such as a candidate website; but he also expanded the social media campaigning strategies that had begun with Howard Dean (Evans, 2008). Barack Obama used prominent social media tools such as Facebook, MySpace, and Twitter to engage and mobilize supporters (Panagopoulos, 2009). The Obama campaign also created a social networking site, *my.barackobama.com*, to accompany Obama's campaign website (Panagopoulos, 2009). My.barackobama.com allowed Obama supporters to "connect to one another, recruit friends, organize meetings, rallies, and fund-raising drives, and offer feedback to the candidate" (Panagopoulos, 2009, p. 10). Many of the features on my.barackobama.com that allow users to connect to the campaign and to each other are characteristic of Web 2.0 and the ability of digital technology to promote participatory democracy.

After his 2008 election, researchers continued to praise Barack Obama for his groundbreaking use of social media. Towner and Dulio (2012) argue that Barack Obama's use of social media in 2008 paved the way for future campaigns, including the 2012 campaign, to build on the "*Obama model*" (111). Scherer (2012) notes that during the 2012 campaign, the Obama team used a new social network called Dashboard to organize over 358.000 offline events, with over 1 million attendees. Another million people downloaded a Facebook application that allowed the Obama campaign to integrate its voter database with supporters' friend networks.

In addition to creating a more participatory political campaign, researchers also argue that President Obama has helped develop a more participatory e-government in the United States (Bertot, Jaeger & Hansen, 2012; Nam, 2012). Nam (2012) commends Obama for creating the Open Government Directive shortly after his 2008 election, which was designed to encourage transparency and citizen participation through social media. Besides the administration's presence on popular social media sites such as Facebook and Twitter, the Open Government Directive produced many "participatory" websites which give citizens access to government information. These sites include Data.gov, which gives citizens the ability to access, download and manipulate hundreds of thousands of agency data sets, and Healthcarerereform.gov, an interactive website with a comment feature that allows people to share stories and ideas about health care (United States White House, 2010).

According to Nam (2012), Obama's use of social media in government has allowed "citizen-sourcing", in which citizens move from "users and choosers of government programs and services to 'makers and shapers' of policies and decisions" (14). Linders (2012) praises Obama for his dedication to "We-Government". He commends innovative participatory social media ventures by the Obama administration, including the "challenge.gov" platform, where the government shares a problem and offers a prize to the citizen who presents the best solution.

Theoretical Framework

Theorists of participatory democracy and the theorists of digital democracy both call for a participatory society in which political participation extends beyond voting; and they both emphasize political debate, discussion, and the formation of community. These aspects of political participation are central to Jurgen Habermas' conception of the public sphere. Habermas' concept of the public sphere emphasizes community and rational-critical debate: the public sphere has been described as "a discursive space in which individuals and groups congregate to discuss matters of mutual interest and, where possible, to reach a common judgment" (Hauser, 1998). Habermas also wrote about the theory of discourse ethics, which explains how individuals engage in communicative rationality (Habermas, 1990).

This study applies a theoretical framework based on *Jurgen Habermas'* concept of the public sphere to explore technoethical considerations within the my.barackobama.com online community. From Habermas' theories on communication, Dahlberg (2001) derives six normative conditions for an online public sphere. These conditions are: autonomy from state and economic power; exchange and critique of criticizable moral-practical validity claims; reflexivity; ideal role taking; sincerity; and discursive inclusion and equality. Dahlberg suggests that the more effectively online media facilitate these conditions, the better they promote the public sphere.

Based on Dahlberg's framework, an ethical participatory democracy would be a place where both citizens and leaders would foster rational-critical debate by engaging in sincere communication, listening to alternative positions, and adopting others' perspectives. The theory of participatory democracy can, therefore, be viewed as an ethical theory that highlights the "rights perspective" of ethics, which examines the protection of individual rights within a social group (Luppicini, 2008). On the other hand, in less ethical approaches to participatory democracy, leaders could bypass rational-critical discussion with individual citizens, choosing instead to exploit digital democracy to promote a candidate or engage in other forms of political marketing (O'Shaughnessy & Henneberg, 2002). In an effort to shed light on the emerging new digital space for participatory democracy, this research examines the ways in which Obama's social networking site, my.barackobama.com, uses or misuses opportunities for creating a healthy participatory democracy.

METHOD

Study Design

This research uses *case study* research methodology. According to Creswell (2007), "case study research is a qualitative approach in which the investigator explores a bounded system (a case) or multiple bounded systems (cases) over time, through detailed, in-depth data collection involving multiple sources of information" (p. 73). This particular study explores a single bounded case, Barack Obama's website (my. barackobama.com) from December 2010 to August 2011. The website was studied during this time frame because the time period followed the 2010 midterm elections for Congress and the Senate, which took place on November 2, 2010 (Gill, 2010) and preceded the 2012 presidential election, which is scheduled to take place on November 6, 2012 (de Voogd, 2011).

Although Barack Obama has a presence on many social networking sites, only the website my.barackobama.com, is studied. This site is significant for two main reasons. Firstly, when my.barackobama.com was launched as a branch of the main Obama campaign website in February 2007 (Arrington, 2007), Obama was not "the only potential candidate to utilize social software," but he was "the first to create a fully fledged social networking site of [his] own" (O'Hear, 2007). Most importantly, however, my.barackobama.com is known for encouraging more active participation in the Obama campaign than any of his other social networking pages. Franklin Hodge, employee of Blue State Digital, which helped set up the database system for Obama's campaign, explains, "Facebook is great for broadcasting yourself to friends, but it's not very action oriented. There are few eatures at my.barackobama.com for broadcasting yourself in the abstract—instead, it's geared to getting people to take action" (Cone, 2008). During Obama's campaign, my.barackobama.com was instrumental not only in attracting supporters, but also in encouraging them to donate, volunteer, and create their own political events (Cone, 2008; Talbot, 2008).

The central research question of this thesis is: How does Barack Obama's website, my.barackobama.com, promote or discourage participatory democracy? Drawing on the theories of participatory democracy (and its emphasis on non-electoral participation, decision-making, and an informed citizenry), and digital democracy, the research attempted to answer the following sub-questions:

RQ1: In what ways do members of my.barackobama.com believe the website has influenced their levels of non-electoral political participation (for example, through such activities as donating or canvassing)?

RQ2: How do members of my.barackobama.com believe the website has influenced their ability to contribute to political decision-making?

RQ3: In what ways do members believe the website has influenced their levels of political knowledge and interest?

RQ4: Drawing on the theory of deliberative democracy, in what ways do members believe the website has provided opportunities for them to discuss and debate ideas with their peers?

RQ5: In what ways do members of my.barackobama.com believe the website has influenced their sense of belonging to a community?

RQ6: How does the written content of my.barackobama.com promote or discourage non-electoral citizen participation in politics?

Data Collection and Analysis

Data collection and analysis relied on interview data and website documentation. In-depth interviews were seen as suitable because the topic of my.barackobama.com and political participation had not yet been researched in detail. The in-depth interviews were semi-structured and conducted by e-mail, due to the fact that the researchers lived in Canada and the members of the groups were residents of the United States or other countries. The interview participants consisted of members of groups on Barack Obama's website, my.barackobama.com. The "group" feature on my.barackobama.com was accessible only to those who joined the website; so the researchers joined the website in order to recruit participants. Twelve interview subjects for this study were chosen through criterion sampling. Respondents were selected from groups on my.barackobama.com whose descriptions identified an interest in one of the significant election issues in the 2008 presidential campaign and/or the 2010 midterm election. "Significant election issues" were determined from two Gallup polls (Saad, 2008; Jones, 2010). The second Gallup poll, conducted between August 27 and 30, 2010, similarly asked respondents, "How important will each of the following issues be for your vote to Congress this year?" (Jones, 2010). From these polls, the researchers created a list of "most important" election issues, with the "most important" 2010 election issues alternating with the "most important" 2008 election issues.

The interviews consisted of open-ended interview questions e-mailed to potential respondents between January 25, 2011, and February 23, 2011 (Appendix). Interview data analysis followed the case study data analysis procedures recommended by Creswell (2007). Creswell (2007) suggests that case study researchers should first read through their data and make initial codes, and then use categorical aggregation to establish themes (p. 157).

The second source of data for this case study was derived from website content. For this study, the sampling units consisted of posts on my.barackobama.com, written by Obama's campaign staff and volunteers. The sampled posts were drawn from the entire population of posts, which consisted of all entries on my.barackobama.com from the launch date of the 2012 campaign website (April 4, 2011) to the date the sample was drawn (July 23, 2011). The researchers used a systematic sampling strategy, which involves selecting units from a population at fixed intervals (Krippendorff, 2004). Systematic sampling "is favored when texts stem from regularly appearing publications, newspapers, television series, interpersonal interaction sequences, or other repetitive or continuous events" (p. 115). Since Obama campaign staff and volunteers posted daily updates on my.barackobama.com, the website's text fell into the category of "repetitive or continuous events." In the last week of July, 2011, the researchers drew a sample of posts from my.barackobama.com. Between April 4, 2011 and July 23, 2011, the Obama campaign staff had posted a total of 320 entries on the website. There were exactly 10 posts on each web page of my.barackobama.com, and the researchers sampled every 5th post, sampling two posts from each web page. To ensure a rich qualitative analysis, the researchers drew a total sample of 75 posts. The sampled entries were posted between April 4, 2011 and July 22, 2011. The second source of data was used in data triangulation to help corroborate findings from the interview data analysis.

FINDINGS AND INTERPRETATIONS

The researchers coded 60 initial themes from the interview data, which were then grouped into five categories, with two or three subcategories each. These were as follows: 1) previous political participation; 2) support for Obama; 3) website's influence on participatory democracy; 4) feelings about democratic participation; and 5) desire to participate in politics. The first category is *previous political participation*. This category covers the answers to the interview question, "Would you say you participated in politics in any way before joining the group?" This category was divided into two subcategories: "was already engaged in political participation beyond voting" and "was politically disengaged." The second category is *support for Obama*. The researchers found that when discussing their political participation, many of the respondents mentioned their support for Obama. References to Barack Obama were so frequent throughout the interviews that the researchers created a separate category for descriptions of support for Obama. This category is divided into three subcategories: "supported Obama," "volunteered for Obama," and "current level of support for Obama." The third category is *website's influence on participatory democracy*. This category addresses respondents' experiences with the website, and is divided into two subcategories and five sub-subcategories. "Website influenced participatory democracy" and "website did not influence participatory democracy" are the two subcategories. The subcategories are further divided into four sub-subcategories each, to describe specific aspects of political participation. These sub-subcategories are: intellectual development; political participation outside of voting; community; and debate and discussion. Next is the category, *feelings about democratic participation*. This category has two subcategories: "has faith in democratic participation" and "disillusioned about democratic participation." The fifth and final category is *desire to participate in politics*.

This category is divided into two subcategories: "desire to contribute ideas," and "desire for administration to listen to people."

After conducting interviews with members of my.barackobama.com, the researchers also performed a content analysis of the website. The main written content on the website was a series of blog-style posts, accompanied by the date of the post and the name of the poster. Often, the contributions were written by campaign staff; but sometimes the posts were written by volunteers and supporters, and sometimes the posts were quotes from Obama's speeches or from the campaign Twitter feed. The researchers drew a sample of these posts; then, as with the interview coding, the researchers coded each sentence according to different categories. These categories were: "stories and quotes from supporters," "stories and quotes from campaign," "encouragement to participate," "motivational rhetoric," "information about campaign activities and website," "news about Obama and the government," and "reasons to support Obama."

From the integration of the interview findings and the content analysis, several key findings emerged which helped answer this study's original research questions. From the interviews and content analysis, four key findings emerged as answers to the research questions. Firstly, the research found that my.barackobama.com generally developed political knowledge and encouraged non-electoral political participation; but secondly, the interviews showed that the website was not likely to promote political discussion or community. In addition, the content analysis echoed the interview findings by showing that Barack Obama's 2012 campaign site appeared to promote community, but was actually mainly focused on promoting campaign participation and informing citizens. Finally, the research findings revealed that a majority of respondents had grown disillusioned about democratic participation. These are discussed in detail below.

According to many of the interview respondents, my.barackobama.com increased their political knowledge and promoted political participation. Six out of twelve interview respondents suggested that their experiences with my.barackobama.com increased their levels of political knowledge, and seven out of twelve respondents said the website had encouraged political participation. In the interviews, one member of my.barackobama.com expressed enthusiasm about both the website's ability to provide information and its ability to promote political participation. This person wrote, "When I called someone on the phone for the campaign and they had a question about something that I didn't know right away, I could usually find the answer in mere moments on the website" (Interviewee 5 excerpt). The same respondent wrote that the website also promoted political participation: "I became very actively involved, starting several groups, organizing events…I continued on to travel to several other states volunteering and working for the campaign before returning home, and became the Regional Field Director for the campaign in my home region" (Interviewee 5 excerpt).

While the interviews with members of my.barackobama.com confirmed the idea that the website was successful at informing visitors and encouraging political involvement, the second key finding from the research was that my.barackobama.com was not so successful at encouraging community or political discussion. The 2008 version of my.barackobama.com did not have a discussion forum, although it did allow users to create blogs and join e-mail discussion groups. Members of the website also had the opportunity to contact the administration: after Barack Obama's election as president, when the researchers first began studying the site, a contact form on the website directed respondents to http://www.whitehouse.gov, the Obama administration's website. In addition, members of the website had the opportunity to build community with each other in e-mail groups, and during the campaign, they had opportunities to meet offline. Nevertheless, out of the seven respondents who answered the follow-up interview questions about political discussion and community, six people said that my.barackobama.com had not created community, and five people indicated that the website had not promoted political discussion.

Three respondents said that that they had not felt part of a community because the website was too focused on campaign spin, and three respondents suggested that the website had promoted online but not offline community. One respondent wrote that other members of the group rejected her because she did not agree with the party line: "I was (still am) a huge proponent of Single Payer health insurance. The members of the group distanced themselves from that idea and me" (Interviewee 6 excerpt). Another respondent wrote, "For me being part of a community involves communication and contact, and I personally prefer physical contact and a vis-à-vis conversation over "impersonal" Internet connections" (Interviewee 13 excerpt).

Respondents also indicated that the website had not promoted political discussion among citizens and also had not helped citizens to communicate with the government. One respondent gave ambivalent responses that were somewhat typical of the other responses about political discussion: "The e-mail group was pretty much a disappointment. Not really much came of that" (Interviewee 4 excerpt). This respondent added, "I don't really think joining barackobama.com changed my participation in political discussions. I did that before and after joining the website. Maybe it developed it slightly" (Interviewee 4 excerpt). Furthermore, respondents said that my.barackobama.com did not help them communicate their opinions to the government. One respondent wrote, "I must say I agree with this concept of being heard, yet I have reservations about the website and its responsiveness...I seriously doubted it was responding to opinion, but merely collecting them" (Interviewee 6 excerpt). Most interview respondents suggested that my.barackobama. com did not really promote interaction among citizens, or interaction between citizens and the campaign (or administration).

These findings are foreshadowed by other studies on campaign websites. Various studies, including Endres and Warnick (2004). Stromer-Galley (2000) found that candidate campaign websites rarely offer opportunities for human-to-human interaction, such as "electronic bulletin boards, chat forums, or e-mail addresses to the candidate" (Stromer-Galley, 2000, p. 122). According to Stromer-Galley (2000), campaigns believe human interactivity is burdensome to manage, it may lead to loss of control, and it may lead to a loss of ambiguity in campaign messages. In fact, the Institute for Politics Democracy and the Internet (IPDI) tells candidates to use caution when implementing user-to-user interactivity on their websites (Endres & Warnick, 2004, p. 325). Fear of risk may explain why my.barackobama.com did not include discussion forums.

While the lack of bulletin boards and chat forums might partly explain why the interview respondents felt that the website did not promote political discussion, members also complained that the website was controlled by campaign spin, which prevented the formation of community. This finding corroborates research by Janack (2006), who found that bloggers on Howard Dean's campaign website "created a self-disciplining system on the campaign site that maintained control over the campaign's message and muted the potential for meaningful online political deliberation and citizen participation." Because campaign sites by their nature are focused on the election or re-election of a candidate, the discussion on campaign websites often becomes a form of "persuasion" to support the candidate. Schneider and Foot (2002) connect the concept of political persuasion to Benjamin Barber's idea of strong democracy: "When political talk is limited to persuasion, it serves what Barber (1984) describes as "thin" democracy; when political talk provides opportunities for individuals to set and control the agenda and express themselves without constraint, "strong" democracy is possible." The interviews in this research study suggest that even when political talk did take place on my.barackobama.com, the discussion was often limited to "persuasion" instead of letting individuals "express themselves without constraint". This reality may have curbed the potential for "strong" or "participatory" democracy on my.barackobama.com.

Campaign spin, however, was not the only reason members of my.barackobama.com felt the website did not promote community. Some members suggested that the website had promoted online community, but not offline community, or they felt that online community did not fit their definition of true community. Although members did have the opportunity to create events such as house parties and attend these events, not everyone took advantage of the opportunity to meet in person. Three out of the seven people who replied to the follow-up questions felt the website did not promote offline community. One person "was suspicious of the meet-ups aspect," while two other people had not met anyone offline, but did not explain why. Further research might be necessary to determine why some respondents did not meet anyone offline. One of these respondents did express a preference for face-to-face interactions, and suggested that online community did not really fit the definition of a "community." The other two respondents indicated that they were ultimately even disappointed by the online community, because the online interactions were controlled by campaign spin and/or never really created a strong community.

A recent study by Cullen and Sommer (2010) suggests that online communities are often not as robust as offline communities. Cullen and Sommer surveyed people who were involved in both online and offline community groups, such as environmental or community action groups. They found that the online respondents had "less sense of cohesion and achievement" (p. 151), and appeared to get less personal fulfillment from their involvement" (p. 151). The offline group of respondents "indicated more positive motivation based on fulfillment, friendship, and their contribution to society" (p. 151), and was "more confident that their contributions to public debate will be acknowledged and taken account of" (p. 152). According to Cullen and Sommer, offline groups are much better at encouraging Robert Putnam's (1995) concept of social capital, "the features of social organization…that facilitate co-ordination and co-operation for mutual benefit" (p. 67). Cullen and Sommer conclude.

> The Internet does not appear to replace normal social intercourse and the interpersonal rewards that citizens gain from participation in civil society, but simply to add another communication channel. This is an important lesson for the government. Politicians and policy makers alike…would be well advised to resist the temptation to shift the majority of their communication, and especially their consultative processes online (p.153).

Cullen and Sommer's warning is notable in light of the reality that campaign websites, like my.barackobama.com, are often expected to include consultative processes.

Campaign websites are expected to encourage consultation between citizens and the campaign because they can enable campaign-to-user and user-to-campaign interactivity (Endres & Warnick, 2004, p. 325). This kind of interactivity "includes any feature that enables campaigns and users to communicate with each other" such as links to volunteer and contribute, and opportunities to sign up for campaign newsletters (Endres & Warnick, 2004, p. 325). Endres and Warnick (2004) evaluated user-to-campaign interactivity on the basis of "responsiveness." Examples of a high degree of responsiveness, according to Endres and Warnick, included contact information for the campaign and opportunities to contact the campaign or candidate through online forms. Schneider and Foot (2002) suggest that these opportunities are a type of "political talk."

Members of my.barackobama.com, however, had access to many links allowing them to volunteer and contribute, and they even had opportunities to contact the Obama administration online; but most members did not feel that these opportunities constituted "responsiveness." In fact, the interview respondents often complained that the administration was unresponsive; one of the recurring themes in the interviews was that the "administration does not listen to the people." Website members felt that the administration did not listen to the people because members did not receive a response to their messages, and because the administration did not make policy changes that would show it took citizens' suggestions into account.

Researchers have noted that campaigns and presidential administrations may simply not have the resources to respond to the flood of responses they receive when they allow online contact forms (Owen & Davis, 2008; Stromer-Galley, 2000). Stromer-Galley (2000) interviewed Bob Dole's web creator and Bill Clinton's website overseer after the 1996 presidential election. Both men explained that the presidential candidates "had neither the staff nor the time to handle the potential bombardment of e-mail messages, which is why they chose not to put an e-mail address up at all…A lack of response to an e-mail message could be more damaging and alienating that not providing an e-mail address at all" (p. 131). The findings from the in-depth interviews in this research study seem to confirm that opinion, since respondents were upset when they communicated their opinions to the government but did not receive a response. The interviews suggest that political supporters see responsiveness as more than the opportunity to click on links and even more than the opportunity to communicate their opinions to government: they want the government to respond to them in tangible ways, using both actions and written words.

Overall, the in-depth interviews with members of my.barackobama.com suggested that the website promoted certain aspects of participatory democracy, such as informing and engaging citizens in political participation; but the website did not promote other aspects, like promoting political dialogue and building community. The interviews discussed the 2008 version of my.barackobama.com, but by the time the researchers performed a content analysis of the site, it had been remodeled and turned into the 2012 campaign site. The lapse in time between the interviews and the content analysis offered a unique opportunity for the researchers to learn whether the 2012 campaign website offered more potential for participatory democracy than the 2008 version of the website.

In fact, the findings from the content analysis turned out to be surprisingly similar to the in-depth interview findings. Although the 2012 version of my.barackobama.com appeared to mainly promote community, it actually primarily encouraged political participation, and secondly offered information. The majority of the content on my.barackobama.com was coded as "stories and quotes from supporters," and "stories from campaign," suggesting that the website promoted community and interaction between the campaign and citizens. The third and fourth most common codes on the website were "encouragement to participate" and "motivational rhetoric," describing content that encouraged political participation. Finally, the rest of the posts were coded as "information about website and campaign events," "news about Obama and the government," and "reasons to support Obama." All these types of posts provided political information. Because the most frequent posts on the website were stories and quotes, they gave the impression that the website mainly promoted discussion and interaction. Nevertheless, a second look at these posts shows that the content of these posts was often similar to the posts encouraging political participation and providing information. For example, the website included several quotes from Obama supporters on Facebook. These quotes included these statements: "What's more wonderful than the amount he's raised so far is that most of the money comes from folks like us in smaller donations, which just goes to prove that big corporations and all their cash can't buy everything they want. Hooray for us!" Another quote from the Facebook page declared: "President Obama 2012! Another 4 yrs, let's get out and vote...i am loving this president." At first glance, these quotes could be taken as a representative set of quotes from the campaign's Facebook page. On a second glance, though, the quotes seem to perform a similar function to many other posts on the website. The first quote mimics motivational rhetoric by suggesting that the campaign belongs to ordinary people; and the second quote encourages participation by urging supporters to vote.

According to Endres and Warnick (2004), the inclusion of different voices on a website is an aspect of text-based interactivity. Text-based interactivity includes dialogic rather than monologic expression, which "manifests a diversity of characteristics, speech styles, and even languages" (pp. 328-329). As an example, Endres and Warnick cite Stan Matsunaka's campaign site, where "one hears the voice of the candidate, his supporters, his father's fellow soldiers, and his father" (p. 329). Since my.barackobama.com includes the "voices" of campaign volunteers and supporters, it incorporates dialogized expression. Endres and Warnick hypothesize that because campaign websites rarely allow discussion boards, text-based interactivity may act as a substitute for actual human interaction "through emulation of dialogue between Web users and members of the campaign" (p. 326). On my.barackobama. com, stories and quotes from supporters might simulate dialogue in the absence of other opportunities for discussion. Since interview respondents said that my.barackobama.com did not promote dialogue or community even in its e-mail groups, though, text-based interactivity seems unlikely to promote the kind of political dialogue or community necessary for participatory democracy.

Furthermore, the stories and quotes on my.barackobama.com seem to encourage political participation or provide reasons to support Obama, rather than promoting open debate. In fact, most of the content on my.barackobama. com implicitly or explicitly encourages political participation or provides information about the campaign. After stories and quotes from supporters and stories from the campaign, the most common code in the content analysis was "encouragement to participate." This content generally involved links to volunteer, donate, or support the campaign in some way. The next most common code, "motivational rhetoric," also encouraged participation more indirectly. The last three codes involved informational content. According to Bimber (2003), information provision is one main goal of campaign websites; the other goal is engaging supporters. Candidates expect their campaign sites to be "providing a variety of means for citizens to become engaged, such as volunteering, donating, or subscribing to e-mail lists" (p. 48). Stromer-Galley (2000) notes that by encouraging citizens to become engaged through donating and volunteering, campaign sites may actually create a façade of interaction with the campaign and the candidate through media interaction (127). Stromer-Galley adds, however, that this type of interaction does not promote democracy: "The real work of democracy...is in human-human interaction...A democratic system in which campaigns close themselves off from engagement with citizens is less democratic" (p. 128).

In light of Stromer-Galley's warning, it is interesting that many of the interview respondents in this study had become disillusioned about democratic participation. Several respondents complained that the website only existed for campaign spin and donations. One interview respondent mentioned both those two common complaints about the website:

I have heard that the ability to raise money in small amounts from many people has made it more difficult for big oil to bribe the media and buy candidates, but after the election, when the website became more directive than interactive, and seemed to be more about spin than substance, I became disaffected. (Interviewee 11 excerpt)

Interview respondents also mentioned that the administration did not listen to the people and only served economic interests and power. One respondent declared: "The political system seeks to control and direct me and harvest my money, not to listen to me or to anyone else who is not rich" (Interviewee 2 excerpt). Overall, since joining my.barackobama.com, respondents seemed to feel more disillusioned than optimistic about democratic participation.

These findings are at odds with the idea that the 2008 Obama campaign successfully engaged voters through the "participatory Internet" (Cogburn & Espinoza-Vasquez, 2011; Levenshaus, 2010). The in-depth interviews with members of my.barackobama.com show that many Obama supporters were less supportive of Obama, and less inclined to participate in politics, than they were before joining the website. There appears to be a gap between the appearance of participatory democracy and the respondents' feelings of frustration with participation. That is, respondents felt that the social media tools on my.barackobama.com failed to live up to their grassroots potential. Instead of helping them participate in politics, respondents felt that the website existed for campaign "spin," to win support, and to gain donations. One respondent called the website "a remarkable way to raise money and fool people into believing Obama was connecting to them on a deeper level" (Interviewee 6 excerpt), while another respondent explained, "It seems like it's really more a way to give the façade of involvement in hopes that people will give money because they feel involved" (Interviewee 4 excerpt).

Both the interviews and content analysis in this research study suggest that, despite its openness and interactivity, the primary goals of my.barackobama.com were reinforcement and mobilization, just like any other campaign site. These main objectives may have created several obstacles to true grassroots participation. Firstly, while the Obama campaign prided itself on empowering supporters, supporters eventually concluded that the real purpose of the website was to win votes and donations, rather than to hear supporters' opinions. Secondly, Bimber notes that reinforcing support on campaign websites sometimes results in campaign spin (2003, p. 49). The interview responses show that campaign spin can frustrate voters who want open debate and unbiased information. Thirdly, because campaign websites are designed mainly for supporters rather than undecided voters, they risk becoming closed communities where only supportive opinions are welcome. Galston (1999) notes that virtual communities in general tend to bring like-minded people together, which may alienate people with differing views. Foot et al. (2007) argue that by merely involving people in the campaign, instead of connecting them to other sites and organizations, campaign websites may become "gated communities." Foot et al. ask, "Will citizens as producers increasingly seek to differentiate themselves and form *ad hoc* communities, or will campaigns and organizations increasingly structure the online democratic activities of citizens?" (p. 102).

Certainly, website members often felt that my.barackobama.com structured their online democratic activities, which contrasts with the Obama campaign's claims that it encouraged grassroots participation. The website members' feelings raise the question of whether campaign websites can ever truly encourage "grassroots" participation. Openness and opportunities for interactivity may benefit both campaigns and citizens in the short term, when the campaign attracts supporters who agree with the candidate and his message; but interactivity by itself does not satisfy citizens when they start to disagree with the candidate's message or politics. When citizens disagree with the campaign message and the campaign does not seem responsive to the citizens' opposing views, citizens will soon begin to feel that the campaign has taken advantage of them to win support.

Lack of responsiveness, in fact, seemed to be the second main reason why website members felt frustrated about democratic participation. Website members were disappointed with the Obama administration because they felt it did not listen to the people and only listened to the rich and powerful. One respondent wrote, "A lot of talk about 'the middle class'...does anyone in the Obama camp really talk to anyone in the middle class?? It is obvious our leaders have no intention of serving the public and would rather remain accountable to economic powers in this country" (Interviewee 6 excerpt). Another respondent wrote, "President Obama's actions since being elected are almost the opposite of the promises he made during the campaign. At least his promises to those who joined mybarackobama.com, perhaps not his promises to the Wall Street billionaires who actually funded his campaign" (Interviewee 12 excerpt). Not only did respondents feel like the administration did not listen to them, but, as the last response above shows, respondents felt as though Barack Obama had betrayed promises he made to them, such as a promise to represent the middle class.

O'Shaughnessy and Henneberg (2002) suggest that modern election campaigns are characterized by a political marketing approach, which involves various election promises. Henneberg (1996) writes that political marketing is an attempt "to establish, maintain, and enhance long-term voter relationships at a profit for society and political parties...This is done by mutual exchange and fulfillment of promises" (p. 777). Farrell and Wortmann (1987) suggest that the promises offered on the political market include more than just specific election promises: "The 'sellers' offer representation to their 'customers' in return for support...They market their particular styles of representation and specific intentions for government as a 'product' which composes party image, leader image, and manifesto proposals or selected images" (p. 298).

The members of my.barackobama.com seemed to feel that they felt they had been offered a "product" in the political market which failed to materialize. One respondent declared, "Since President Obama has shown nothing but contempt for the people who elected him and violated so many campaign promises, my attitude towards the U.S. government has deteriorated considerably since joining the website" (Interviewee 12 excerpt). The interview respondents in this study may have felt betrayed by the Obama administration for all these reasons. For example, respondents may have hoped to "feel the effects" of Obama's economic policies, and may have been disappointed when the economy did not substantially improve after Obama took office. One respondent wrote, "I am less engaged in politics as the country is not changing much under this administration in the ways I had hoped for" (Interviewee 7 excerpt).

The main reason for respondents' dissatisfaction with the Obama administration seems to be consistent with Naurin's view that "The notion of election promises can be used to denote a broader kind of responsibility of the representatives. Common visions or goals of society can also be defined as election promises" (2011, p. 151). In other words, more than just fulfilling specific electoral promises, citizens may expect their representatives to realize a particular vision of society. This idea is particularly relevant to an explanation of disillusionment with Barack Obama's administration. During his campaign, Obama presented a vision of democracy in which communal participation in politics could transform America (Jenkin & Cos, 2010; Sweet & McCue, 2010). According to Sweet and McCue (2010), "Obama's rhetoric is an attempt to reconstitute the U.S. electorate's understanding of 'the people' and of their collective agency as citizens capable of self-governance" (p. 603). Levenshaus (2010) notes that my.barackobama.com offered the same "you-centred" discourse of citizen participation that Obama presented in his speeches. In a personal interview, one of Obama's campaign staff explained, "In everything we did, the narrative, the underlying expectation or message was that you have the power to affect the course of the campaign" (Levenshaus, 2010, p. 325). The Obama campaign thus offered citizens vague but optimistic promises that if they united in self-government, they could create a new future for America. In his speeches and on his website, therefore, Barack Obama presented a discourse of participatory democracy. Not only did he tell his supporters that they were "agents of change" (Jenkins & Cos, 2010, p. 191) who could collectively take charge of America's destiny, but he also suggested that if they did so, they could help fulfill the American Dream (Jenkins & Cos, 2010). Although this message was obviously not a specific campaign promise, it was still an implied promise that support for Obama could help create a participatory democracy that would lead to national transformation and renewal.

Moreover, Obama portrayed himself as a co-collaborator in the process of transforming America. Jenkins and Cos (2010) explain that in his speeches, Obama presented himself as the authorial persona and Americans as the "second persona" (p. 191); but he also suggested that he was united with Americans: "Obama touched on his authorial persona and that of the American people as agents of change, the second persona, to describe the 'whole' of America…Thus, it is clear that 'he' was not in this battle alone" (Jenkins & Cos, 2010, p. 191). Jenkins and Cos (2010) add that Obama connected with Americans by sharing personal stories: "Obama presented himself as an average American whose story was part of the larger American story" (p. 195). Obama thus presented himself as a representative for all Americans, portraying himself as "one of them", and someone who could participate with them in the work of self-governance. Naurin (2011) explains that, "when they accept being a representative, politicians somehow also accept carrying the unrealized wishes and hopes for all society, and these unrealized wishes and hopes can be summarized under the expression 'broken election promises' when citizens give words to them" (p. 152). In his speeches, in the content of his website, and in his website's opportunities for interactivity, Barack Obama took Americans' hopes on his shoulders more fully than many other political candidates. He marketed himself as the representative who would help ordinary, middle class Americans realize their unspoken hopes and wishes.

Not surprisingly, when Barack Obama failed to make life demonstrably better for the middle class, some of his supporters felt that he had broken faith with them. One respondent declared,

If I could change anything about the website, I would change it to be about a candidate who is actually honest and believes in the American people, not just the corporations, the military industrial complex and the international bankers…I am thoroughly disgusted by Barack Obama, his marketing, his lies and everything about him. (Interviewee 12 excerpt)

Recent researchers have echoed this study's research findings, which showed citizen dissatisfaction with the way Obama favoured the status quo (Cohen, 2012; Dawson, 2012). In Souls: A Critical *Journal* of Black Politics, Culture, and Society, Cathy Cohen (2012) argues that Obama's support of neoliberalism causes him to develop policies that benefit certain marginalized communities, such as the gay and lesbian community, while ignoring the black community. Michael Dawson (2012) writes, "Although he promised no real left agenda, Barack Obama was elected by hopeful voters who projected onto him a deep desire for a truly transformative and progressive presidency. Instead, the Obama administration expanded US imperialism abroad and economic policies that benefited the wealthiest of individuals and corporations domestically" (117).

Members of my.barackobama.com were more than just disillusioned with Barack Obama, however. In his rhetoric and on my.barackobama.com, Obama had promised that ordinary Americans could make a difference and participate in the work of governing America. When his supporters noticed that the website seemed to exist for campaign spin and donations, they felt as though the campaign did not genuinely care about their participation. Moreover, supporters watched the Obama administration after the election for signs that it would continue to promote participatory democracy. Naurin (2011) notes that as part of citizens' broader understanding of electoral promises, citizens see representation as a more ongoing process than just winning an election: "Elections do not get the same attention in citizens' perceptions of politicians as they do in the scholarly discussion about representative democracy…the process of representation is seen as a continuous process rather than as elected terms linked together in a chain" (p. 153). After the election, when the Obama administration failed to respond to the people or incorporate their suggestions into policy, Obama's supporters felt disillusioned about the ideal of participatory democracy and even democratic participation in general. One respondent said that the experience with the website had "soured me completely on the Democratic Party and I know [sic] longer believe there is any purpose in engaging in electoral politics at the national level…I no longer believe there is any chance they will take my views into account" (Interviewee 12 excerpt).

Nevertheless, some supporters still felt that citizens could make a difference in politics, even if the Obama administration was unresponsive. Another member of the website acknowledged disappointment with my.barackobama.com, yet remained optimistic about democratic participation:

It became clear quickly that MyBarackObama was NOT interested in participatory democracy… Since this was billed as the brave new experiment in participatory democracy, I concluded that there would not actually be any such thing in my lifetime…so if I wanted to try to make myself useful I would need to figure out how to do so despite the system rather than through it. (Interviewee 2 excerpt)

This quote expresses a common theme in the interviews: although my.barackobama.com had failed to realize supporters' hopes for participatory democracy, some of them still hoped to make a difference in politics.

Pulling together key findings from the analysis, Table 1 highlights general normative conditions of the online public sphere found within the My.barackobama.com community.

Recent studies agree that despite widespread optimism about participatory government, "participatory" tools like the ones stemming from Obama's Open Government Initiative have little actual effect on democratic engagement (Dahlberg, 2011; Perez, 2013). Perez (2013) blames this in part on the digital tools themselves. He notes that it is difficult to replicate the vitality of face-to-face discussions online, and it is also difficult to link online discussions with offline meetups. Perez explains that, paradoxically, "Web-based democratization seems to work best when it draws on spontaneous and non-hierarchical social processes…Once the project is institutionalized….to generate a more systemic deliberation and consultation process, it seems much more difficult to create the kind of motivation and enthusiasm that are found in the non-coordinated, civic projects (p. 80)."

The findings from this case study of my.barackobama.com certainly demonstrate that work needs to be done to better nurture normative conditions to promote participatory democracy within an online public sphere. Drawing on the respondents' feelings of both optimism and discouragement with democratic participation, the conclusion of this research study explores the lessons that can be learned from my.barackobama.com and present some recommendations for the future.

Table 1. Normative conditions of the online public sphere found within My.barackobama.com

Limited	Moderate	Strong
Autonomy from state and economic power	X	
Critique of moral-practical validity claims	X	
Reflexivity	X	
Ideal role taking	X	
Sincerity	X	
Discursive inclusion and equality	X	

CONCLUSION

The research resulted in four main research findings: that the website often promoted political knowledge and political participation; that the website rarely promoted political discussion or community; that the 2012 campaign site appeared to promote political discussion and community, but actually promoted political knowledge and involvement; and that many website members were disillusioned about democratic participation. These research findings show that my.barackobama.com was successful at promoting certain aspects of participatory democracy, including political knowledge and non-electoral political participation. However, like other political websites in the past, my.barackobama.com failed to promote political discussion and community. This failure contributed to members' eventual disillusionment with democratic participation.

The study also highlights the role of political leaders in promoting or discouraging participatory democracy. This study suggests that in 2008, Barack Obama and his campaign team created such high hopes of participatory democracy that it was almost impossible for the administration to live up to the expectations it had created. When members of my.barackobama. com realized that the administration had not lived up to its implied promise of participatory democracy, they ended up losing faith in democratic participation.

While the research sheds light on the way my.barackobama.com influenced participatory democracy, it also illuminates the relationship between participatory democracy and political digital media in general. Many scholarly books and articles claim that digital media will promote participatory democracy, but the findings of this study suggest that this may not happen easily. Firstly, as writers like Mansbridge (1983) admit, implementing true "participatory democracy" with direct democratic participation by the people is nearly impossible in large nation-states. Because of their size, nation-states like the United States use representative systems, and in representative democracies, digital media cannot implement direct democracy; it can only encourage greater democratic participation. If the leaders of representative democracies suggest that ordinary people will have more influence in political decision-making than is possible, the people will undoubtedly be disappointed and disillusioned by the reality.

Secondly, this research study illustrates the importance of human interaction in participatory democracy. While political websites often do a good job of providing political information and offering opportunities to participate in politics, they may be less successful at promoting interaction among citizens and between citizens and the government; yet for citizens, human interaction is essential to the concept of participatory democracy. Citizens want a place where they can discuss and debate political issues without feeling ostracized by others for having different views; and they also want opportunities to share their ideas with their political leaders and representatives. For citizens, "being heard" by government means more than the opportunity to express their opinions to the government; it means receiving an acknowledgement that the government has really listened to their opinions, and that leaders will take their opinions into account when making decisions.

As Bimber and Davis (2003) explain, "Campaigning is a process of communication... the fabric holding [campaign strategies] together is communication" (p. 44). Web 2.0 has been praised for facilitating communication, not just from government to citizens, but from citizens to government and from citizens to other citizens. A healthy participatory democracy can only emerge when both candidates and citizens use these communicative tools not just to serve their own interests, but to understand and serve the interests of each other. Used effectively, the participatory Internet has the potential to enhance democracy by providing more opportunities for citizens to participate in government, and by giving governments more opportunities to listen to citizens.

REFERENCES

Bertot, J. C., Jaeger, P. T., & Hansen, D. (2012). The impact of policies of government social media usage: Issues, challenges, and recommendations. *Government Information Quarterly*, *29*(1), 30–40. doi:10.1016/j.giq.2011.04.004

Cohen, C. J. (2012). Obama, neoliberalism, and the 2012 election. *Souls: A Critical Journal of Black Politics. Cultura e Scuola*, *14*(1-2), 19–27.

Dahlberg, L. (2011). Reconstructing digital democracy: An outline of four 'positions'. *New Media & Society*, *13*, 855–866. doi:10.1177/1461444810389569

Dawson, M. C. (2012). Beyond 2012. *Souls: A Critical Journal of Black Politics. Cultura e Scuola*, *14*(1-2), 117–122.

Linders, D. (2012). From e-government to we-government: Defining a typology for citizen coproduction in the age of social media. *Government Information Quarterly*, *29*(4), 446–454. doi:10.1016/j.giq.2012.06.003

Nam, T. (2012). Suggesting frameworks of citizen-sourcing via Government 2.0. *Government Information Quarterly*, *29*(1), 12–20. doi:10.1016/j.giq.2011.07.005

Perez, O. (2013). Open government, technological innovation, and the politics of democratic disillusionment: (E-)democracy from Socrates to Obama. *I/S: A Journal of Law and Policy for the Information Society*, *9*(1), 61-138.

Scherer, M. (2012, November). Exclusive: Obama's 2012 digital fundraising outperformed 2008. *Time Magazine*. Retrieved from http://swampland.time.com/2012/11/15/exclusive-obamas-2012-digital-fundraising-outperformed-2008/

Towner, T. L., & Dulio, D. A. (2012). New media and political marketing in the United States: 2012 and beyond. *Journal of Political Marketing*, *11*, 95–119. doi:10.1080/15377857.2012.642748

United States White House. (2010). *The Obama administration's commitment to an open government: A status report*. Retrieved from http://www.whitehouse.gov/sites/default/files/opengov_report.pdf

ADDITIONAL READING

Baumgartner, J., & Morris, J. (2010). MyFaceTube politics: Social networking web sites and political engagement of young adults. *Social Science Computer Review*, *28*(1), 24–44. doi:10.1177/0894439309334325

Bennett, W. The personalization of politics: Political identity, social media, and changing patterns of participation. *The Annals of the American Academy of Political and Social Science*, *644*(1), 20–39. doi:10.1177/0002716212451428

Chua, A. Y. K., Goh, D. H., & Ang, R. P. Web 2.0: Applications in government web sites. *Online Information Review*, *36*(2), 175–195. doi:10.1108/14684521211229020

Coleman, S. (1999). Cutting out the middle man: From virtual representation to direct deliberation. In B. N. Hague, & B. Loader (Eds.), *Digital democracy: Discourse and decision making in the information age* (pp. 195–210). New York, NY: Routledge.

Davis, R. (1999). *The web of politics: The internet's impact on the American political system*. New York: Oxford University Press.

Dawson, M. C. (2012). Beyond 2012. *Souls*, *14*(2), 117–122. doi:10.1080/10999949.2012.722788

Foot, K. A., Schneider, S. M., & Dougherty, M. (2007). Online structure for political action in the 2004 U.S. congressional electoral web sphere. In R. Kluver, N.W. Jankowski, A. Foot, & S.M. Schneider. The internet and national elections: A comparative study of web campaigning (pp. 92-104). New York (NY).

Fountain, J. E. (2002). Toward a theory of federal bureaucracy in the 21st century. In E. C. Kamarck, & J. S. Nye (Eds.), *Governance.com: Democracy in the information age* (pp. 117–140). Washington, D.C.: Brookings Institution Press.

Garvy, H. (2007). *Rebels with a cause: A collective memoir of the hopes, rebellions and repression of the 1960s*. Los Gatos, California: Shire Press.

Hong, S., & Nadler, D. (2012). Which candidates do the public discuss online in an election campaign?: The use of social media by 2012 presidential candidates and its impact on candidate salience. *Government Information Quarterly*, *29*(4), 455–461. doi:10.1016/j.giq.2012.06.004

Lilleker, D. G., & Jackson, N. A. (2010). Towards a more participatory style of election campaigning: The impact of web 2.0 on the UK 2010 general election. *Policy & Internet*, *2*(3), 69–98. doi:10.2202/1944-2866.1064

Linders, D. (2012). From e-government to we-government: Defining a typology for citizen coproduction in the age of social media. *Government Information Quarterly*, *29*(4), 446–454. doi:10.1016/j.giq.2012.06.003

Lindlof, T. R., & Taylor, B. C. (2002). *Qualitative communication research methods*. Thousand Oaks, California: Sage.

Malina, A. (1999). Perspectives on citizen democratization and alienation in the virtual public sphere. In B. N. Hague, & B. Loader (Eds.), *Digital democracy: Discourse and decision making in the Information Age* (pp. 23–38). New York, NY: Routledge.

McMillan, J., & Buhle, P. (2003). *The new left revisited*. Philadelphia, PA: Temple University Press.

Park, H. M., & Perry, J. L. (2008). Does internet use really facilitate civic engagement? Empirical evidence from the American national election studies. In K. Yang, & E. Bergrud (Eds.), *Civic engagement in a network society* (pp. 237–269). Charlotte, North Carolina: Information Age Publishing.

Park, H. M., & Perry, J. L. (2009). Do campaign web sites really matter in electoral civic engagement? Empirical evidence from the 2004 and 2006 post-election internet tracking survey. In C. Panagopoulos (Ed.), *Politicking online: The transformation of election campaign communications* (pp. 101–123). New Brunswick, NJ: Rutgers University Press.

Putnam, R. D. (1995). Bowling alone: America's declining social capital. *Journal of Democracy*, *6*(1), 65–78. doi:10.1353/jod.1995.0002

Quiring, O. (2009). What do users associate with 'interactivity'? A qualitative study on user schemata. *New Media & Society*, *11*(6), 899–920. doi:10.1177/1461444809336511

Selwyn, N., & Robson, K. (1998). Using e-mail as a research tool. *Social Research Update 21*. Retrieved from http://sru.soc.surrey.ac.uk/SRU21.html

Talbot, D. (2008, September/October). How Obama really did it: Social technology helped bring him to the brink of the presidency. *Technology Review*, *111*(5), 78–83.

Towner, T. L., & Duleo, D. A. (2012). New media and political marketing in the United States: 2012 and beyond. *Journal of Political Marketing*, *11*(1-2), 95–119. doi:10.1080/15377857.2012.642748

Unsworth, K., & Townes, A. (2012). Social media and E-Government: A case study assessing Twitter use in the implementation of the open government directive. *Proceedings of the American Society for Information Science and Technology*, *49*(1), 1–3. doi:10.1002/meet.14504901298

Williams, C. B., & Gordon, J. A. (2003). Dean Meetups: Using the Internet for grassroots organizing. Retrieved from http://www.deanvolunteers.org/Survey/Dean_Meetup_Analysis.htm.

Williams, C. B., & Gultai, G. J. (2009). The political impact of Facebook: Evidence from the 2006 elections and the 2008 nomination contest. In C. Panagopoulos (Ed.), *Politicking online: The transformation of election campaign communications* (pp. 249–271). New Brunswick, NJ: Rutgers University Press.

Williams, J. (2012). Change you can believe in, You better not believe it. *Critical Sociology*, *38*(5), 747–769. doi:10.1177/0896920511426804

KEY TERMS AND DEFINITIONS

Case Study: Case study is a qualitative research approach where the investigator explores a bounded system (a case) or multiple bounded systems.

Digital Media: Digital media are media in a digital format.

E-Democracy: E-democracy refers to the use of information and communication technologies (ICT's) to broaden political participation among citizens.

Participatory Democracy: Participatory democracy refers to a democratic system involving intensive active engagement of citizens in the self-governing process.

Public Sphere: Public sphere refers to a discursive space where individuals and groups rationally discuss matters of mutual interest to reach consensus where possible.

Chapter 15
Ethics for eLearning:
Two Sides of the Ethical Coin

Deb Gearhart
Ohio University, USA

ABSTRACT

Among the top concerns an eLearning program administrator faces are ethical concerns for eLearning, which develop both internally and externally. This chapter is a review of some ethical concerns facing eLearning administrators and looks at two sides of the ethical coin. The first side of the coin looks at internal ethical issues, which have brought about quality concerns for eLearning programs and which partially led to five new federal regulations facing Institutions of Higher Education (IHE). The flip side of the coin looks at ethical concerns coming from outside the program by way of unethical behaviors from students and how eLearning program administrators can deal with these unethical practices.

INTRODUCTION

Before discussing the ethical dilemmas facing eLearning administrators, it is important to understand the basis for ethical concerns related to technology, known as technoethics. Technoethics is defined as the study of moral, legal, and social issues involving technology. Technoethics examines the impact of technology on our social, legal and moral systems, and evaluates the social policies and laws that have been framed in response to issues generated by its development and use (Tavani, 2004).

An early look at ethics in eLearning demonstrated a lack of a national code of ethics for higher education (Gearhart, 2000). Early research on young adults entering higher education found students are not prepared to be self-directed nor able to understand the complications of moral solutions facing students during their post-secondary educational experiences (Burgan, 1996). Gearhart (2005) discussed the "psychological distant" student who, through the use of technology for distance delivery, when not facing others face to face, saw no problems with unethical behaviors, not comprehending that actions were hurting

DOI: 10.4018/978-1-4666-6122-6.ch015

others; "out of sight, out of mind" so to speak. As early as 1990, informal polls showed that as many as three quarters of students on campuses admit to some sort of academic fraud (Gearhart, 2000), which is increasing through eLearning. Such academic fraud and unethical behaviors include harassment, defamation, infringing on intellectual property rights, hacking, plagiarism, and cheating.

According to Strike and Ternasky (1993) ethics can be applied to education in three principal ways: assisting in educational policy making, assessing the institutions' roles as moral educators, and informing standards to govern the conduct of educators. Many institutions of higher education have chosen to develop and enforce policy dealing with the unethical behaviors that have caused concerns for eLearning. However, some egregious unethical practices by institutions, particularly in the area of eLearning; have led to the Department of Education taking action to correct these practices.

When looking at what approach to take with the issues of ethics related to eLearning, two different aspects of the issues were most evident. Two areas need to be addressed – a discussion of the new federal regulations which have come about from institutional unethical behavior or a discussion of the unethical behaviors demonstrated by students in relation to use of technology used in eLearning courses. Both are extremely important to eLearning program administrators; this chapter will deal with both sides of the ethical coin.

THE ETHICAL THEORY BASIS FOR THE FIRST SIDE OF THE COIN: MASON'S FOUR ETHICAL ISSUES FOR THE INFORMATION AGE: THEN AND NOW

Mason's paper in 1986 addressed four ethical issues for the information age – privacy, accuracy, property, and accessibility (PAPA) and are just as true, if not more true as issues today. Society has accepted some forms of privacy invasion through the use of social media, like FaceBook. Many people put themselves on the Internet in a very public way. However, since 9/11 an individual's privacy has been invaded more than any other time in the information age with enhanced capabilities for surveillance of the members of our society. Accuracy of information has changed with Wikipedia and openness to information with the ability of individuals to easily change it. Property, who owns it? Individuals putting information, research, on the Internet have an expectation of owning their works. However, as with accuracy, when information can be publically maintained on websites, ownership is a joint effort. Accessibility - there is still a digital divide and issues for those you need accommodations for disabilities. As long as our societies have those who do not have the resources for the technology or need accommodations to the technology accessibility will continue to drive a wedge into each society. Freeman and Peace (2005)reexamined Mason's four ethical issues ethics years later and found they hold true today as a basis for reviewing concerns today such as hacking, identity theft, software piracy, viruses and worms among others.

Crowell, Narvaez, and Gomberg (2005) developed the Four Component Model, based on Mason's ethical concerns, to review moral psychology and information ethics. Their model represents the internal processes necessary for a moral act to ensue and the model is comprised of:

- **Ethical sensitivity:** Perceiving the relevant elements in the situation and constructing an interpretation of those elements. In this component one looks at what actions are possible, who and what might be affected by the action and how the involved parties might be affected by the action.
- **Ethical judgment:** Relates to reasoning about the possible actions and deciding which is most moral or ethical.

- **Ethical motivation:** Involves prioritizing what is considered to be the most moral or ethical action over all others and being intent upon following that course.
- **Ethical action:** Combines the strength of will with the social and psychological skills necessary to carry out the intended course of action. This component is dependent both on having the requisite shills and on persisting in the face of any obstacles or challenges to the action that may arise (pg. 22).

An example of this is computer-mediated communication which is less socially constrained than traditional forms of interpersonal interaction. In this way the technological medium creates a kind of "psychological distance" between the communication and the audience. This form of communication lends itself to being more open and frank than their traditional counterparts. It also leads to more hurtful types of communication because social inhibition is decreased (Crowell, Narvaez, and Gomberg, 2005, pg. 25).

Miller, Urbaczewski, and Salisbury (2005) discuss information privacy and public documents online. They contend that technological advances pushes the concerns reflected in the Privacy Act of 1974. Information, as a matter of public record, is not new however the use of technology to obtain and store information has placed a concern on individual privacy. This reflects on the use of public record information as a means of student authentication for online courses. It is meant to be used only by the student to identify the student however obtaining and storing the data is considered by some an infringement on privacy.

Schultz (2006) noted that part of the ethical issues surrounding information technologies (IT) deals with technology that is not yet conceived. To deal with the ever changing IT world Schultz looks at several ethical theories – theory of right; theory of value; and theory of justice. The theory of right consists of three kinds of theory: intuitionist, end-based, and duty-based (Schultz, 2006). Although there are no definitive answers on what is wrong or what is right, one can have intuitive feelings about right or wrong. The end-based theory of right considers that in the end, right will prevail. Duty-based theory of right considers the concept that one has a duty to be ethical. The theory of value works from the concept that everyone develops a set of values to base ethical practices upon. The theory of justice is designed around the assessment of justice based on an individual's concepts of right and wrong.

Keep these theories and information in mind while reviewing the federal regulations from the first side of the coin.

FIRST SIDE OF THE COIN

Institutions of Higher Education (IHE) have been engaged in meeting the state authorization rules. There are, however, a number of other federal regulations that affect eLearning and most IHE have not been paying close enough attention to these regulations and to their ramifications.

All involved with eLearning are familiar with the HEOA passed in 2008 and in particular to H.R.3746 Part H and S.1642 Part G—Program Integrity:

(ii) the agency or association requires that an institution that offers distance education programs to have processes by which it establishes that the student who registers in a distance education course or program is the same student who participates, completes academic work, and receives academic credit;.

There are five other federal regulations over the past several years that eLearning program administrators need to pay careful attention to as some have come about from unethical practices in the field. The five include:

- Definition of a "Credit Hour."
- Prohibition against "Incentive Compensation."
- Definition of "Misrepresentation."
- *Clarification* of "Last Day of Attendance."
- Definition of "Gainful Employment."

Definition of a "Credit Hour"

This federal regulation describes a credit hour as the amount of work represented in intended learning outcomes verified by quantifiable evidence of student achievement, as measured against the standard of the Carnegie Unit. It is the institution's responsibility to determine credit for the courses delivered in its programs but at the same time, the institution responsibility to meet regulatory standards. Credit is delivered in a variety of delivery formats at different levels among all institutions. An institution needs to have proper assessment mechanisms in place to demonstrate students are meeting the learning outcomes of their academic programs, in line with the expectations of the institution's regulatory agency. Failure to do so can lead to sanctions to the institution, including the loss of federal financial aid to the institution (Goldstein, 2011).

Prohibition Against "Incentive Compensation"

Unethical recruiting practices that have surfaced over the past several years has led to this federal regulation which basically states that an institution will not provide any commission, bonus, or other incentive payment based directly or indirectly on success in securing enrollments or financial aid to an persons or entities engaged in any student recruiting or admission activities or in making decisions regarding the award of student financial assistance, except for the recruitment of international students (Goldstein, 2011).

Definition of "Misrepresentation"

Misrepresentation is the federal regulation that eLearning programs should be most worried about, whether there is blatant unethical practices or just mistakenly providing incorrect information to a potential or current student. Misrepresentation is any false, erroneous, or misleading statement made to a student, prospective student or any member of the public, or to an accrediting agency, a state agency or Department of Education by the institution, or one of its representatives or persons with whom an institution has an agreement to provide educational programs or marketing, advertising, recruiting, or admissions services (Goldstein, 2011). Goldstein (2011) goes on to describe "substantial misrepresentation" as a misrepresentation on which the person to whom it was made could reasonably be expected to rely, or has relied on, to that person's detriment. When so many unethical practices in eLearning have come to light over the past several years, misrepresentation should be the greatest concern to eLearning program administrators. The public is very wary of eLearning programs and administrators need to be forthright about all information going to the public on their eLearning programs.

Some of the areas where eLearning program administrators need to be most accurate about when providing information include: accreditation status, requirements for specialized accreditation to practice in the field, prospect of employment, accepting and transferability of credits, false, erroneous or misleading statement about completion of program leading to licensure or certification (Goldstein, 2011). Key to remember is that any institutional communication is subject to the rule, which means if you use third party vendors, like a call center, statements made by the third-party vendor on your institution's behalf establishes institutional liability.

Clarification of Last Day of Attendance (LDA)

Often noted as the "sleeper" issue for eLearning programs, LDA is the date an institution uses to calculate the amount of Title IV refund a student receives when withdrawing from a program. What eLearning programs have used in the past was the last day a student logged into the learning management system. The new regulation is looking for academic engagement in the eLearning environment and just logging into the course does not constitute engagement in the course.

What the Inspector General has now issued is that if an institution's eLearning program lacks sufficient engagement it would be classified as correspondence study and may cause the institution to be ineligible to participate in Title IV programs (Goldstein, 2011). There are two-fold ramifications of this federal regulation. First, the institution needs to provide clear information to the student on withdrawal policies and procedures and the instructor needs to be able to provide the LDA based on student engagement and not just the last log in day to the LMS. Secondly, institutions now must clearly demonstrate what student engagement means for their eLearning courses to prevent the loss of Title IV eligibility for eLearning activity.

Definition of Gainful Employment

Since 1965, gainful employment has been defined, for a post-secondary institution, as at least one year of training that leads to a certificate or degree that prepares students for gainful employment in a recognized occupation. A recognized occupation had derived from the Standard Occupation Classification code (Goldstein, 2011). Since June 30, 2011, this has changed for institutions to now meeting a new metrics dealing with debt-to-income ratios and using loan repayment rates. This new definition and regulation is applicable for public and non-profit institutions.

These five new federal regulations are applicable to all institutions, not only for for-profit institutions and not only for totally online institutions. So what does this mean for eLearning programs and the ethical dilemmas they face? First there is a lesson to all institutions that unethical behaviors where students and other stakeholders are concerned can cause major ramifications to the institution. It demonstrates that the federal government will step in when institutions do not police themselves. eLearning programs have been viewed under a microscope for decades. Naysayers, within the academic community and outside it, question with quality of education which is only delivered in an alternative format. Sometimes the concerns have been warranted, but most of the time institutions do an excellent job of running their programs, in-class and online. However, this is only one side of the ethical coin, the unethical practices of institutions. The second part of this article deals with the unethical behaviors of students use of technology and of unethical practices in eLearning courses.

THE OPPOSITE SIDE OF THE COIN

Federal regulations and ethical issues have long been under scrutiny. Wax and Cassell (1979) edited a book looking at the ethical issues of that time – informed consent; harm and benefit to the subjects; privacy; many of these issues continue, some in new forms related to the technology used today.

Most IHE began addressing the HEOA Program integrity rule by working with what their eLearning programs already had in place, secure login to the LMS and proctored exams. Many institutions look to stricter proctoring policies to both strengthen student authentication and curb cheating concerns; technology based proctoring methods to address both concerns. The real issue that surfaced is the lack of understanding of ethical concerns by students. For decades surveys of students have demonstrated that students admit to cheating. Some common statistics include:

- 80% of "high-achieving" high school students admit to cheating.
- 51% of high school students did not believe cheating was wrong.
- 95% of cheating high school students said that they had not been detected.
- 75% of college students admitted cheating.
- 90% of college students didn't believe cheaters would be caught.
- 72% of students reported one or more instances of serious cheating on written work.
- Almost 85% of college students said cheating was necessary to get ahead (Cheating Statistics, 2001).

A Look at Student Authentication

Bailie and Jortberg (2009) reported on a study using early forms of student authentication for online proctored exams in order to meet the federal rules on student authentication for online learning. Their study demonstrated the successful use of technology-based authentication for exam proctoring. This was driven by the institution's need to meet regional accreditation, in this study's case the policy established by the Higher learning Commission (HLC). Providing a secure login and password and authentication for proctored exams will not be sufficient as institutions move forward. The impetus will be driven to authenticate students at the time they are accepted into online programs. Some Student Information Systems are moving in the direction of providing for a photo id in the student record.

Cost of technology is an issue. Bailie and Jortberg mention in their article that technology-based forms of authentication will reduce administrative costs to institutions, however, since 2009, when this article was published the cost to an institution providing authentication to meet regional accreditation and federal regulations has been growing and is often passed onto the student. In the new SACS COC standard 4.8.3 asks that the following be documented: "has a written procedure distributed at the time of registration or enrollment that notifies students of any projected additional student charges associated with verification of student identity. (SACS COC, 2012)."

Building a Culture of Academic Integrity in Today's Students

What the federal regulations fail to take into account and what IHE's need to consider the ever developing global era of collaboration where learners from around the world work in groups or communities and that shared resources and work will redefine academic dishonesty. Roberts and Shalin (2009) commented that values can be socially created and embedded in a culture. What we consider cheating is more endemic in these collective cultures where it is often considered an honor to represent another's work as your own.

Roberts and Shalin (2009) discussed the mixed messages learners receive from faculty on their views on academic dishonesty and their actual practice. Many faculty do not support University policy or their own policy in their syllabi. Roberts and Shalin presented the Ten Principles of Academic Integrity established by McCabe and Pavela:

1. Affirm the importance of academic integrity.
2. Foster a love of learning.
3. Treat students as ends in themselves.
4. Promote an environment of trust in the classroom.
5. Encourage student responsibility for academic integrity.
6. Clarify expectations for students.
7. Develop fair and relevant forms of assessment.
8. Reduce opportunities to engage in academic dishonesty.

9. Challenge academic dishonesty when it occurs.
10. Help define and support campus-wide academic integrity standards (McCabe and Pavela in Roberts and Shalin, 2009, pg. 185).

When students feel that they will not be caught if they cheat, or do not feel that "borrowing" information from others or on the Web is cheating, then the faculty members do need to follow through with the policies and procedures of their institution and in their syllabi to help students understand that there are ramifications to unethical behavior. This can be done in a number of ways.

The following list includes the common categories of academic dishonesty from a UTTC study:

- **Plagiarism:** Using another's words or ideas without appropriate attribution or without following citation conventions.
- **Fabrication:** Making up data, results, information or numbers, and recording and recording and reporting them.
- **Falsification:** Manipulating research, data, or results to inaccurately portray information in reports (research, financial, or other) or academic assignments.
- **Misrepresentation:** Falsely representing oneself, efforts, or abilities.
- **Misbehavior:** Acting in ways that are not overly misconduct but are counter to prevailing behavioral experience (Gallant in McNabb & Olmstead, 2009).

UTTC study found that faculty members see no significant differences in face-to-face classroom cheating and that now done online. The article provided some strategies for creating communities of integrity in online courses: course design incorporating policing, prevention and virtue, communication between faculty and students and between students, and collaboration in coursework (McNabb & Olmstead, 2009).

Course design with the three approaches to minimizing online cheating was also an emphasis of Olt (2002). The virtues approach seeks to develop students who do not want to cheat. With the prevention approach students find reduced pressure to cheat through reduced opportunities. Through policing, the goal is which is most often used, catching and punishing cheaters (Olt, 2002).

Ethics education, through the virtues approach, was the emphasis of a survey of business school deans noting an increased emphasis on ethics education over the past 20 years. This emphasis not only addressed the concerns in higher education, but also to deal with the corporate and governmental ethical scandals (Heisler, Westfall, & Kitahara, 2012).

Some Easy Deterrents for Cheating

Many students have never been taught about what cheating is, especially when it comes to plagiarism. Also students feel that doing something online is not cheating. Commonly known as "psychological distance", students do understand they are doing harm because they do not see the ramifications of their actions. No matter the reasoning to their cheating, when students do understand the ramifications behaviors will change. A few examples of helping students understand, through honor codes, netiquette and course design are easy ways faculty can curb unethical behaviors in the eLearning environment.

In a blog on ethics in online education the blogger (Gyanfinder, 2012) commented that students today are more "technology wired" and that technological connections influence students behavior, value systems and code of conduct. Many students feel that the Ctrl+C – Ctrl+V phenomenon is ok, however this is seen as plagiarism and concerns grow. Where should these ethical practices be learned? The blogger pointed out it is important to learn ethical practices in the classroom. That is stance of this author. Faculty are often too quick to accuse students of plagiarism rather than provide a teachable moment on the ethical practices based around educational use of technology.

LoSchiavo and Shatz (2011) reviewed three studies conducted on online introductory psychology courses and student self-reporting of cheating on online quizzes. In all three studies, most students reported they cheat on at least one of the quizzes, mostly by reviewing the textbook. However, in the study where students were randomly selected to sign an honor code, cheating was reduced.

Mintu-Wimsatt, Kerne, and Lozada (2010) discuss the need for establishing "ground rules" for online course discussion commonly known as netiquette. According to the authors, many netiquette guidelines and policies are based on institutional policies within student codes of conduct, academic honesty policies or university/student codes of ethics.

Olt (2009) provides seven strategies for helping design and implement discussion boards to prevent student plagiarism in online courses:

1. Ensure that discussion questions encourage higher-order thinking skills.
2. Relate discussion questions to the course as a whole.
3. Rotate the curriculum.
4. Encourage interactivity.
5. Ensure that as an instructor you take an active role in online discussions.
6. Ensure that the workload is manageable.
7. Assess discussions and provide feedback.

This review of literature on ethical practices provides a guide for eLearning program administrators to deal with the ethical concerns for both sides of the ethical coin.

Where Are We Heading: Ethics and Mobile Learning

As a student, instructor, or eLearning program administrator, are there ethical issues in mobile learning? If so, are they the same as ones one might expect in eLearning? Smith Nash (2006) to a look early on at several of the more worrisome ethical issues that could accompany mobile learning and suggests approaches to raise awareness. The goal is to avoid ethically problematic design or behaviors. Some of the other issues are not as easily addressed.

1. **Privacy Issues:** Mobile devices can invade privacy. Guidelines need to be set. They need to be clear and they should be enforced.
2. **Uniformity of Access:** Ethical constructs that deal with justice and the administration of justice suggest that all individuals who participate in an activity should be able to do so with equal chances of success; which is to say they should be on a level playing field. The digital divide is still with us and in some ways is growing.
3. **Language Barriers:** Colleges ethically obligated to provide training, mentoring and support to learners who may not have the background or language skills to succeed in mobile learning, as well as legally obligated.
4. **Learning Preferences:** Some students who sign up for mobile learning may not realize that they do not actually learn or retain content as well through audio as via other modalities. For example, they may be a spatial learner who needs to organize text. Courses need to be designed for multiple modalities and the institution needs to determine how that fits into its mobile learning initiatives.
5. **Posting and Other Concerns:** Netiquette notwithstanding, impulse control is often lessened in an environment where one feels safe and fairly anonymous. One way to combat rudeness in discussion board is to attach a real identity and impose social control.
6. **Recognizing Consequences of Actions in an Impersonal and Sometimes Invisible World (Psychological Distance):** It is not always easy to be aware the consequences of one's actions when the action takes place "off

the screen" and in the real world (rather than the virtual world). In the future, m-learning will help bring the "real world" more readily into the virtual world by means of video, images, and sound captured on location.

7. **Cyber-Bullying and Cyber-Stalking:** According to utilitarian ethics, what matters most is the amount of discomfort one causes another. This puts the burden on the course creator or provider to assure an environment where individual learners are not caused discomfort or anguish through the actions of others. Harassing instant messaging, posts, comment-spam, etc. are definitely off-limits. (Smith Nash, 2006).

Frameworks for Developing Ethical Practices with Informational Technology

The concerns in this chapter go beyond just dealing with distance education courses and needs to move to the heart of developing ethical practices though one's education into the workplace. Information technology has been changing ethical practices that have been long established. This section of the chapter deals with building an ethical framework for information technology

Building an Ethical Framework for Information Technology

West (2001) defines ethics as a philosophical system, or code of morals that serves as a standard of conduct and moral judgment. West goes on to comment that technologies have the potential to create greater autonomy and freedom for humans. A discussion of values and ethics, West notes, is needed both within the information technology industry as well as in the communities in which they are deployed. There is a need for awareness and consciousness about the systems that are being built and the uses which they are being implemented. Community participation is necessary to assure that we create the kinds of societies and cultures we desire. The six levels of active participation for change are resistance, local value influence, and technological democratization, development of a code or codes of ethics, philosophical leadership, and a philosophical paradigm shift.

1. **Resistance:** It is normal to have resistance to, particularly to technology. Technology is changing our lives and our society. Resistance comes in all forms, passively to aggressive. While resistance can be effective, it most likely will not be after the technology is released.
2. **Local Value Influence:** Unlike resistance which is after the technology is released, local value influence takes place before the technology is released. Technology becomes more effective if the local community has an active part in shaping the way technology will be used in prior to its implementation.
3. **Technological Democratization:** There are five criteria used for creating democratic technology: 1. promotion of technological practices within the community; 2. promotion of technological practices for the workplace; 3. technological practices that help enable disadvantaged individuals and other groups to participate in society; 4. people should resist the distribution of technologies that have potentially adverse social or environmental consequences; and 5. technologies and technological practices need to be created with greater participatory design (West, 2001, pgs. 144-145).
4. **Toward a Code of Ethics:** The first step might be a common set of ethical principles for the industry. Although individual moral codes should be sufficient there are many reasons why that is not enough. Common moral standards include respect, being fait to others, being truthful among others. In the technology world it is also about honoring intellectual property, copyrights, not

honoring others creations and crediting as your own.

5. **Philosophical Leadership:** Developed over the concept that community leaders choose to make sure that we lead the technology and that the technology doesn't lead us. "It is only through philosophical leadership that the industry will promote more positive values (West, 2001, pg. 148)."
6. **Philosophical Paradigm Shift:** According to West (2001), any changes in attitude in dealing with technology, in and out of the industry, will take place within a broader philosophical context. Our societies must create how we ethically adopt technologies which will support value to the environment and human life. Technology should improve the human condition.

Another framework was developed by Spinello (2003) from an ethical analysis of a series of case studies in information technology. Spinello presents two broad categories of modern ethical theories: teleological, the ethics of ends and deontological, the ethics of duty. Teleological theories give priority to the good over the right and evaluative actions by the goal or consequences that they attain. Deontological theories, on the other hand, argue priority of the right over the good or the independence of the right from the good (Spinello, 2003, pg. 4). These theories are often stated as so the "ends justify the means or the means justify the ends." For most, ethical values are a combination of the two.

Spinello discusses the need for critical analysis of the moral problems that arise in information technology. He provides a three step ethical analysis:

- Three steps for ethical analysis – intuition, critical normative evaluation, public policy implications
    - What are your first impressions or thoughtful reactions to the ethical issues trigged by the situation – in other words, what does your *moral intuition* say about the action or policy you are reviewing: is it right or wrong?
    - Consider the issue from the viewpoint of *ethical theory* and review the following questions in order to develop a coherent rationale that defends your normative conclusion about the situation: Does the action (or policy) optimize the consequences and generate the greatest net expectable utility for all parties involved? Does the course of action violate any ethical duties or infringe upon any basic human rights? If moral duties are in conflict, what is the higher duty? Is this conduct consistent with the norms of justice?
    - Finally, consider the possible public policy implications of this situation and of your normative conclusion. Should the recommended behavior be prescribed through legislation or specific regulations? (Spinello, 2003, pg.18)

SUMMARY

Shelton (2011) discussed Khan's eight dimensions of eLearning framework which includes an ethical dimension. These ethical considerations of eLearning relate to social and political influence, cultural diversity, bias, geographical diversity, learner diversity, the digital divide, etiquette, and legal issues. Add to Khan's dimensions accessibility concerns of the disabled. Online learning reaches such a vast and diverse population and designing course and reaching this diverse population can bring about ethical issues not affecting the traditional classroom. Appropriate use of technology and best practices in ethics may have diverted many of the federal regulation put into place because inappropriate and unethical practices.

Ethics for eLearning is a conundrum for IHE. Ethical issues bombard institutions from all sides. This chapter reviewed some of the ethical issues facing eLearning from two sides of the coin. First, was the review of the new federal regulations facing IHE, in particular eLearning program administrators. An institution violating any of the federal regulations can lead to the loss of Title IV benefits, federal financial aid. This has already happened. The flip side of the coin is the concern about unethical behaviors of students in eLearning and with increased technology use for course delivery the issues compound. Although cheating does appear widespread, no matter how the courses are delivered, the majority of the students are doing the right thing. However, eLearning program administrators spend extensive resources to control ethical issues dealing with students who chose to cheat. Many issues have been discussed in this article and ethical frameworks for technology have been presented. The important take-aways from this article would be two-fold:

- IHE are responsible for the actions of all employees and third-party vendors in relation to all the information passed on to prospective and current students and other stakeholders. Ethical behavior is warranted at all times.
- IHE are responsible for providing students the expectations of the institution and the faculty their expectations related to ethical behavior in the classroom and for the academic program.

Failure to do either as IHE leads to consequences IHE do not expect and hurt institution, faculty, students and stakeholders.

REFERENCES

Bailie, J. L., & Jortberg, M. A. (2009). Online learner authentication: Verifying the identity of online learners. *MERLOT Journal of Online Learning and Teaching, 5*(2).

Burgan, M. (1996). Teaching the subject: Developmental identities in teaching. In *Ethical dimensions of college and university teaching: Understanding and honoring the special relationship between teachers and students*. San Francisco, CA: Jossey-Bass Publishers.

Cheating Statistics. (2001). *Caveon test security.* Retrieved from http://www.caveon.com/resources/cheating-statistics/

Crowell, C. R., Narvaez, D., & Gomberg, A. (2005). Moral psychology and information ethics: Psychological distance and the components of moral behavior in a digital world. In *Information ethics: Privacy and intellectual property*. Hershey, PA: Information Science Publishing.

Freeman, L. A., & Peace, A. G. (2005). Revisiting Mason: The last 18 years and onward. In *Information ethics: Privacy and intellectual property.* Hershey, PA: Information Science Publishing.

Gearhart, D. (2000). *Ethics in distance education: Are distance educators and administrators following ethical practices?* Unpublished Manuscript.

Gearhart, D. (2005). *The ethical use of technology and the Internet in research and learning*. Paper presented at the Center of Excellence in Computer Information Systems 2005 Spring Symposium. New York, NY.

Goldstein, M. B. (2011). *Living with the (rest of the) new program integrity regulations in an online environment.* Paper presented at the 23rd Annual WCET Conference. Denver, CO.

Gyanfinder. (2012). *Ethics in online education.* Retrieved from http://www.gyanfinder.com/blogs/1/42/ethics-in-online-education

Heisler, W., Westfall, F., & Kitahara, R. (2012). Technological approaches to maintaining academic integrity in management education. In *Handbook of research on teaching ethics in business and management education*. Hershey, PA: IGI Global.

LoSchiavo, F. M., & Shatz, M. A. (2011). The impact of an honor code on cheating in online courses. *MERLOT Journal of Online Learning and Teaching, 7*(2).

Mason, R. O. (1986). Four ethical issues of the information age. *Management Information Systems Quarterly, 10*(1). doi:10.2307/248873

McNabb, L., & Olmstead, A. (2009). Communities of integrity in online courses: Faculty member beliefs and strategies. *MERLOT Journal of Online Learning and Teaching, 5*(2).

Miller, D. W., Urbaczewski, A., & Salisbury, W. D. (2005). Does public access imply ubiquitous or immediate? Issues surrounding public documents online. In *Information ethics: Privacy and intellectual property*. Hershey, PA: Information Science Publishing.

Mintu-Wimsatt, A., Kernek, C., Lozada, H. R. (2010). Netiquette: Make it part of your syllabus. *MERLOT Journal of Online Learning and Teaching, 6*(1).

Olt, M. R. (2002). Ethics and distance education: Strategies for minimizing academic dishonesty in online assessment. *Online Journal of Distance Learning Administration, 5*(3).

Olt, M. R. (2009). Seven strategies for plagiarism-proofing discussion threads in online courses. *MERLOT Journal of Online Learning and Teaching, 5*(2).

Roberts, C. J., & Shalin, H. (2009). Issues of academic integrity: an online course for students Addressing academic dishonesty. *MERLOT Journal of Online Learning and Teaching, 5*(2).

SACS COC. (2012). *The principles of accreditation: Foundations for quality enhancement*. Decatur, GA: Southern Association of Colleges and Schools Commission on Colleges.

Schultz, R. A. (2006). *Contemporary issues in ethics and information technology*. Hershey, PA: IRM Press.

Shelton, K. (2011). A review of paradigms for evaluating the quality of online education programs. *The Online Journal for Distance Learning Administrators, 4*(1).

Smith Nash, S. (2006). *Ethics and mobile learning: Should we worry?* Retrieved from http://elearnqueen.blogspot.com/2006/09/ethics-and-mobile-learning-should-we.html

Spinello, R. A. (2003). *Case studies in information technology ethics* (2nd ed.). Upper Saddle River, NJ: Prentice Hall.

Strike, K. A., & Ternasky, P. L. (1991). *Ethics for professional in education: Perspectives for preparation and practice*. New York, NY: Teachers College Press, Columbia University.

Tavani, H. T. (2004). *Ethics and technology: Ethical issues in an age of information and communication technology*. Hoboken, NJ: John Wiley & Sons.

Wax, M. L., & Cassell, J. (1979). *Federal regulations ethical issues and social research*. Boulder, CO: Westview Press, Inc.

West, C. K. (2001). *Techno-human MESH: The growing power of information technologies*. Westport, CT: Quorum Books.

KEY TERMS AND DEFINITIONS

Academic Integrity: An ethical academic policy which deals with ethical concerns like cheating, plagiarism, and honesty.

Administration: Management of the affairs of an organization, in this case of a distance education organization.

ELearning: Often known as online learning is a course delivery methodology using web-based instruction.

Ethical Framework: Is a structure of principles that address important ethical responsibilities.

Ethics: An area of study that deals with ideas about what is good and bad behavior.

Quality: In this case specifically quality assurance, is defined as a program for the systematic monitoring and evaluation of the various aspects of a project, service, or facility to ensure that standards of quality are being met.

Technoethics: Technoethics is defined as the study of moral, legal, and social issues involving technology.

Chapter 16
Privacy vs. Security:
Smart Dust and Human Extinction

Mark Walker
New Mexico State University, USA

ABSTRACT

This chapter bridges the dilemma created by intrusive surveillance technologies needed to safeguard people's security and the potential negative consequences such technologies might have on individual privacy. It begins by highlighting recent tensions between concerns for privacy and security. Next, it notes the increasing threat to human life posed by emerging technologies (e.g., genetic engineering and nanotechnology). The chapter then turns to a potential technological means to mitigate some of this threat, namely ubiquitous microscopic sensors. One consequence of the deployment of such technology appears to be an erosion of personal privacy on a scale hitherto unimaginable. It is then argued that many details of an individual's private life are actually irrelevant for security purposes and that it may be possible to develop technology to mask these details in the data gleaned from surveillance devices. Such a development could meet some, perhaps many, of the concerns about privacy. It is also argued that if it is possible to use technology to mask personal information, this may actually promote the goal of security, since it is conjectured that the public is likely to be more willing to accept invasive technology if it is designed to mask such details. Finally, some applications to society's current uses of surveillance technology are drawn. Policy recommendations for surveillance organizations such as the National Security Agency are briefly canvassed.

DOI: 10.4018/978-1-4666-6122-6.ch016

INTRODUCTION

Edward Snowden's revelation of the extent of surveillance of U.S. and foreign citizens' personal communication by the National Security Agency (NSA) and other agencies has been descried as "one of the most significant leaks in U.S. political history" (Greenwald, MacAskill, & Poitras, 2013). The revelation has prompted what one author calls a "long overdue" discussion of privacy and security (Bird, 2013).

The dilemma that pits privacy versus security is familiar enough. Those concerned with privacy are troubled by the extent to which the NSA and other agencies use technology to monitor emails, phone calls and other personal communication between private citizens. Snowden's disclosure of the invasion of privacy of private citizens has evoked comparisons between private citizens in the U.S. and Orwell's dystopian novel *1984* (Bird, 2013). Yet, a case for the NSA's activities can also be made in the name of security. If surveillance can help us avoid a terrorist attack such as the destruction of the World Trade Center on September 11th, 2001, then the moral payoff in terms of saving of lives and property is clear.

A little armchair anthropology suggests the dilemma between privacy and security is probably at least as old as humanity itself.[1] It is not hard to imagine that our pre-civilization ancestors, who lived in groups of several hundred people, faced the same dilemma. Imagine a couple from one pre-civilization tribe thought they were having a private conversation when they found another member eavesdropping. The couple protests to the chieftain that their privacy was violated; the eavesdropper defends his actions on the basis that he wanted to make sure that no secrets about their impending attack on a neighboring tribe were being discussed. And thus, according to our little "just so" story, the dilemma between privacy and security was born.

So, although the dilemma has a long history, technological developments are heightening the tension. Consider first the relationship between technology and security. The number of persons that a single individual can kill in half an hour has risen dramatically in recent history. It was only a few centuries ago that the best one could do in this respect was to dispatch a couple of dozen or so people in this time frame with a sword. Such numbers would depend crucially on the skill and the strength of the swordsman. For over a century now, mechanized weaponry such as machine guns has increased this number into the hundreds, chemical weaponry into the thousands, and nuclear technology into the millions. And no longer do assailants require great physical prowess or skill: these technologies leverage the power of our minds, not our bodies. As horrifying as this is, developments in technology promise to only increase this number. In a small secret lab set up for a few thousand dollars, terrorists could—even as you read this— be working on a virus with the potential to wipe out humanity. Much of the knowledge is freely available online. For example, research scientists have published genetic information about the Spanish flu that killed millions of people in 1918 (Tumpey et al., 2005). You could set up a small genetic lab in your basement or garage for a few tens of thousands of dollars (equipment can be ordered online for your convenience) and start constructing the next pandemic virus.[2] Looking just a little further down the road, a number of technologists have theorized that self-replicating nanobots could potentially be deployed to destroy humanity.[3]

It is horrifying when we hear about innocents killed by terrorists, or a gunman's victims in shooting rampages in high schools and universities; but it is absolutely unthinkable that individuals might be able to, in the foreseeable future, turn this same destructive rage against all of humanity. The primary worry here, of course, is that it is difficult enough to get whole nations to work peacefully and co-operatively, how can we possibly think that we could do this for all living individuals? That is, even if we could get every government to foreswear the developing technologies that could potentially put the whole world at risk—that could wipe out every last single human—how will we ensure that individuals will not do so? As intimated, the problem is that there is some possibility that such weapons could be developed in secrecy by a single person or a few individuals. Detecting a small genetics lab is not like looking for evidence that Iran or North Korea is building atomic weapons: there is no large-scale infrastructure that is visible from spy satellites in orbit. The problem is more like this: how do you know that your neighbor is restoring a 1969 Ford Mustang in his garage, as he claims, rather than using the space for a genetics lab? Fortunately, the number of people bent on destroying many or all humans is fairly small, but the fact that there are some is in itself a significant worry. Even if we attribute a very small probability to human extinction, the ramifications of it are so profound that developments in technology force us to take very seriously the threat to our security.

Technology has also greatly changed the nature and scope of concerns about privacy. The main threats to the privacy of our ancestors probably came from eavesdroppers and spies. With the development of the written word came the threat of interception and reading of private communication. Recent developments have taken the need for a human physical presence right out of the equation: we are familiar with cameras that monitor our streets and our transactions at ATMs.

As cameras have become smaller and cheaper, some have theorized that it will be possible to develop what is sometimes called 'smart dust.' Universal surveillance through smart dust is a prediction that microscopic video, auditory and other sensors could be manufactured for next-to-nothing, and dispersed throughout the world. Much like the 'dumb dust' that continually settles on our household furnishings, smart dust would be dispersed by air currents and would be equally pervasive in our environment. The information from the individual smart dust sensors could be sent to the Internet, since each individual mote of smart dust would have two-way communication with nearby motes allowing a wireless connection to the Internet. In this scenario, everyone with Internet access could see what everyone else is doing.

True, smart dust is only on the drawing board at this time, and so this may sound like mere science fiction, but the technology is much closer than one might think.[4] For example, RFID chips have been miniaturized to an amazing extent in the last decade or so. The period at the end of this sentence is about the size of an RFID chip created by Hitachi that was used on an admission ticket in 2005. Hitachi has recently announced that they have succeeded in creating an RFID chip with the same memory capacity that is 1/64 the size. Of course these RFID tags don't quite have the capabilities projected for smart dust (they are basically a 128 bit ROM), but here too, miniaturization is proceeding at breakneck speed, both in academia and private industry (Warneke et al., 2001; Van Nedervelde, 2006).

The application of smart dust to concerns about security is perhaps obvious. If smart dust were deployed globally it would be impossible or at least much less likely that individuals or a small group of individuals could succeed in their destructive plans. Think about Ted Kaczynski, also known as the 'Unabomber'. He killed three and injured three others by mailing bombs to his victims. He was able to evade capture for over two decades despite a small army of FBI agents attempting to discover his identity. With smart dust it seems likely that Ted Kaczynski would have been apprehended much earlier, since it would have been possible to view his bomb-making activities in his remote Montana cabin on the Internet.

While universal surveillance used in service of security may not completely eliminate the possibility of widespread or global extinction, it has some promise within an overall defensive strategy. What role it might actually play in a defense strategy is probably too difficult to say at this point. In part, it will depend on a host of factors like cost and effectiveness, and whether there are better technological and non-technological alternatives in our defensive strategy. Let us not forget that compromise, concession and other tactics of diplomacy may increase our security as effectively as any technological development. Nevertheless, to the extent that it is a promising defensive strategy, we should see at least prima facie moral value in universal surveillance. Contrariwise, the main objection to smart dust is the moral one mentioned above: it impinges upon our privacy in a manner that is without precedent: even the dense network of spies working for Stalin during the 1930's in the Soviet Union were not nearly as invasive to privacy as smart dust would be.

Thus, it is clear, then, that technological developments are exacerbating the moral dilemma of how to weigh the values of privacy and security. The most appropriate intellectual discipline to study the problem is technoethics, for the ethical issue of human extinction is a direct result of these recent developments in technology (Luppicini & Adell, 2008).

In pursuing this tension, it will help to say a bit more about the crucial terms 'security' and 'privacy'. The term 'security' we will understand as the absence of violence and threats of violence. It should be apparent from this understanding that security can be had or undermined in varying degrees, e.g., the Cuban Missile Crisis presented more of a threat to our security than the malicious activities of the Unabomber. This understanding of security is broader than the traditional notion that understands security in terms of absence of military threats or attacks to the nation (Haftendorn, 1991; Miller, 2001). Our understanding includes the phenomena covered by the traditional definition—war and the threat of war is covered by our understanding—but also the absence of violence, or the threat of violence by non-military persons, e.g., terrorist organizations or lone individuals bent on destruction. Conversely, our definition is narrower than the expanded concept of security that has recently developed, particularly since the end of the cold war, which includes freedom from threats of disease, natural disasters and environmental degradation (Ullman, 1983; Miller, 2001). The reason for holding to a narrower understanding of 'security' is simply that it best serves our present purposes. There is no suggestion that other understandings of 'security' might not be more appropriate for different purposes.[5]

The concept of 'privacy' is also contested. Legal and moral reasoning about privacy includes interests in:

1. Determining what information others may have about oneself.
2. Determining what access others have to one's physical and mental life.
3. Self-determination with respect to family, friends and lifestyle choices (DeCew 1997).

For our purposes, we will understand privacy narrowly as the capacity to determine the content, extent and timing of information about oneself that others may have (Westin, 1967; Parent, 1983), that is in line with (1). The reason for this narrower understanding of 'privacy' is, again, that it best serves our present purposes. There is no suggestion that other understandings of 'privacy' might not be more appropriate for different purposes.[6]

Sometimes it is suggested that people may not mind or at least would get used to universal surveillance because it has a much more democratic or egalitarian element to it than traditional methods of surveillance (Brin, 1998; van Nedervelde, 2006). For as indicated, the suggestion is that everyone would have access to information about what everyone else is doing—everyone would suffer a similar erosion of privacy. This then points to a symmetrical exchange of information so unlike that which existed under Stalin, when information was collected and concentrated in the hands of the few, and used against the many.[7] David Brin (1998) has argued that our basic choice is between governments and other authorities collecting information on the citizenry; or, the mutual collection of information by governments on citizens and the citizenry collecting information on authorities. The latter is more democratic, and provides more freedom and security. As radical as Brin's thesis seems to many, Brin too draws a line at how far information gathering should invade upon privacy. He says, "I surely won't give up essential privacy: of home, hearth, and the intimacy that one shares with just a few" (1998, p. 26, p. 334). Ubiquitous smart dust poses a threat to what Brin terms 'essential privacy,' for security concerns do not stop at the door to home and hearth: murderous plots can be nurtured at home just as well, if not better, than anywhere else.

We will not directly enter the debate here about the moral dilemma created by a concern for security and privacy, that is, the question of how we should weigh the value of security against privacy. Rather, for the purposes of this paper we shall *dialectically assume* that the value of security trumps that of the value of privacy given the reasoning: the magnitude of the risk that we face with the misuse of genetic technologies and nanotechnology demonstrates that it is good and wise to renounce our privacy in favor of security. If one thinks it is wrong (or even crazy) that we should give up privacy, especially what Brin terms the essential privacy of home and hearth, for the sake of security, the argument here is still congenial with this view. It will be argued that valuing security leads to valuing at least a certain amount of privacy. So, the argument is that whether we take security or privacy as more important, it turns out that we should value a certain amount of privacy. The argument of this paper is directed against the extreme security position: we ought not to be concerned at all with privacy. The argument, in other words, is that the extreme security position is self-refuting: valuing security requires us to value at least some privacy.

Part of the reasoning here is the difficulty of convincing enough people to appreciate the urgency of renouncing privacy in favor of security. The rationalization for ranking the value of privacy over that of security is likely to be multifarious. Some will not believe in the potential destructive power of these technologies and so will not be willing to give up their privacy. Others will think that even if these technologies do have such destructive powers, the innate goodness of every human being is such that they would never turn such technology against humanity. Some will see (quite rightly, perhaps) that universal surveillance will weaken their position of power. It is, after all, potentially a radical democratization of the power of information. Many will find it so restrictive of their liberty that they'll be willing to risk global extinction. Others may be simply too modest: they would rather risk global annihilation than suffer the embarrassment of being seen naked or worse.

This is not to suggest that any of these are good reasons: as indicated we are assuming for the purposes of the argument that it would be good and wise to trade our privacy for security if we could. Rather, the claim is that (rightly or wrongly) a large number of people will object to universal surveillance. It does not matter that we may think that it is the height of immorality and stupidity to renounce the security that smart dust might provide simply because, for example, one is afraid of others looking at them naked. This does not change the fact that for some this will be a reason, and so too will large numbers of people resist smart dust for the reasons adumbrated.

Now it may be thought that such resistance will not matter since smart dust will be ubiquitous just like dumb dust. Resistance will be futile. However, dumb dust can be resisted and so too, can smart dust. For example, if you are interested in keeping your privacy, you might send out a large electromagnetic pulse into your living space to destroy the electronics in the smart dust. Once you destroy all the smart dust, and keep positive pressure in your domicile, you keep more smart dust from entering. The diabolical might also try to hack the smart dust network, e.g., a misanthropist might reprogram the smart dust around his garage to send back auditory and visual input suggesting that a '69 Mustang is being restored while in fact he is using the space to develop biological weapons. It is true that one might be able to design smart dust to be impervious to such attacks; but this will drive up the cost of the smart dust, and lead to an escalating arms race between those who seek their privacy and those interested in universal surveillance. There are, after all, a number of avenues that one might use to attack smart dust, including pressure, heat, radiation and chemicals, in an effort to maintain a 'clean room' free of smart dust. So, a sufficiently high dose of ultrasound might destroy at least some smart dust designs. One might try high intensity infrared light as a means to cook the smart dust, x-rays or a chemical bath for the room. In this sort of battle, it is generally easier to make a small breach in the defense than to ensure universal coverage of the defense net. This would be another instance of what is sometimes referred to as 'information warfare.' So long as there is significant opposition to smart dust and a continuing battle between those resisting universal surveillance and those promoting it, it will be enough for those who would destroy humanity to simply get lost amongst a large number of those seeking personal privacy for non-security threatening reasons. For example, suppose, on a monitor, we observe via the smart dust network someone come home and flick on a switch that sends out a large electromagnetic pulse destroying all the smart dust in the room, and so thwarting our attempt to observe her further. One thought, of course, would be that this person was seeking privacy in order to construct some humanity-destroying weapon. However there is a greater probability that the person is simply seeking her privacy for more domestic matters.

This suggests an additional reason some may resist smart dust: a version of the 'tragedy of the commons' (Hardin, 1968). That is, some might hope to opt out of universal surveillance because security is a 'common good,' and so individuals will have little incentive to sacrifice their personal privacy to promote the common good. Of course, the problem is that many may reason this way, and thus reduce the effectiveness of universal surveillance. One might hope that governmental regulations would make such electronic pulse and other attacks on the smart dust network illegal. But as noted, universal surveillance promises to democratize power, so many political leaders may be reluctant to pass such legislation. Of course, if this is to be an effective deterrent against those individuals, organizations or nations who seek to destroy humanity, then such legislation would have to be pretty well universal amongst the nations of the earth. The likelihood of this happening in the near-term does not seem high. This will require a wholesale change in societal attitudes. For those worried about security, the problem here, echoing Malthus, is that technology changes geometrically while societal attitudes and behavior change arithmetically. To illustrate with a single example: it has taken at least four decades for social policy and behaviors concerning recycling to catch on. Competent observers could see that it was the right thing to do in the 1960s, yet it is only within the last decade or so that many North American cities have offered public facilities for recycling. Think of all the technologies that have moved to the mainstream in the intervening forty year period: microwaves, fax machines, Walkman's, personal computers, the Internet, email, cell phones, laptops, scanners, digital cameras, tablets, IPod's and so on. Public attitudes on recycling have been slow to change, but technology speeds along.

Despite the foregoing somewhat pessimistic line of reasoning, it will be suggested that there may be a way that we can keep all or most of the benefits of universal surveillance while assuaging many, if not most, of the concerns about loss of privacy. The argument is that we can develop software that will protect the privacy of individuals while still allowing us to obtain sufficient information about potential global catastrophic risks. The suggestion is for 'farsighted surveillance': to view the world, not the personal details of peoples' lives. One way this could be done is to use computer software to remove people from our smart dust view of the world. Presently, I am writing this in a hotel room. If there was smart dust in this room with me at this moment it would show me working on my laptop. Imagine a computer program were devised to remove me from the picture of the room. Although I am sitting in a chair, the image provided would show the chair empty—a computer program would extrapolate what the chair would look like if no one were sitting in it. Those observing the hotel room on their monitor would see an empty room, but not an unchanging room. One would occasionally see a coffee cup moving, the keys on the computer moving, the shower running, etc. Still, one would not be able to determine my identity from these facts alone, nor whether I was here by myself, with my wife or engaged in some extramarital affair. This begins to touch on at least some of the worries people have about privacy: under this scenario no one will see me naked, having sex (with others or myself), relieving myself, and so on. On the other hand, the information available should be sufficient to see whether I am plotting to destroy the world. If the table in the hotel room was festooned with components necessary to make a bomb, and one could see the pieces slowly coming together, it would be possible to see the emerging threat without knowing the identity of the person creating the threat. And there are at least two ways to learn the identity of the person and thwart the threat. The first being that one could simply inform the police and have the individual arrested. So, for example, suppose smart dust in the Unabomber's remote cabin allowed us to see the bombs being made. We could inform the authorities and learn about his identity upon his arrest. The second is that we might allow the computer code, which blocks humans from appearing on the screen, to be overridden in some cases. Here is one of many different possible protocols that might be developed to support such a position: suppose you see on your computer screen something like a bomb or a lab where a virus is being created in some distant part of the globe. You inform one or more privacy committees that determine whether the computer code that protects individual privacy should be broken. A committee is composed of three weapons experts and three experts on individual rights. They meet and agree that there is a threat, and plug in their security over-ride codes, which in turn reveal the identity of the individual. Since the activities of the committee themselves are open to inspection, at least some of the risk of a centralization of surveillance would be mitigated.

This is termed 'farsighted surveillance' for two reasons. First, we should want our surveillance to be blurry about the close-up details of individuals' personal lives. If our goal is security, then the details of peoples' personal lives—who they are sleeping with, what they are smoking, etc.—are actually noise in the system. How blurry the details of one's personal life should be will clearly be a matter of some dispute. For example, we could think of ways to further protect the privacy of individuals such as developing software to blank out what is being written on a computer screen. This seems quite possible and would protect the privacy of secret lovers, for example. On the other hand, people could be hatching plans for universal destruction on their computers, and so it would seem important to monitor a person's computer activities. A compromise might be reached by sophisticated software that can determine whether the material is personal in nature or whether it is potentially useful in genocidal activities. Certain spy agencies allegedly monitor Internet traffic at present with the aid of software designed to search for terrorist communication. Presumably this same software could easily be adapted to read computer screens with the aid of an OCR system. Likewise, we can imagine less privacy than the scenario sketched. Images of persons could be replaced by 'stick figures'; thus, the picture of the hotel room might show a stick figure sitting at a desk writing this paper. You could not determine identity from the stick figure, but you could determine, for example, how many people were in the room. What can be done here to protect the identity of individuals will depend in part on the sophistication of computer software. In effect, we might imagine this as a 'volume control' on privacy: a control that might be raised or lowered depending on our needs and the sophistication of our technology. Obviously, how much intrusion into privacy ought to be permitted is an important question, but no specific recommendations will be made here, in part, for fear that it will deflect from the more general point: we may be able to meet most of our security needs without compromising too much in the way of personal privacy.

The second reason for calling this 'farsighted' is that it makes universal surveillance a much more realistic possibility. The reason is based on the conjecture that the opposition to universal surveillance will be directly proportional to the degree that surveillance intrudes on our personal lives. By our 'personal lives' something very broad is intended, specifically, our activities that are not directed toward murder, genocide or human extinction. Admittedly, blocking out personal information with software may allow further scope for those attempting to destroy humanity to evade detection. However, the suggestion is that this will be more than compensated for by greater acceptance of surveillance. So, for example, if there is no privacy software, that is, if the privacy volume is set at zero, then we can imagine that tens of thousands if not millions of people might attempt to disrupt the smart dust network. If much of our privacy is maintained, then this number (it is conjectured) should fall drastically. Investigating a few hundred or a few thousand attempts to evade the smart dust network is a more manageable number. Of course not all of those attempting to evade the smart dust network will be bent on creating civilization-destroying weapons, but, if there are such individuals, they will almost certainly be among the number attempting to evade the smart dust network. Figure 1 summarizes the conjectured relation.

In terms of security, arguably the best possibility seems to be the upper right hand corner of the graph where 100% of the population has no privacy, that is, where surveillance obtains extensive personal information. The worst security outcome is where no one is under surveillance—anywhere on the left hand side of the graph. As intimated, a second conjecture here is that sophisticated software will mean that it will be possible to filter out personal information with little loss in security. That is, 100% surveillance with no privacy will not provide significantly more security than 100% surveillance with many protections of privacy. If this is correct, and if security is our number one priority, then we should consider attempting to assuage resistance to universal surveillance by building in safeguards to privacy.

Just to be clear: the argument is not that we ought to balance the value of individual privacy against the ideal of security. To the contrary, the argument is this: suppose that security trumps all consideration of the value of privacy. How would we best realize this? The answer is to develop some compromise to allow for privacy, that is, to make surveillance farsighted. The reason again is that it is entirely foreseeable that many will not see such a commitment to security as wise or just. They will not fully appreciate the risk to civilization that modern technologies pose, or they may be more concerned with the discovery of their extramarital affair than the potential end of humanity, etc. So long as people do not place security as the number one goal there is the real possibility that they will resist smart dust and other surveillance methods both in their personal lives and at the level of public policy. The thought then, is that even if there is some loss of information entailed by farsighted surveillance, this will be more than compensated for by less resistance to ubiquitous surveillance. In other words, the view that we ought to be concerned exclusively with security—perhaps because of the extinction

Figure 1. Information obtained through surveillance

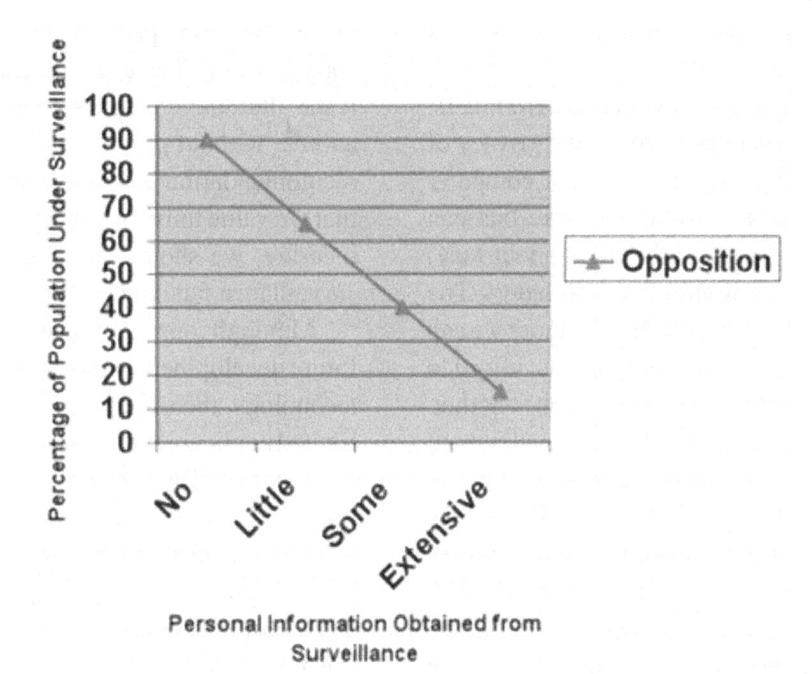

risks mentioned—is self-refuting: if we hope to increase security through increased surveillance then the best means to achieve this end is to accommodate at least some privacy concerns, that is, to make surveillance farsighted. It is worth emphasizing again that there is no suggestion here that farsighted surveillance would completely eliminate all attempts to evade surveillance. Obviously would-be terrorists will have motivation to attempt to do so, as will hackers and others—if only because it is a challenge. At best, the hope with farsighted surveillance is that such attempts to evade universal surveillance will be *reduced* to a manageable level, not that they will be completely eliminated.

It is worth noting too that the argument shows only the necessity, not the sufficiency, of farsighted measures as a means to ensure that attempts to increase security through universal surveillance succeed. The reason it may not be sufficient is that it is difficult to tell at this stage whether people would be willing to accept farsighted surveillance. As argued, the less information obtained about personal lives the more willing people may be to accept universal surveillance. The question is whether the resistance to farsighted surveillance will be sufficiently low that instituting universal farsighted surveillance will actually increase security. After all, some attempts at increasing security may actually decrease security, e.g., it is conceivable that using racial profiling in an attempt to catch terrorists could actually increase terrorists' attacks. Similarly, before attempting to deploy farsighted surveillance for security reasons we would need to be reasonably confident that it would not actually undermine efforts to increase security. Acknowledging that farsighted surveillance is not sufficient to guarantee increased security is consistent with our earlier argument that it is necessary: without farsighted measures we can be reasonably confident that attempts to improve security through (non-farsighted) universal surveillance will be less effective.[8]

Developing software to protect privacy may have a good and unintended consequence: it should provide additional impetus to develop sophisticated software for detecting would-be mass killers. For example, if we want to protect the privacy of someone writing a love letter on their computer then we will need to develop software that can distinguish love letters from letters about (say) biological weapons disguised as love letters. The more sophisticated our software, the more we can protect privacy, that is, the more we can afford to keep our personal lives private from one another. On the security side, such software development would be a major advantage. It would be pretty difficult for humans to continually monitor every other human without computer assistance. Presumably the vast majority of Kaczynski's time was spent in non-terrorist related activities, e.g., eating, sleeping and so on. The amount of time he actually spent working on his bombs may have been a very small part of his routine. Software that could wade through the minutiae of people's lives to look for tools and components for weapons would be extremely useful. If we attempted to do so without software help, it would take up much of our time. Moreover, without some incentive to the contrary, if humans could see into private lives, most people would peer into the lives of intimates and celebrities, while little time would be spent peering into the lives of those who are most likely to attempt genocide. More importantly perhaps, software may be much better at recognizing threats than most, if not all, people: two terrorists could appear to be writing love letters and exchanging gifts and we might not be able to detect that they were writing in code and exchanging parts for weapons. Sophisticated software may have a much better chance at detecting such activities. A similar point applies to the actual creation of weapons: sophisticated software might be better than most humans at distinguishing between someone using a soldering iron for benign purposes, like fixing a computer, as opposed to making a bomb. So the development of such software is important for security reasons. It seems too that those advocating privacy would have strong reason to promote the development of such software if the alternative is to have humans monitor one another without privacy software. So, the policy recommendation is simple enough: to the extent that we value universal surveillance as a security measure, we should develop software to make surveillance farsighted.

Although our argument was predicated on future developments in weapons and surveillance technology, the same line of reasoning has some limited applicability to our present situation. That is, even if one finds the possibilities of human extinction or universal surveillance too implausible to merit serious consideration, still the lessons learnt from magnifying the problem in this way can be applied to our current practices. As noted, it is no secret that many of our public spaces are presently under surveillance, even if we are often not mindful of this fact. (Ironically, after speaking about smart dust to a philosophy class, a student pointed out something I had never noticed after two semesters: our classroom had a security camera neatly and unobtrusively mounted in a back corner of the room). We might ask whether we should seek to turn up the privacy volume on current surveillance practices. We certainly have the basic technological building blocks to do so: software that discriminates human faces from other objects is ubiquitous in digital cameras. So, the basic idea seems plausible enough: face-recognition technology could be used to increase the privacy volume by blurring-out the identity of individual faces. For example, many banks have security cameras that seem to often unnecessarily invade our privacy: almost always there is no need for banks to have a visual record of the facial identity of those using the bank. Farsighted surveillance suggests that our faces could, and perhaps should, be blurred out of the record. Only when there are actual signs of maleficence would it be necessary to override the program to identify suspected criminals. A record of when and why the privacy software is overridden should go some small way to making sure that surveillance is used for legitimate purposes, and help assuage fears that it is being used for illegitimate purposes.

As a policy recommendation to the NSA and other surveillance organizations, further developing farsighted surveillance ought to be a top priority, if not *the* top priority. The reason is obvious: if these agencies' activities are curtailed in the name of privacy and these agencies are correct that surveillance is necessary for our security, then our security will be compromised. As noted above, security needs can be met as well, if not better, by the use of farsighted surveillance, which should also assuage some of the concerns of advocates for privacy.

In short, we should always ask about any surveillance practice: To what extent can we maintain both privacy and security with surveillance? That is, can the privacy volume be raised while preserving security? In many cases, the answer appears to be 'yes.' Admittedly, only a rough outline of how this might be possible at present has been provided. Providing a more definitive answer as to what is possible in the way of farsighted surveillance requires a better understanding of current and prospective developments in technology—something that is beyond the author's competence. So, it is hoped that those with the relevant expertise to evaluate the technological aspects will weigh in on this proposal.

REFERENCES

Bird, S. J. (2013). Security and Privacy: Why Privacy Matters. *Science and Engineering Ethics*, *19*(3), 669–671. doi:10.1007/s11948-013-9458-z PMID:23893336

Brin, D. (1998). *The transparent society*. Reading, MA: Addison-Wesley.

Decew, J. (1997). *In pursuit of privacy: Law, ethics, and the rise of technology*. Ithaca, NY: Cornell University Press.

Decew, J. (2013). Privacy. *The Stanford Encyclopedia of Philosophy*. Retrieved from http://plato.stanford.edu/archives/fall2013/entries/privacy/

Freitas, R. (2001). Some limits to global ecophagy by biovorous nanoreplicators, with public policy recommendations. *Foresight Institute*. Retrieved July 26, 2012, from http://www.foresight.org/nano/Ecophagy.html

Greenwald, G., MacAskill, E., & Poitras, L. (2013). Edward Snowden: The whistleblower behind the NSA surveillance revelations. *The Guardian*, p. 9.

Haftendorn, H. (1991). The security puzzle: Theory building and discipline building in international security. *International Studies Quarterly*, *35*(1), 3–17. doi:10.2307/2600386

Hardin, G. (1968). The tragedy of the commons. *Science*, *162*, 1243–1248. doi:10.1126/science.162.3859.1243 PMID:5699198

Joy, B. (2000). Why the future doesn't need us. *Wired*. Retrieved July 26, 2012, from http://www.wired.com/wired/archive/8.04/joy_pr.html

Luppicini, R., & Adell, R. (Eds.). (2008). *Handbook of research on technoethics*. Hershey, PA: Idea Group. doi:10.4018/978-1-60566-022-6

Mann, S., Nolan, J., & Wellman, B. (2003). Sousveillance: Inventing and using wearable computing devices for data collection in surveillance environments. *Surveillance & Society*, *1*(3), 331–355.

Miller, B. (2001). The concept of security: Should it be redefined? *The Journal of Strategic Studies*, *24*(2), 13–42. doi:10.1080/01402390108565553

Parent, W. (1983). Privacy, morality and the law. *Philosophy & Public Affairs*, *12*, 269–288.

Rees, M. (2003). *Our final hour: A scientist's warning: How terror, error, and environmental disaster threaten humankind's future in this century—On earth and beyond*. New York, NY: Basic Book.

Tumpey, T. M., Basler, C. F., Aguilar, P. V., Zeng, H., Solórzano, A., & Swayne, D. E. (2005). Character- ization of the reconstructed 1918 Spanish influenza pandemic virus. *Science, 310*, 77–80. doi:10.1126/science.1119392 PMID:16210530

Ullman, R. (1983). Redefining security. *International Security, 8*(1), 129–150. doi:10.2307/2538489

Van Nedervelde, P. (2006). Lifeboat foundation security preserver. *Lifeboat Foundation*. Retrieved July 26, 2012, from http://lifeboat.com/ex/security.preserver

Warneke, B., Last, M., Liebowitz, B., & Pister, K. S. J. (2001). Smart dust: Communicating with a cubic-millimeter computer. *Computer, 34*, 44–51. doi:10.1109/2.895117

Wenger, A., & Wollenmann, R. (Eds.). (2007). *Bio-terrorism: Confronting a complex threat*. Boulder, CO: Lynne Rienner.

Westin, A. (1967). *Privacy and freedom*. New York, NY: Atheneum.

KEY TERMS AND DEFINITIONS

Human Extinction: Refers to the eradication of all, or almost all, human life along with advanced technological civilization.

Farsighted Surveillance: Refers to surveillance techniques that purposely obfuscate the close-up details of individuals' personal lives in favor of information that is relevant to security, e.g., a computer program that "blurs out" individual faces while still permitting surveillance for bombs and other threats to security.

Nanobots: Refers to the prospective development of molecular-size robots; often nanobots are imagined to have self-replication capacities.

Privacy: Refers to the capacity to determine the content, extent and timing of information about oneself that others may have.

Security: Refers to (i) the absence of violence and (ii) threats of violence.

Smart Dust: Refers to the prediction that microscopic video, auditory and other sensors could be manufactured for next-to-nothing, and dispersed throughout the world.

Terrorism: Refers to large scale acts (or threats) of violence by non-state actors (primarily individuals or groups) to influence policy or societal behavior.

Universal Surveillance: Refers to the possibility of continuous and pervasive surveillance in both public and traditionally private spaces like individual homes.

ENDNOTES

1. An informal understanding of 'privacy' and 'security' is sufficient for the present. Below I will discuss these concepts in more depth.
2. Popular accounts of the threat from biological agents can be found in Joy (2000) and Rees (2003).Academic literature includes, Bostrom (2002), Bostrom and Ćirković (2008), and Wenger and Wollenmann (2007).
3. See Freitas (2000), Bostrom (2002), and several of the papers in Bostrom and Ćirković (2008).
4. A very good overview of the current state of the art and possible future developments is offered by van Nedervelde (2006).
5. Thus, there is no intention here to join the debate about the proper use of 'security' (Miller 2001). Furthermore, it seems likely that the ends of an expanded notion of se-

curity such as that offered by Ullman noted (which includes an absence of threats from disease, natural disaster and environmental degradation) might also be served by universal surveillance. However, we shall leave these possible connections for a different occasion.

6 See DeCew (1997, 2013) for review and criticism of the narrower understanding.

7 This idea of 'watching the watchers' is sometimes described as 'sousveillance'as watching from 'underneath' whereas 'surveillance' is the watching by authorities from 'above' (Mann, Noolan, & Wellman, 2003). I shall use 'surveillance' throughout in a more traditional sense that does not distinguish between 'top down' and 'bottom up' observation.

8 Obviously there is at least the possibility that non-farsighted surveillance might work if backed up with a Stalinistic police force and system of informants. I think it is safe to say that proponents of increased security through universal surveillance are not advocating universal surveillance backed up by such a police force as a means to increase security.

Related References

To continue our tradition of advancing information science and technology research, we have compiled a list of recommended IGI Global readings. These references will provide additional information and guidance to further enrich your knowledge and assist you with your own research and future publications.

Aayeshah, W., & Bebawi, S. (2014). The Use of Facebook as a Pedagogical Platform for Developing Investigative Journalism Skills. In G. Mallia (Ed.), *The Social Classroom: Integrating Social Network Use in Education* (pp. 83–99). Hershey, PA: Information Science Reference.

Adi, A., & Scotte, C. G. (2013). Barriers to Emerging Technology and Social Media Integration in Higher Education: Three Case Studies. In M. Pătruţ, & B. Pătruţ (Eds.), *Social Media in Higher Education: Teaching in Web 2.0* (pp. 334–354). Hershey, PA: Information Science Reference. doi:10.4018/978-1-4666-2970-7.ch017

Agazzi, E. (2012). How Can the Problems of An Ethical Judgment on Science and Technology Be Correctly Approached? In R. Luppicini (Ed.), *Ethical Impact of Technological Advancements and Applications in Society* (pp. 30–38). Hershey, PA: Information Science Reference. doi:10.4018/978-1-4666-1773-5.ch003

Agina, A. M., Tennyson, R. D., & Kommers, P. (2013). Understanding Children's Private Speech and Self-Regulation Learning in Web 2.0: Updates of Vygotsky through Piaget and Future Recommendations. In P. Ordóñez de Pablos, H. Nigro, R. Tennyson, S. Gonzalez Cisaro, & W. Karwowski (Eds.), *Advancing Information Management through Semantic Web Concepts and Ontologies* (pp. 1–53). Hershey, PA: Information Science Reference.

Ahrens, A., Bassus, O., & Zaščerinska, J. (2014). Enterprise 2.0 in Engineering Curriculum. In M. Cruz-Cunha, F. Moreira, & J. Varajão (Eds.), *Handbook of Research on Enterprise 2.0: Technological, Social, and Organizational Dimensions* (pp. 599–617). Hershey, PA: Business Science Reference.

Akputu, O. K., Seng, K. P., & Lee, Y. L. (2014). Affect Recognition for Web 2.0 Intelligent E-Tutoring Systems: Exploration of Students' Emotional Feedback. In J. Pelet (Ed.), *E-Learning 2.0 Technologies and Web Applications in Higher Education* (pp. 188–215). Hershey, PA: Information Science Reference.

Al-Hajri, S., & Tatnall, A. (2013). A Socio-Technical Study of the Adoption of Internet Technology in Banking, Re-Interpreted as an Innovation Using Innovation Translation. In A. Tatnall (Ed.), *Social and Professional Applications of Actor-Network Theory for Technology Development* (pp. 207–220). Hershey, PA: Information Science Reference.

Al Hujran, O., Aloudat, A., & Altarawneh, I. (2013). Factors Influencing Citizen Adoption of E-Government in Developing Countries: The Case of Jordan. [IJTHI]. *International Journal of Technology and Human Interaction*, *9*(2), 1–19. doi:10.4018/jthi.2013040101

Alavi, R., Islam, S., Jahankhani, H., & Al-Nemrat, A. (2013). Analyzing Human Factors for an Effective Information Security Management System. [IJSSE]. *International Journal of Secure Software Engineering*, *4*(1), 50–74. doi:10.4018/jsse.2013010104

Altun, N. E., & Yildiz, S. (2013). Effects of Different Types of Tasks on Junior ELT Students' Use of Communication Strategies in Computer-Mediated Communication. [IJCALLT]. *International Journal of Computer-Assisted Language Learning and Teaching*, *3*(2), 17–40. doi:10.4018/ijcallt.2013040102

Amaldi, P., & Smoker, A. (2013). An Organizational Study into the Concept of "Automation Policy" in a Safety Critical Socio-Technical System. [IJSKD]. *International Journal of Sociotechnology and Knowledge Development*, *5*(2), 1–17. doi:10.4018/jskd.2013040101

An, I. S. (2013). Integrating Technology-Enhanced Student Self-Regulated Tasks into University Chinese Language Course. [IJCALLT]. *International Journal of Computer-Assisted Language Learning and Teaching*, *3*(1), 1–15. doi:10.4018/ijcallt.2013010101

Andacht, F. (2013). The Tangible Lure of the Technoself in the Age of Reality Television. In R. Luppicini (Ed.), *Handbook of Research on Technoself: Identity in a Technological Society* (pp. 360–381). Hershey, PA: Information Science Reference.

Anderson, A., & Petersen, A. (2012). Shaping the Ethics of an Emergent Field: Scientists' and Policymakers' Representations of Nanotechnologies. In R. Luppicini (Ed.), *Ethical Impact of Technological Advancements and Applications in Society* (pp. 219–231). Hershey, PA: Information Science Reference. doi:10.4018/978-1-4666-1773-5.ch017

Anderson, J. L. (2014). Games and the Development of Students' Civic Engagement and Ecological Stewardship. In J. Bishop (Ed.), *Gamification for Human Factors Integration: Social, Education, and Psychological Issues* (pp. 199–215). Hershey, PA: Information Science Reference. doi:10.4018/978-1-4666-5071-8.ch012

Ann, O. C., Lu, M. V., & Theng, L. B. (2014). A Face Based Real Time Communication for Physically and Speech Disabled People. In Assistive Technologies: Concepts, Methodologies, Tools, and Applications (pp. 1434-1460). Hershey, PA: Information Science Reference. doi: doi:10.4018/978-1-4666-4422-9.ch075

Aricak, O. T., Tanrikulu, T., Siyahhan, S., & Kinay, H. (2013). Cyberbullying: The Bad and the Ugly Side of Information Age. In M. Pătruţ, & B. Pătruţ (Eds.), *Social Media in Higher Education: Teaching in Web 2.0* (pp. 318–333). Hershey, PA: Information Science Reference. doi:10.4018/978-1-4666-2970-7.ch016

Ariely, G. (2011). Boundaries of Socio-Technical Systems and IT for Knowledge Development in Military Environments. [IJSKD]. *International Journal of Sociotechnology and Knowledge Development*, *3*(3), 1–14. doi:10.4018/jskd.2011070101

Ariely, G. (2013). Boundaries of Socio-Technical Systems and IT for Knowledge Development in Military Environments. In J. Abdelnour-Nocera (Ed.), *Knowledge and Technological Development Effects on Organizational and Social Structures* (pp. 224–238). Hershey, PA: Information Science Reference.

Arjunan, S., Kumar, D. K., Weghorn, H., & Naik, G. (2014). Facial Muscle Activity Patterns for Recognition of Utterances in Native and Foreign Language: Testing for its Reliability and Flexibility. In Assistive Technologies: Concepts, Methodologies, Tools, and Applications (pp. 1462-1480). Hershey, PA: Information Science Reference. doi: doi:10.4018/978-1-4666-4422-9.ch076

Arling, P. A., Miech, E. J., & Arling, G. W. (2013). Comparing Electronic and Face-to-Face Communication in the Success of a Long-Term Care Quality Improvement Collaborative. [IJRQEH]. *International Journal of Reliable and Quality E-Healthcare*, *2*(1), 1–10. doi:10.4018/ijrqeh.2013010101

Asghari-Oskoei, M., & Hu, H. (2014). Using Myoelectric Signals to Manipulate Assisting Robots and Rehabilitation Devices. In Assistive Technologies: Concepts, Methodologies, Tools, and Applications (pp. 970-990). Hershey, PA: Information Science Reference. doi: doi:10.4018/978-1-4666-4422-9.ch049

Aspradaki, A. A. (2013). Deliberative Democracy and Nanotechnologies in Health. [IJT]. *International Journal of Technoethics*, *4*(2), 1–14. doi:10.4018/jte.2013070101

Asselin, S. B. (2014). Assistive Technology in Higher Education. In Assistive Technologies: Concepts, Methodologies, Tools, and Applications (pp. 1196-1208). Hershey, PA: Information Science Reference. doi: doi:10.4018/978-1-4666-4422-9.ch062

Auld, G., & Henderson, M. (2014). The Ethical Dilemmas of Social Networking Sites in Classroom Contexts. In G. Mallia (Ed.), *The Social Classroom: Integrating Social Network Use in Education* (pp. 192–207). Hershey, PA: Information Science Reference.

Awwal, M. A. (2012). Influence of Age and Genders on the Relationship between Computer Self-Efficacy and Information Privacy Concerns. [IJTHI]. *International Journal of Technology and Human Interaction*, *8*(1), 14–37. doi:10.4018/jthi.2012010102

Ballesté, F., & Torras, C. (2013). Effects of Human-Machine Integration on the Construction of Identity. In R. Luppicini (Ed.), *Handbook of Research on Technoself: Identity in a Technological Society* (pp. 574–591). Hershey, PA: Information Science Reference. doi:10.4018/978-1-4666-4607-0.ch063

Baporikar, N. (2014). Effective E-Learning Strategies for a Borderless World. In J. Pelet (Ed.), *E-Learning 2.0 Technologies and Web Applications in Higher Education* (pp. 22–44). Hershey, PA: Information Science Reference.

Bardone, E. (2011). Unintended Affordances as Violent Mediators: Maladaptive Effects of Technologically Enriched Human Niches. [IJT]. *International Journal of Technoethics*, *2*(4), 37–52. doi:10.4018/jte.2011100103

Basham, R. (2014). Surveilling the Elderly: Emerging Demographic Needs and Social Implications of RFID Chip Technology Use. In M. Michael, & K. Michael (Eds.), *Uberveillance and the Social Implications of Microchip Implants: Emerging Technologies* (pp. 169–185). Hershey, PA: Information Science Reference.

Bates, M. (2013). The Ur-Real Sonorous Envelope: Bridge between the Corporeal and the Online Technoself. In R. Luppicini (Ed.), *Handbook of Research on Technoself: Identity in a Technological Society* (pp. 272–292). Hershey, PA: Information Science Reference.

Bauer, K. A. (2012). Transhumanism and Its Critics: Five Arguments against a Posthuman Future. In R. Luppicini (Ed.), *Ethical Impact of Technological Advancements and Applications in Society* (pp. 232–242). Hershey, PA: Information Science Reference. doi:10.4018/978-1-4666-1773-5.ch018

Bax, S. (2011). Normalisation Revisited: The Effective Use of Technology in Language Education. [IJCALLT]. *International Journal of Computer-Assisted Language Learning and Teaching*, *1*(2), 1–15. doi:10.4018/ijcallt.2011040101

Baya'a, N., & Daher, W. (2014). Facebook as an Educational Environment for Mathematics Learning. In G. Mallia (Ed.), *The Social Classroom: Integrating Social Network Use in Education* (pp. 171–190). Hershey, PA: Information Science Reference.

Bayerl, P. S., & Janneck, M. (2013). Professional Online Profiles: The Impact of Personalization and Visual Gender Cues on Online Impression Formation. [IJSKD]. *International Journal of Sociotechnology and Knowledge Development*, *5*(3), 1–16. doi:10.4018/ijskd.2013070101

Bell, D., & Shirzad, S. R. (2013). Social Media Business Intelligence: A Pharmaceutical Domain Analysis Study. [IJSKD]. *International Journal of Sociotechnology and Knowledge Development*, *5*(3), 51–73. doi:10.4018/ijskd.2013070104

Bergmann, N. W. (2014). Ubiquitous Computing for Independent Living. In Assistive Technologies: Concepts, Methodologies, Tools, and Applications (pp. 679-692). Hershey, PA: Information Science Reference. doi: doi:10.4018/978-1-4666-4422-9.ch033

Bertolotti, T. (2011). Facebook Has It: The Irresistible Violence of Social Cognition in the Age of Social Networking. [IJT]. *International Journal of Technoethics*, *2*(4), 71–83. doi:10.4018/jte.2011100105

Berzsenyi, C. (2014). Writing to Meet Your Match: Rhetoric and Self-Presentation for Four Online Daters. In H. Lim, & F. Sudweeks (Eds.), *Innovative Methods and Technologies for Electronic Discourse Analysis* (pp. 210–234). Hershey, PA: Information Science Reference.

Best, L. A., Buhay, D. N., McGuire, K., Gurholt, S., & Foley, S. (2014). The Use of Web 2.0 Technologies in Formal and Informal Learning Settings. In G. Mallia (Ed.), *The Social Classroom: Integrating Social Network Use in Education* (pp. 1–22). Hershey, PA: Information Science Reference.

Bhattacharya, S. (2014). Model-Based Approaches for Scanning Keyboard Design: Present State and Future Directions. In Assistive Technologies: Concepts, Methodologies, Tools, and Applications (pp. 1497-1515). Hershey, PA: Information Science Reference. doi: doi:10.4018/978-1-4666-4422-9.ch078

Bibby, S. (2011). Do Students Wish to 'Go Mobile'?: An Investigation into Student Use of PCs and Cell Phones. [IJCALLT]. *International Journal of Computer-Assisted Language Learning and Teaching*, *1*(2), 43–54. doi:10.4018/ijcallt.2011040104

Bishop, J. (2014). The Psychology of Trolling and Lurking: The Role of Defriending and Gamification for Increasing Participation in Online Communities Using Seductive Narratives. In J. Bishop (Ed.), *Gamification for Human Factors Integration: Social, Education, and Psychological Issues* (pp. 162–179). Hershey, PA: Information Science Reference. doi:10.4018/978-1-4666-5071-8.ch010

Bishop, J., & Goode, M. M. (2014). Towards a Subjectively Devised Parametric User Model for Analysing and Influencing Behaviour Online Using Neuroeconomics. In J. Bishop (Ed.), *Gamification for Human Factors Integration: Social, Education, and Psychological Issues* (pp. 80–95). Hershey, PA: Information Science Reference. doi:10.4018/978-1-4666-5071-8.ch005

Biswas, P. (2014). A Brief Survey on User Modelling in Human Computer Interaction. In Assistive Technologies: Concepts, Methodologies, Tools, and Applications (pp. 102-119). Hershey, PA: Information Science Reference. doi: doi:10.4018/978-1-4666-4422-9.ch006

Black, D. (2013). The Digital Soul. In R. Luppicini (Ed.), *Handbook of Research on Technoself: Identity in a Technological Society* (pp. 157–174). Hershey, PA: Information Science Reference.

Blake, S., Winsor, D. L., Burkett, C., & Allen, L. (2014). iPods, Internet and Apps, Oh My: Age Appropriate Technology in Early Childhood Educational Environments. In K-12 Education: Concepts, Methodologies, Tools, and Applications (pp. 1650-1668). Hershey, PA: Information Science Reference. doi: doi:10.4018/978-1-4666-4502-8.ch095

Boghian, I. (2013). Using Facebook in Teaching. In M. Pătruţ, & B. Pătruţ (Eds.), *Social Media in Higher Education: Teaching in Web 2.0* (pp. 86–103). Hershey, PA: Information Science Reference. doi:10.4018/978-1-4666-2970-7.ch005

Boling, E. C., & Beatty, J. (2014). Overcoming the Tensions and Challenges of Technology Integration: How Can We Best Support our Teachers? In K-12 Education: Concepts, Methodologies, Tools, and Applications (pp. 1504-1524). Hershey, PA: Information Science Reference. doi: doi:10.4018/978-1-4666-4502-8.ch087

Bonanno, P. (2014). Designing Learning in Social Online Learning Environments: A Process-Oriented Approach. In G. Mallia (Ed.), *The Social Classroom: Integrating Social Network Use in Education* (pp. 40–61). Hershey, PA: Information Science Reference.

Bongers, B., & Smith, S. (2014). Interactivating Rehabilitation through Active Multimodal Feedback and Guidance. In Assistive Technologies: Concepts, Methodologies, Tools, and Applications (pp. 1650-1674). Hershey, PA: Information Science Reference. doi: doi:10.4018/978-1-4666-4422-9.ch087

Bottino, R. M., Ott, M., & Tavella, M. (2014). Serious Gaming at School: Reflections on Students' Performance, Engagement and Motivation. [IJGBL]. *International Journal of Game-Based Learning*, *4*(1), 21–36. doi:10.4018/IJGBL.2014010102

Brad, S. (2014). Design for Quality of ICT-Aided Engineering Course Units. [IJQAETE]. *International Journal of Quality Assurance in Engineering and Technology Education*, *3*(1), 52–80. doi:10.4018/ijqaete.2014010103

Braman, J., Thomas, U., Vincenti, G., Dudley, A., & Rodgers, K. (2014). Preparing Your Digital Legacy: Assessing Awareness of Digital Natives. In G. Mallia (Ed.), *The Social Classroom: Integrating Social Network Use in Education* (pp. 208–223). Hershey, PA: Information Science Reference.

Bratitsis, T., & Demetriadis, S. (2013). Research Approaches in Computer-Supported Collaborative Learning. [IJeC]. *International Journal of e-Collaboration*, *9*(1), 1–8. doi:10.4018/jec.2013010101

Brick, B. (2012). The Role of Social Networking Sites for Language Learning in UK Higher Education: The Views of Learners and Practitioners. [IJCALLT]. *International Journal of Computer-Assisted Language Learning and Teaching*, *2*(3), 35–53. doi:10.4018/ijcallt.2012070103

Burke, M. E., & Speed, C. (2014). Knowledge Recovery: Applications of Technology and Memory. In M. Michael, & K. Michael (Eds.), *Uberveillance and the Social Implications of Microchip Implants: Emerging Technologies* (pp. 133–142). Hershey, PA: Information Science Reference.

Burton, A. M., Liu, H., Battersby, S., Brown, D., Sherkat, N., Standen, P., & Walker, M. (2014). The Use of Motion Tracking Technologies in Serious Games to Enhance Rehabilitation in Stroke Patients. In J. Bishop (Ed.), *Gamification for Human Factors Integration: Social, Education, and Psychological Issues* (pp. 148–161). Hershey, PA: Information Science Reference. doi:10.4018/978-1-4666-5071-8.ch009

Burusic, J., & Karabegovic, M. (2014). The Role of Students' Personality Traits in the Effective Use of Social Networking Sites in the Educational Context. In G. Mallia (Ed.), *The Social Classroom: Integrating Social Network Use in Education* (pp. 224–243). Hershey, PA: Information Science Reference.

Busch, C. D., Lorenzo, A. M., Sánchez, I. M., González, B. G., García, T. P., Riveiro, L. N., & Loureiro, J. P. (2014). In-TIC for Mobile Devices: Support System for Communication with Mobile Devices for the Disabled. In Assistive Technologies: Concepts, Methodologies, Tools, and Applications (pp. 345-356). Hershey, PA: Information Science Reference. doi: doi:10.4018/978-1-4666-4422-9.ch017

Bute, S. J. (2013). Integrating Social Media and Traditional Media within the Academic Environment. In M. Pătruţ, & B. Pătruţ (Eds.), *Social Media in Higher Education: Teaching in Web 2.0* (pp. 75–85). Hershey, PA: Information Science Reference. doi:10.4018/978-1-4666-2970-7.ch004

Butler-Pascoe, M. E. (2011). The History of CALL: The Intertwining Paths of Technology and Second/Foreign Language Teaching. [IJCALLT]. *International Journal of Computer-Assisted Language Learning and Teaching*, *1*(1), 16–32. doi:10.4018/ijcallt.2011010102

Cabrera, L. (2012). Human Implants: A Suggested Framework to Set Priorities. In R. Luppicini (Ed.), *Ethical Impact of Technological Advancements and Applications in Society* (pp. 243–253). Hershey, PA: Information Science Reference. doi:10.4018/978-1-4666-1773-5.ch019

Cacho-Elizondo, S., Shahidi, N., & Tossan, V. (2013). Intention to Adopt a Text Message-based Mobile Coaching Service to Help Stop Smoking: Which Explanatory Variables? [IJTHI]. *International Journal of Technology and Human Interaction*, *9*(4), 1–19. doi:10.4018/ijthi.2013100101

Caldelli, R., Becarelli, R., Filippini, F., Picchioni, F., & Giorgetti, R. (2014). Electronic Voting by Means of Digital Terrestrial Television: The Infrastructure, Security Issues and a Real Test-Bed. In Assistive Technologies: Concepts, Methodologies, Tools, and Applications (pp. 905-915). Hershey, PA: Information Science Reference. doi: doi:10.4018/978-1-4666-4422-9.ch045

Camacho, M. (2013). Making the Most of Informal and Situated Learning Opportunities through Mobile Learning. In M. Pătruţ, & B. Pătruţ (Eds.), *Social Media in Higher Education: Teaching in Web 2.0* (pp. 355–370). Hershey, PA: Information Science Reference. doi:10.4018/978-1-4666-2970-7.ch018

Camilleri, V., Busuttil, L., & Montebello, M. (2014). MOOCs: Exploiting Networks for the Education of the Masses or Just a Trend? In G. Mallia (Ed.), *The Social Classroom: Integrating Social Network Use in Education* (pp. 348–366). Hershey, PA: Information Science Reference.

Campos, P., Noronha, H., & Lopes, A. (2013). Work Analysis Methods in Practice: The Context of Collaborative Review of CAD Models. [IJSKD]. *International Journal of Sociotechnology and Knowledge Development*, *5*(2), 34–44. doi:10.4018/jskd.2013040103

Cao, G. (2013). A Paradox Between Technological Autonomy and Ethical Heteronomy of Philosophy of Technology: Social Control System. [IJT]. *International Journal of Technoethics*, *4*(1), 52–66. doi:10.4018/jte.2013010105

Carofiglio, V., & Abbattista, F. (2013). BCI-Based User-Centered Design for Emotionally-Driven User Experience. In M. Garcia-Ruiz (Ed.), *Cases on Usability Engineering: Design and Development of Digital Products* (pp. 299–320). Hershey, PA: Information Science Reference. doi:10.4018/978-1-4666-4046-7.ch013

Carpenter, J. (2013). Just Doesn't Look Right: Exploring the Impact of Humanoid Robot Integration into Explosive Ordnance Disposal Teams. In R. Luppicini (Ed.), *Handbook of Research on Technoself: Identity in a Technological Society* (pp. 609–636). Hershey, PA: Information Science Reference.

Carroll, J. L. (2014). Wheelchairs as Assistive Technology: What a Special Educator Should Know. In Assistive Technologies: Concepts, Methodologies, Tools, and Applications (pp. 623-633). Hershey, PA: Information Science Reference. doi: doi:10.4018/978-1-4666-4422-9.ch030

Casey, L. B., & Williamson, R. L. (2014). A Parent's Guide to Support Technologies for Preschool Students with Disabilities. In Assistive Technologies: Concepts, Methodologies, Tools, and Applications (pp. 1340-1356). Hershey, PA: Information Science Reference. doi: doi:10.4018/978-1-4666-4422-9.ch071

Caviglione, L., Coccoli, M., & Merlo, A. (2013). On Social Network Engineering for Secure Web Data and Services. In L. Caviglione, M. Coccoli, & A. Merlo (Eds.), *Social Network Engineering for Secure Web Data and Services* (pp. 1–4). Hershey, PA: Information Science Reference. doi:10.4018/978-1-4666-3926-3.ch001

Chadwick, D. D., Fullwood, C., & Wesson, C. J. (2014). Intellectual Disability, Identity, and the Internet. In Assistive Technologies: Concepts, Methodologies, Tools, and Applications (pp. 198-223). Hershey, PA: Information Science Reference. doi: doi:10.4018/978-1-4666-4422-9.ch011

Chao, L., Wen, Y., Chen, P., Lin, C., Lin, S., Guo, C., & Wang, W. (2012). The Development and Learning Effectiveness of a Teaching Module for the Algal Fuel Cell: A Renewable and Sustainable Battery. [IJTHI]. *International Journal of Technology and Human Interaction*, *8*(4), 1–15. doi:10.4018/jthi.2012100101

Charnkit, P., & Tatnall, A. (2013). Knowledge Conversion Processes in Thai Public Organisations Seen as an Innovation: The Re-Analysis of a TAM Study Using Innovation Translation. In A. Tatnall (Ed.), *Social and Professional Applications of Actor-Network Theory for Technology Development* (pp. 88–102). Hershey, PA: Information Science Reference.

Chen, E. T. (2014). Challenge and Complexity of Virtual Team Management. In E. Nikoi, & K. Boateng (Eds.), *Collaborative Communication Processes and Decision Making in Organizations* (pp. 109–120). Hershey, PA: Business Science Reference. doi:10.4018/978-1-4666-4979-8.ch062

Chen, R., Xie, T., Lin, T., & Chen, Y. (2013). Adaptive Windows Layout Based on Evolutionary Multi-Objective Optimization. [IJTHI]. *International Journal of Technology and Human Interaction*, *9*(3), 63–72. doi:10.4018/jthi.2013070105

Chen, W., Juang, Y., Chang, S., & Wang, P. (2012). Informal Education of Energy Conservation: Theory, Promotion, and Policy Implication. [IJTHI]. *International Journal of Technology and Human Interaction*, *8*(4), 16–44. doi:10.4018/jthi.2012100102

Chino, T., Torii, K., Uchihira, N., & Hirabayashi, Y. (2013). Speech Interaction Analysis on Collaborative Work at an Elderly Care Facility. [IJSKD]. *International Journal of Sociotechnology and Knowledge Development*, *5*(2), 18–33. doi:10.4018/jskd.2013040102

Chiu, M. (2013). Gaps Between Valuing and Purchasing Green-Technology Products: Product and Gender Differences. [IJTHI]. *International Journal of Technology and Human Interaction*, *8*(3), 54–68. doi:10.4018/jthi.2012070106

Chivukula, V., & Shur, M. (2014). Web-Based Experimentation for Students with Learning Disabilities. In Assistive Technologies: Concepts, Methodologies, Tools, and Applications (pp. 1156-1172). Hershey, PA: Information Science Reference. doi: doi:10.4018/978-1-4666-4422-9.ch060

Coakes, E., Bryant, A., Land, F., & Phippen, A. (2011). The Dark Side of Technology: Some Sociotechnical Reflections. [IJSKD]. *International Journal of Sociotechnology and Knowledge Development*, *3*(4), 40–51. doi:10.4018/IJSKD.2011100104

Cole, I. J. (2013). Usability of Online Virtual Learning Environments: Key Issues for Instructors and Learners. In C. Gonzalez (Ed.), *Student Usability in Educational Software and Games: Improving Experiences* (pp. 41–58). Hershey, PA: Information Science Reference.

Colombo, B., Antonietti, A., Sala, R., & Caravita, S. C. (2013). Blog Content and Structure, Cognitive Style and Metacognition. [IJTHI]. *International Journal of Technology and Human Interaction*, *9*(3), 1–17. doi:10.4018/jthi.2013070101

Constantinides, M. (2011). Integrating Technology on Initial Training Courses: A Survey Amongst CELTA Tutors. [IJCALLT]. *International Journal of Computer-Assisted Language Learning and Teaching*, *1*(2), 55–71. doi:10.4018/ijcallt.2011040105

Cook, R. G., & Crawford, C. M. (2013). Addressing Online Student Learning Environments and Socialization Through Developmental Research. In M. Khosrow-Pour (Ed.), *Cases on Assessment and Evaluation in Education* (pp. 504–536). Hershey, PA: Information Science Reference.

Corritore, C. L., Wiedenbeck, S., Kracher, B., & Marble, R. P. (2012). Online Trust and Health Information Websites. [IJTHI]. *International Journal of Technology and Human Interaction*, *8*(4), 92–115. doi:10.4018/jthi.2012100106

Covarrubias, M., Bordegoni, M., Cugini, U., & Gatti, E. (2014). Supporting Unskilled People in Manual Tasks through Haptic-Based Guidance. In Assistive Technologies: Concepts, Methodologies, Tools, and Applications (pp. 947-969). Hershey, PA: Information Science Reference. doi: doi:10.4018/978-1-4666-4422-9.ch048

Coverdale, T. S., & Wilbon, A. D. (2013). The Impact of In-Group Membership on e-Loyalty of Women Online Shoppers: An Application of the Social Identity Approach to Website Design. [IJEA]. *International Journal of E-Adoption*, *5*(1), 17–36. doi:10.4018/jea.2013010102

Crabb, P. B., & Stern, S. E. (2012). Technology Traps: Who Is Responsible? In R. Luppicini (Ed.), *Ethical Impact of Technological Advancements and Applications in Society* (pp. 39–46). Hershey, PA: Information Science Reference. doi:10.4018/978-1-4666-1773-5.ch004

Crespo, R. G., Martíne, O. S., Lovelle, J. M., García-Bustelo, B. C., Díaz, V. G., & Ordoñez de Pablos, P. (2014). Improving Cognitive Load on Students with Disabilities through Software Aids. In Assistive Technologies: Concepts, Methodologies, Tools, and Applications (pp. 1255-1268). Hershey, PA: Information Science Reference. doi: doi:10.4018/978-1-4666-4422-9.ch066

Croasdaile, S., Jones, S., Ligon, K., Oggel, L., & Pruett, M. (2014). Supports for and Barriers to Implementing Assistive Technology in Schools. In Assistive Technologies: Concepts, Methodologies, Tools, and Applications (pp. 1118-1130). Hershey, PA: Information Science Reference. doi: doi:10.4018/978-1-4666-4422-9.ch058

Cucchiarini, C., & Strik, H. (2014). Second Language Learners' Spoken Discourse: Practice and Corrective Feedback through Automatic Speech Recognition. In H. Lim, & F. Sudweeks (Eds.), *Innovative Methods and Technologies for Electronic Discourse Analysis* (pp. 169–189). Hershey, PA: Information Science Reference.

Dafoulas, G. A., & Saleeb, N. (2014). 3D Assistive Technologies and Advantageous Themes for Collaboration and Blended Learning of Users with Disabilities. In Assistive Technologies: Concepts, Methodologies, Tools, and Applications (pp. 421-453). Hershey, PA: Information Science Reference. doi: doi:10.4018/978-1-4666-4422-9.ch021

Dai, Z., & Paasch, K. (2013). A Web-Based Interactive Questionnaire for PV Application. [IJSKD]. *International Journal of Sociotechnology and Knowledge Development*, *5*(2), 82–93. doi:10.4018/jskd.2013040106

Daradoumis, T., & Lafuente, M. M. (2014). Studying the Suitability of Discourse Analysis Methods for Emotion Detection and Interpretation in Computer-Mediated Educational Discourse. In H. Lim, & F. Sudweeks (Eds.), *Innovative Methods and Technologies for Electronic Discourse Analysis* (pp. 119–143). Hershey, PA: Information Science Reference.

Davis, B., & Mason, P. (2014). Positioning Goes to Work: Computer-Aided Identification of Stance Shifts and Semantic Themes in Electronic Discourse Analysis. In H. Lim, & F. Sudweeks (Eds.), *Innovative Methods and Technologies for Electronic Discourse Analysis* (pp. 394–413). Hershey, PA: Information Science Reference.

Dogoriti, E., & Pange, J. (2014). Considerations for Online English Language Learning: The Use of Facebook in Formal and Informal Settings in Higher Education. In G. Mallia (Ed.), *The Social Classroom: Integrating Social Network Use in Education* (pp. 147–170). Hershey, PA: Information Science Reference.

Donegan, M. (2014). Features of Gaze Control Systems. In Assistive Technologies: Concepts, Methodologies, Tools, and Applications (pp. 1055-1061). Hershey, PA: Information Science Reference. doi: doi:10.4018/978-1-4666-4422-9.ch054

Douglas, G., Morton, H., & Jack, M. (2012). Remote Channel Customer Contact Strategies for Complaint Update Messages. [IJTHI]. *International Journal of Technology and Human Interaction*, *8*(2), 43–55. doi:10.4018/jthi.2012040103

Drake, J. R., & Byrd, T. A. (2013). Searching for Alternatives: Does Your Disposition Matter? [IJTHI]. *International Journal of Technology and Human Interaction*, *9*(1), 18–36. doi:10.4018/jthi.2013010102

Driouchi, A. (2013). ICTs and Socioeconomic Performance with Focus on ICTs and Health. In ICTs for Health, Education, and Socioeconomic Policies: Regional Cases (pp. 104-125). Hershey, PA: Information Science Reference. doi: doi:10.4018/978-1-4666-3643-9.ch005

Driouchi, A. (2013). Social Deficits, Social Cohesion, and Prospects from ICTs. In ICTs for Health, Education, and Socioeconomic Policies: Regional Cases (pp. 230-251). Hershey, PA: Information Science Reference. doi: doi:10.4018/978-1-4666-3643-9.ch011

Driouchi, A. (2013). Socioeconomic Reforms, Human Development, and the Millennium Development Goals with ICTs for Coordination. In ICTs for Health, Education, and Socioeconomic Policies: Regional Cases (pp. 211-229). Hershey, PA: Information Science Reference. doi: doi:10.4018/978-1-4666-3643-9.ch010

Drula, G. (2013). Media and Communication Research Facing Social Media. In M. Pătruţ, & B. Pătruţ (Eds.), *Social Media in Higher Education: Teaching in Web 2.0* (pp. 371–392). Hershey, PA: Information Science Reference. doi:10.4018/978-1-4666-2970-7.ch019

Druzhinina, O., Hvannberg, E. T., & Halldorsdottir, G. (2013). Feedback Fidelities in Three Different Types of Crisis Management Training Environments. [IJSKD]. *International Journal of Sociotechnology and Knowledge Development*, *5*(2), 45–62. doi:10.4018/jskd.2013040104

Eason, K., Waterson, P., & Davda, P. (2013). The Sociotechnical Challenge of Integrating Telehealth and Telecare into Health and Social Care for the Elderly. [IJSKD]. *International Journal of Sociotechnology and Knowledge Development*, *5*(4), 14–26. doi:10.4018/ijskd.2013100102

Edenius, M., & Rämö, H. (2011). An Office on the Go: Professional Workers, Smartphones and the Return of Place. [IJTHI]. *International Journal of Technology and Human Interaction*, *7*(1), 37–55. doi:10.4018/jthi.2011010103

Eke, D. O. (2011). ICT Integration in Nigeria: The Socio-Cultural Constraints. [IJTHI]. *International Journal of Technology and Human Interaction*, *7*(2), 21–27. doi:10.4018/jthi.2011040103

Evett, L., Ridley, A., Keating, L., Merritt, P., Shopland, N., & Brown, D. (2014). Designing Serious Games for People with Disabilities: Game, Set, and Match to the Wii. In J. Bishop (Ed.), *Gamification for Human Factors Integration: Social, Education, and Psychological Issues* (pp. 97–105). Hershey, PA: Information Science Reference. doi:10.4018/978-1-4666-5071-8.ch006

Evmenova, A. S., & Behrmann, M. M. (2014). Communication Technology Integration in the Content Areas for Students with High-Incidence Disabilities: A Case Study of One School System. In Assistive Technologies: Concepts, Methodologies, Tools, and Applications (pp. 26-53). Hershey, PA: Information Science Reference. doi: doi:10.4018/978-1-4666-4422-9.ch003

Evmenova, A. S., & King-Sears, M. E. (2014). Technology and Literacy for Students with Disabilities. In Assistive Technologies: Concepts, Methodologies, Tools, and Applications (pp. 1269-1291). Hershey, PA: Information Science Reference. doi: doi:10.4018/978-1-4666-4422-9.ch067

Ewais, A., & De Troyer, O. (2013). Usability Evaluation of an Adaptive 3D Virtual Learning Environment. [IJVPLE]. *International Journal of Virtual and Personal Learning Environments*, *4*(1), 16–31. doi:10.4018/jvple.2013010102

Farrell, H. J. (2014). The Student with Complex Education Needs: Assistive and Augmentative Information and Communication Technology in a Ten-Week Music Program. In K-12 Education: Concepts, Methodologies, Tools, and Applications (pp. 1436-1472). Hershey, PA: Information Science Reference. doi: doi:10.4018/978-1-4666-4502-8.ch084

Fathulla, K. (2012). Rethinking Human and Society's Relationship with Technology. [IJSKD]. *International Journal of Sociotechnology and Knowledge Development*, *4*(2), 21–28. doi:10.4018/jskd.2012040103

Fidler, C. S., Kanaan, R. K., & Rogerson, S. (2011). Barriers to e-Government Implementation in Jordan: The Role of Wasta. [IJTHI]. *International Journal of Technology and Human Interaction, 7*(2), 9–20. doi:10.4018/jthi.2011040102

Fischer, G., & Herrmann, T. (2013). Socio-Technical Systems: A Meta-Design Perspective. In J. Abdelnour-Nocera (Ed.), *Knowledge and Technological Development Effects on Organizational and Social Structures* (pp. 1–36). Hershey, PA: Information Science Reference.

Foreman, J., & Borkman, T. (2014). Learning Sociology in a Massively Multi-Student Online Learning Environment. In J. Bishop (Ed.), *Gamification for Human Factors Integration: Social, Education, and Psychological Issues* (pp. 216–224). Hershey, PA: Information Science Reference. doi:10.4018/978-1-4666-5071-8.ch013

Fornaciari, F. (2013). The Language of Technoself: Storytelling, Symbolic Interactionism, and Online Identity. In R. Luppicini (Ed.), *Handbook of Research on Technoself: Identity in a Technological Society* (pp. 64–83). Hershey, PA: Information Science Reference.

Fox, J., & Ahn, S. J. (2013). Avatars: Portraying, Exploring, and Changing Online and Offline Identities. In R. Luppicini (Ed.), *Handbook of Research on Technoself: Identity in a Technological Society* (pp. 255–271). Hershey, PA: Information Science Reference.

Fox, W. P., Binstock, J., & Minutas, M. (2013). Modeling and Methodology for Incorporating Existing Technologies to Produce Higher Probabilities of Detecting Suicide Bombers. [IJORIS]. *International Journal of Operations Research and Information Systems, 4*(3), 1–18. doi:10.4018/joris.2013070101·

Franchi, E., & Tomaiuolo, M. (2013). Distributed Social Platforms for Confidentiality and Resilience. In L. Caviglione, M. Coccoli, & A. Merlo (Eds.), *Social Network Engineering for Secure Web Data and Services* (pp. 114–136). Hershey, PA: Information Science Reference. doi:10.4018/978-1-4666-3926-3.ch006

Frigo, C. A., & Pavan, E. E. (2014). Prosthetic and Orthotic Devices. In Assistive Technologies: Concepts, Methodologies, Tools, and Applications (pp. 549-613). Hershey, PA: Information Science Reference. doi: doi:10.4018/978-1-4666-4422-9.ch028

Fuhrer, C., & Cucchi, A. (2012). Relations Between Social Capital and Use of ICT: A Social Network Analysis Approach. [IJTHI]. *International Journal of Technology and Human Interaction, 8*(2), 15–42. doi:10.4018/jthi.2012040102

Galinski, C., & Beckmann, H. (2014). Concepts for Enhancing Content Quality and eAccessibility: In General and in the Field of eProcurement. In Assistive Technologies: Concepts, Methodologies, Tools, and Applications (pp. 180-197). Hershey, PA: Information Science Reference. doi: doi:10.4018/978-1-4666-4422-9.ch010

Galván, J. M., & Luppicini, R. (2012). The Humanity of the Human Body: Is Homo Cybersapien a New Species? [IJT]. *International Journal of Technoethics, 3*(2), 1–8. doi:10.4018/jte.2012040101

García-Gómez, A. (2013). Technoself-Presentation on Social Networks: A Gender-Based Approach. In R. Luppicini (Ed.), *Handbook of Research on Technoself: Identity in a Technological Society* (pp. 382–398). Hershey, PA: Information Science Reference.

Gill, L., Hathway, E. A., Lange, E., Morgan, E., & Romano, D. (2013). Coupling Real-Time 3D Landscape Models with Microclimate Simulations. [IJEPR]. *International Journal of E-Planning Research, 2*(1), 1–19. doi:10.4018/ijepr.2013010101

Godé, C., & Lebraty, J. (2013). Improving Decision Making in Extreme Situations: The Case of a Military Decision Support System. [IJTHI]. *International Journal of Technology and Human Interaction, 9*(1), 1–17. doi:10.4018/jthi.2013010101

Griol, D., Callejas, Z., & López-Cózar, R. (2014). Conversational Metabots for Educational Applications in Virtual Worlds. In Assistive Technologies: Concepts, Methodologies, Tools, and Applications (pp. 1405-1433). Hershey, PA: Information Science Reference. doi: doi:10.4018/978-1-4666-4422-9.ch074

Griol Barres, D., Callejas Carrión, Z., Molina López, J. M., & Sanchis de Miguel, A. (2014). Towards the Use of Dialog Systems to Facilitate Inclusive Education. In Assistive Technologies: Concepts, Methodologies, Tools, and Applications (pp. 1292-1312). Hershey, PA: Information Science Reference. doi: doi:10.4018/978-1-4666-4422-9.ch068

Groba, B., Pousada, T., & Nieto, L. (2014). Assistive Technologies, Tools and Resources for the Access and Use of Information and Communication Technologies by People with Disabilities. In Assistive Technologies: Concepts, Methodologies, Tools, and Applications (pp. 246-260). Hershey, PA: Information Science Reference. doi: doi:10.4018/978-1-4666-4422-9.ch013

Groß, M. (2013). Personal Knowledge Management and Social Media: What Students Need to Learn for Business Life. In M. Pătruţ, & B. Pătruţ (Eds.), *Social Media in Higher Education: Teaching in Web 2.0* (pp. 124–143). Hershey, PA: Information Science Reference. doi:10.4018/978-1-4666-2970-7.ch007

Gu, L., Aiken, M., Wang, J., & Wibowo, K. (2011). The Influence of Information Control upon Online Shopping Behavior. [IJTHI]. *International Journal of Technology and Human Interaction, 7*(1), 56–66. doi:10.4018/jthi.2011010104

Hainz, T. (2012). Value Lexicality and Human Enhancement. [IJT]. *International Journal of Technoethics, 3*(4), 54–65. doi:10.4018/jte.2012100105

Harnesk, D., & Lindström, J. (2014). Exploring Socio-Technical Design of Crisis Management Information Systems. In Crisis Management: Concepts, Methodologies, Tools and Applications (pp. 514-530). Hershey, PA: Information Science Reference. doi: doi:10.4018/978-1-4666-4707-7.ch023

Hicks, D. (2014). Ethics in the Age of Technological Change and its Impact on the Professional Identity of Librarians. In *Technology and Professional Identity of Librarians: The Making of the Cybrarian* (pp. 168–187). Hershey, PA: Information Science Reference.

Hicks, D. (2014). Technology, Profession, Identity. In *Technology and Professional Identity of Librarians: The Making of the Cybrarian* (pp. 1–20). Hershey, PA: Information Science Reference.

Hirata, M., Yanagisawa, T., Matsushita, K., Sugata, H., Kamitani, Y., Suzuki, T., et al. (2014). Brain-Machine Interface Using Brain Surface Electrodes: Real-Time Robotic Control and a Fully Implantable Wireless System. In Assistive Technologies: Concepts, Methodologies, Tools, and Applications (pp. 1535-1548). Hershey, PA: Information Science Reference. doi: doi:10.4018/978-1-4666-4422-9.ch080

Hodge, B. (2014). Critical Electronic Discourse Analysis: Social and Cultural Research in the Electronic Age. In H. Lim, & F. Sudweeks (Eds.), *Innovative Methods and Technologies for Electronic Discourse Analysis* (pp. 191–209). Hershey, PA: Information Science Reference.

Hoey, J., Poupart, P., Boutilier, C., & Mihailidis, A. (2014). POMDP Models for Assistive Technology. In Assistive Technologies: Concepts, Methodologies, Tools, and Applications (pp. 120-140). Hershey, PA: Information Science Reference. doi: doi:10.4018/978-1-4666-4422-9.ch007

Hogg, S. (2014). An Informal Use of Facebook to Encourage Student Collaboration and Motivation for Off Campus Activities. In G. Mallia (Ed.), *The Social Classroom: Integrating Social Network Use in Education* (pp. 23–39). Hershey, PA: Information Science Reference.

Holmqvist, E., & Buchholz, M. (2014). A Model for Gaze Control Assessments and Evaluation. In Assistive Technologies: Concepts, Methodologies, Tools, and Applications (pp. 332-343). Hershey, PA: Information Science Reference. doi: doi:10.4018/978-1-4666-4422-9.ch016

Hsiao, S., Chen, D., Yang, C., Huang, H., Lu, Y., & Huang, H. et al. (2013). Chemical-Free and Reusable Cellular Analysis: Electrochemical Impedance Spectroscopy with a Transparent ITO Culture Chip. [IJTHI]. *International Journal of Technology and Human Interaction*, *8*(3), 1–9. doi:10.4018/jthi.2012070101

Hsu, M., Yang, C., Wang, C., & Lin, Y. (2013). Simulation-Aided Optimal Microfluidic Sorting for Monodispersed Microparticles. [IJTHI]. *International Journal of Technology and Human Interaction*, *8*(3), 10–18. doi:10.4018/jthi.2012070102

Huang, W. D., & Tettegah, S. Y. (2014). Cognitive Load and Empathy in Serious Games: A Conceptual Framework. In J. Bishop (Ed.), *Gamification for Human Factors Integration: Social, Education, and Psychological Issues* (pp. 17–30). Hershey, PA: Information Science Reference. doi:10.4018/978-1-4666-5071-8.ch002

Huseyinov, I. N. (2014). Fuzzy Linguistic Modelling in Multi Modal Human Computer Interaction: Adaptation to Cognitive Styles using Multi Level Fuzzy Granulation Method. In Assistive Technologies: Concepts, Methodologies, Tools, and Applications (pp. 1481-1496). Hershey, PA: Information Science Reference. doi: doi:10.4018/978-1-4666-4422-9.ch077

Hwa, S. P., Weei, P. S., & Len, L. H. (2012). The Effects of Blended Learning Approach through an Interactive Multimedia E-Book on Students' Achievement in Learning Chinese as a Second Language at Tertiary Level. [IJCALLT]. *International Journal of Computer-Assisted Language Learning and Teaching*, *2*(1), 35–50. doi:10.4018/ijcallt.2012010104

Iglesias, A., Ruiz-Mezcua, B., López, J. F., & Figueroa, D. C. (2014). New Communication Technologies for Inclusive Education in and outside the Classroom. In Assistive Technologies: Concepts, Methodologies, Tools, and Applications (pp. 1675-1689). Hershey, PA: Information Science Reference. doi: doi:10.4018/978-1-4666-4422-9.ch088

Inghilterra, X., & Ravatua-Smith, W. S. (2014). Online Learning Communities: Use of Micro Blogging for Knowledge Construction. In J. Pelet (Ed.), *E-Learning 2.0 Technologies and Web Applications in Higher Education* (pp. 107–128). Hershey, PA: Information Science Reference.

Ionescu, A. (2013). Cyber Identity: Our Alter-Ego? In R. Luppicini (Ed.), *Handbook of Research on Technoself: Identity in a Technological Society* (pp. 189–203). Hershey, PA: Information Science Reference.

Jan, Y., Lin, M., Shiao, K., Wei, C., Huang, L., & Sung, Q. (2013). Development of an Evaluation Instrument for Green Building Literacy among College Students in Taiwan. [IJTHI]. *International Journal of Technology and Human Interaction*, *8*(3), 31–45. doi:10.4018/jthi.2012070104

Jawadi, N. (2013). E-Leadership and Trust Management: Exploring the Moderating Effects of Team Virtuality. [IJTHI]. *International Journal of Technology and Human Interaction*, *9*(3), 18–35. doi:10.4018/jthi.2013070102

Jiménez-Castillo, D., & Fernández, R. S. (2014). The Impact of Combining Video Podcasting and Lectures on Students' Assimilation of Additional Knowledge: An Empirical Examination. In J. Pelet (Ed.), *E-Learning 2.0 Technologies and Web Applications in Higher Education* (pp. 65–87). Hershey, PA: Information Science Reference.

Jin, L. (2013). A New Trend in Education: Technoself Enhanced Social Learning. In R. Luppicini (Ed.), *Handbook of Research on Technoself: Identity in a Technological Society* (pp. 456–473). Hershey, PA: Information Science Reference.

Johansson, L. (2012). The Functional Morality of Robots. In R. Luppicini (Ed.), *Ethical Impact of Technological Advancements and Applications in Society* (pp. 254–262). Hershey, PA: Information Science Reference. doi:10.4018/978-1-4666-1773-5.ch020

Johansson, L. (2013). Robots and the Ethics of Care. [IJT]. *International Journal of Technoethics, 4*(1), 67–82. doi:10.4018/jte.2013010106

Johri, A., Dufour, M., Lo, J., & Shanahan, D. (2013). Adwiki: Socio-Technical Design for Mananging Advising Knowledge in a Higher Education Context. [IJSKD]. *International Journal of Sociotechnology and Knowledge Development, 5*(1), 37–59. doi:10.4018/jskd.2013010104

Jones, M. G., Schwilk, C. L., & Bateman, D. F. (2014). Reading by Listening: Access to Books in Audio Format for College Students with Print Disabilities. In Assistive Technologies: Concepts, Methodologies, Tools, and Applications (pp. 454-477). Hershey, PA: Information Science Reference. doi: doi:10.4018/978-1-4666-4422-9.ch022

Kaba, B., & Osei-Bryson, K. (2012). An Empirical Investigation of External Factors Influencing Mobile Technology Use in Canada: A Preliminary Study. [IJTHI]. *International Journal of Technology and Human Interaction, 8*(2), 1–14. doi:10.4018/jthi.2012040101

Kampf, C. E. (2012). Revealing the Socio-Technical Design of Global E-Businesses: A Case of Digital Artists Engaging in Radical Transparency. [IJSKD]. *International Journal of Sociotechnology and Knowledge Development, 4*(4), 18–31. doi:10.4018/jskd.2012100102

Kandroudi, M., & Bratitsis, T. (2014). Classifying Facebook Usage in the Classroom or Around It. In G. Mallia (Ed.), *The Social Classroom: Integrating Social Network Use in Education* (pp. 62–81). Hershey, PA: Information Science Reference.

Kidd, P. T. (2014). Social Networking Technologies as a Strategic Tool for the Development of Sustainable Production and Consumption: Applications to Foster the Agility Needed to Adapt Business Models in Response to the Challenges Posed by Climate Change. In Sustainable Practices: Concepts, Methodologies, Tools and Applications (pp. 974-987). Hershey, PA: Information Science Reference. doi: doi:10.4018/978-1-4666-4852-4.ch054

Kirby, S. D., & Sellers, D. M. (2014). The Live-Ability House: A Collaborative Adventure in Discovery Learning. In Assistive Technologies: Concepts, Methodologies, Tools, and Applications (pp. 1626-1649). Hershey, PA: Information Science Reference. doi: doi:10.4018/978-1-4666-4422-9.ch086

Kitchenham, A., & Bowes, D. (2014). Voice/Speech Recognition Software: A Discussion of the Promise for Success and Practical Suggestions for Implementation. In Assistive Technologies: Concepts, Methodologies, Tools, and Applications (pp. 1005-1011). Hershey, PA: Information Science Reference. doi: doi:10.4018/978-1-4666-4422-9.ch051

Konrath, S. (2013). The Empathy Paradox: Increasing Disconnection in the Age of Increasing Connection. In R. Luppicini (Ed.), *Handbook of Research on Technoself: Identity in a Technological Society* (pp. 204–228). Hershey, PA: Information Science Reference.

Koutsabasis, P., & Istikopoulou, T. G. (2013). Perceived Website Aesthetics by Users and Designers: Implications for Evaluation Practice. [IJTHI]. *International Journal of Technology and Human Interaction*, *9*(2), 39–52. doi:10.4018/jthi.2013040103

Kraft, E., & Wang, J. (2012). An Exploratory Study of the Cyberbullying and Cyberstalking Experiences and Factors Related to Victimization of Students at a Public Liberal Arts College. In R. Luppicini (Ed.), *Ethical Impact of Technological Advancements and Applications in Society* (pp. 113–131). Hershey, PA: Information Science Reference. doi:10.4018/978-1-4666-1773-5.ch009

Kulman, R., Stoner, G., Ruffolo, L., Marshall, S., Slater, J., Dyl, A., & Cheng, A. (2014). Teaching Executive Functions, Self-Management, and Ethical Decision-Making through Popular Videogame Play. In Assistive Technologies: Concepts, Methodologies, Tools, and Applications (pp. 771-785). Hershey, PA: Information Science Reference. doi: doi:10.4018/978-1-4666-4422-9.ch039

Kunc, L., Míkovec, Z., & Slavík, P. (2013). Avatar and Dialog Turn-Yielding Phenomena. [IJTHI]. *International Journal of Technology and Human Interaction*, *9*(2), 66–88. doi:10.4018/jthi.2013040105

Kuo, N., & Dai, Y. (2012). Applying the Theory of Planned Behavior to Predict Low-Carbon Tourism Behavior: A Modified Model from Taiwan. [IJTHI]. *International Journal of Technology and Human Interaction*, *8*(4), 45–62. doi:10.4018/jthi.2012100103

Kurt, S. (2014). Accessibility Issues of Educational Web Sites. In Assistive Technologies: Concepts, Methodologies, Tools, and Applications (pp. 54-62). Hershey, PA: Information Science Reference. doi: doi:10.4018/978-1-4666-4422-9.ch004

Kuzma, J. (2013). Empirical Study of Cyber Harassment among Social Networks. [IJTHI]. *International Journal of Technology and Human Interaction*, *9*(2), 53–65. doi:10.4018/jthi.2013040104

Kyriakaki, G., & Matsatsinis, N. (2014). Pedagogical Evaluation of E-Learning Websites with Cognitive Objectives. In D. Yannacopoulos, P. Manolitzas, N. Matsatsinis, & E. Grigoroudis (Eds.), *Evaluating Websites and Web Services: Interdisciplinary Perspectives on User Satisfaction* (pp. 224–240). Hershey, PA: Information Science Reference. doi:10.4018/978-1-4666-5129-6.ch013

Lee, H., & Baek, E. (2012). Facilitating Deep Learning in a Learning Community. [IJTHI]. *International Journal of Technology and Human Interaction*, *8*(1), 1–13. doi:10.4018/jthi.2012010101

Lee, W., Wu, T., Cheng, Y., Chuang, Y., & Sheu, S. (2013). Using the Kalman Filter for Auto Bit-rate H.264 Streaming Based on Human Interaction. [IJTHI]. *International Journal of Technology and Human Interaction*, *9*(4), 58–74. doi:10.4018/ijthi.2013100104

Li, Y., Guo, N. Y., & Ranieri, M. (2014). Designing an Online Interactive Learning Program to Improve Chinese Migrant Children's Internet Skills: A Case Study at Hangzhou Minzhu Experimental School. In Z. Yang, H. Yang, D. Wu, & S. Liu (Eds.), *Transforming K-12 Classrooms with Digital Technology* (pp. 249–265). Hershey, PA: Information Science Reference.

Lin, C., Chu, L., & Hsu, H. (2013). Study on the Performance and Exhaust Emissions of Motorcycle Engine Fuelled with Hydrogen-Gasoline Compound Fuel. [IJTHI]. *International Journal of Technology and Human Interaction*, *8*(3), 69–81. doi:10.4018/jthi.2012070107

Lin, L. (2013). Multiple Dimensions of Multitasking Phenomenon. [IJTHI]. *International Journal of Technology and Human Interaction*, *9*(1), 37–49. doi:10.4018/jthi.2013010103

Lin, T., Li, X., Wu, Z., & Tang, N. (2013). Automatic Cognitive Load Classification Using High-Frequency Interaction Events: An Exploratory Study. [IJTHI]. *International Journal of Technology and Human Interaction*, *9*(3), 73–88. doi:10.4018/jthi.2013070106

Lin, T., Wu, Z., Tang, N., & Wu, S. (2013). Exploring the Effects of Display Characteristics on Presence and Emotional Responses of Game Players. [IJTHI]. *International Journal of Technology and Human Interaction*, *9*(1), 50–63. doi:10.4018/jthi.2013010104

Lin, T., Xie, T., Mou, Y., & Tang, N. (2013). Markov Chain Models for Menu Item Prediction. [IJTHI]. *International Journal of Technology and Human Interaction*, *9*(4), 75–94. doi:10.4018/ijthi.2013100105

Lin, X., & Luppicini, R. (2011). Socio-Technical Influences of Cyber Espionage: A Case Study of the GhostNet System. [IJT]. *International Journal of Technoethics*, *2*(2), 65–77. doi:10.4018/jte.2011040105

Linek, S. B., Marte, B., & Albert, D. (2014). Background Music in Educational Games: Motivational Appeal and Cognitive Impact. In J. Bishop (Ed.), *Gamification for Human Factors Integration: Social, Education, and Psychological Issues* (pp. 259–271). Hershey, PA: Information Science Reference. doi:10.4018/978-1-4666-5071-8.ch016

Lipschutz, R. D., & Hester, R. J. (2014). We Are the Borg! Human Assimilation into Cellular Society. In M. Michael, & K. Michael (Eds.), *Uberveillance and the Social Implications of Microchip Implants: Emerging Technologies* (pp. 366–407). Hershey, PA: Information Science Reference.

Liu, C., Zhong, Y., Ozercan, S., & Zhu, Q. (2013). Facilitating 3D Virtual World Learning Environments Creation by Non-Technical End Users through Template-Based Virtual World Instantiation. [IJVPLE]. *International Journal of Virtual and Personal Learning Environments*, *4*(1), 32–48. doi:10.4018/jvple.2013010103

Liu, F., Lo, H., Su, C., Lou, D., & Lee, W. (2013). High Performance Reversible Data Hiding for Mobile Applications and Human Interaction. [IJTHI]. *International Journal of Technology and Human Interaction*, *9*(4), 41–57. doi:10.4018/ijthi.2013100103

Liu, H. (2012). From Cold War Island to Low Carbon Island: A Study of Kinmen Island. [IJTHI]. *International Journal of Technology and Human Interaction*, *8*(4), 63–74. doi:10.4018/jthi.2012100104

Lixun, Z., Dapeng, B., & Lei, Y. (2014). Design of and Experimentation with a Walking Assistance Robot. In Assistive Technologies: Concepts, Methodologies, Tools, and Applications (pp. 1600-1605). Hershey, PA: Information Science Reference. doi: doi:10.4018/978-1-4666-4422-9.ch084

Low, R., Jin, P., & Sweller, J. (2014). Instructional Design in Digital Environments and Availability of Mental Resources for the Aged Subpopulation. In Assistive Technologies: Concepts, Methodologies, Tools, and Applications (pp. 1131-1154). Hershey, PA: Information Science Reference. doi: doi:10.4018/978-1-4666-4422-9.ch059

Luczak, H., Schlick, C. M., Jochems, N., Vetter, S., & Kausch, B. (2014). Touch Screens for the Elderly: Some Models and Methods, Prototypical Development and Experimental Evaluation of Human-Computer Interaction Concepts for the Elderly. In Assistive Technologies: Concepts, Methodologies, Tools, and Applications (pp. 377-396). Hershey, PA: Information Science Reference. doi: doi:10.4018/978-1-4666-4422-9.ch019

Luor, T., Lu, H., Johanson, R. E., & Yu, H. (2012). Minding the Gap Between First and Continued Usage of a Corporate E-Learning English-language Program. [IJTHI]. *International Journal of Technology and Human Interaction, 8*(1), 55–74. doi:10.4018/jthi.2012010104

Luppicini, R. (2013). The Emerging Field of Technoself Studies (TSS). In R. Luppicini (Ed.), *Handbook of Research on Technoself: Identity in a Technological Society* (pp. 1–25). Hershey, PA: Information Science Reference.

Magnani, L. (2012). Material Cultures and Moral Mediators in Human Hybridization. In R. Luppicini (Ed.), *Ethical Impact of Technological Advancements and Applications in Society* (pp. 1–20). Hershey, PA: Information Science Reference. doi:10.4018/978-1-4666-1773-5.ch001

Maher, D. (2014). Learning in the Primary School Classroom using the Interactive Whiteboard. In K-12 Education: Concepts, Methodologies, Tools, and Applications (pp. 526-538). Hershey, PA: Information Science Reference. doi: doi:10.4018/978-1-4666-4502-8.ch031

Manolache, M., & Patrut, M. (2013). The Use of New Web-Based Technologies in Strategies of Teaching Gender Studies. In M. Pătruț, & B. Pătruţ (Eds.), *Social Media in Higher Education: Teaching in Web 2.0* (pp. 45–74). Hershey, PA: Information Science Reference. doi:10.4018/978-1-4666-2970-7.ch003

Manthiou, A., & Chiang, L., & Liang (Rebecca) Tang. (2013). Identifying and Responding to Customer Needs on Facebook Fan Pages. [IJTHI]. *International Journal of Technology and Human Interaction, 9*(3), 36–52. doi:10.4018/jthi.2013070103

Marengo, A., Pagano, A., & Barbone, A. (2013). An Assessment of Customer's Preferences and Improve Brand Awareness Implementation of Social CRM in an Automotive Company. [IJTD]. *International Journal of Technology Diffusion, 4*(1), 1–15. doi:10.4018/jtd.2013010101

Martin, I., Kear, K., Simpkins, N., & Busvine, J. (2013). Social Negotiations in Web Usability Engineering. In M. Garcia-Ruiz (Ed.), *Cases on Usability Engineering: Design and Development of Digital Products* (pp. 26–56). Hershey, PA: Information Science Reference. doi:10.4018/978-1-4666-4046-7.ch002

Martins, T., Carvalho, V., & Soares, F. (2014). An Overview on the Use of Serious Games in Physical Therapy and Rehabilitation. In Assistive Technologies: Concepts, Methodologies, Tools, and Applications (pp. 758-770). Hershey, PA: Information Science Reference. doi: doi:10.4018/978-1-4666-4422-9.ch038

Mathew, D. (2013). Online Anxiety: Implications for Educational Design in a Web 2.0 World. In M. Pătruţ, & B. Pătruţ (Eds.), *Social Media in Higher Education: Teaching in Web 2.0* (pp. 305–317). Hershey, PA: Information Science Reference. doi:10.4018/978-1-4666-2970-7.ch015

Mazzanti, I., Maolo, A., & Antonicelli, R. (2014). E-Health and Telemedicine in the Elderly: State of the Art. In Assistive Technologies: Concepts, Methodologies, Tools, and Applications (pp. 693-704). Hershey, PA: Information Science Reference. doi: doi:10.4018/978-1-4666-4422-9.ch034

Mazzara, M., Biselli, L., Greco, P. P., Dragoni, N., Marraffa, A., Qamar, N., & de Nicola, S. (2013). Social Networks and Collective Intelligence: A Return to the Agora. In L. Caviglione, M. Coccoli, & A. Merlo (Eds.), *Social Network Engineering for Secure Web Data and Services* (pp. 88–113). Hershey, PA: Information Science Reference. doi:10.4018/978-1-4666-3926-3.ch005

McColl, D., & Nejat, G. (2013). A Human Affect Recognition System for Socially Interactive Robots. In R. Luppicini (Ed.), *Handbook of Research on Technoself: Identity in a Technological Society* (pp. 554–573). Hershey, PA: Information Science Reference. doi:10.4018/978-1-4666-4607-0.ch015

McDonald, A., & Helmer, S. (2011). A Comparative Case Study of Indonesian and UK Organisational Culture Differences in IS Project Management. [IJTHI]. *International Journal of Technology and Human Interaction*, *7*(2), 28–37. doi:10.4018/jthi.2011040104

McGee, E. M. (2014). Neuroethics and Implanted Brain Machine Interfaces. In M. Michael, & K. Michael (Eds.), *Uberveillance and the Social Implications of Microchip Implants: Emerging Technologies* (pp. 351–365). Hershey, PA: Information Science Reference.

McGrath, E., Lowes, S., McKay, M., Sayres, J., & Lin, P. (2014). Robots Underwater! Learning Science, Engineering and 21st Century Skills: The Evolution of Curricula, Professional Development and Research in Formal and Informal Contexts. In K-12 Education: Concepts, Methodologies, Tools, and Applications (pp. 1041-1067). Hershey, PA: Information Science Reference. doi: doi:10.4018/978-1-4666-4502-8.ch062

Meissonierm, R., Bourdon, I., Amabile, S., & Boudrandi, S. (2012). Toward an Enacted Approach to Understanding OSS Developer's Motivations. [IJTHI]. *International Journal of Technology and Human Interaction*, *8*(1), 38–54. doi:10.4018/jthi.2012010103

Melius, J. (2014). The Role of Social Constructivist Instructional Approaches in Facilitating Cross-Cultural Online Learning in Higher Education. In J. Keengwe, G. Schnellert, & K. Kungu (Eds.), *Cross-Cultural Online Learning in Higher Education and Corporate Training* (pp. 253–270). Hershey, PA: Information Science Reference. doi:10.4018/978-1-4666-5023-7.ch015

Melson, G. F. (2013). Building a Technoself: Children's Ideas about and Behavior toward Robotic Pets. In R. Luppicini (Ed.), *Handbook of Research on Technoself: Identity in a Technological Society* (pp. 592–608). Hershey, PA: Information Science Reference. doi:10.4018/978-1-4666-4607-0.ch068

Mena, R. J. (2014). The Quest for a Massively Multiplayer Online Game that Teaches Physics. In T. Connolly, T. Hainey, E. Boyle, G. Baxter, & P. Moreno-Ger (Eds.), *Psychology, Pedagogy, and Assessment in Serious Games* (pp. 292–316). Hershey, PA: Information Science Reference.

Meredith, J., & Potter, J. (2014). Conversation Analysis and Electronic Interactions: Methodological, Analytic and Technical Considerations. In H. Lim, & F. Sudweeks (Eds.), *Innovative Methods and Technologies for Electronic Discourse Analysis* (pp. 370–393). Hershey, PA: Information Science Reference.

Millán-Calenti, J. C., & Maseda, A. (2014). Telegerontology®: A New Technological Resource for Elderly Support. In Assistive Technologies: Concepts, Methodologies, Tools, and Applications (pp. 705-719). Hershey, PA: Information Science Reference. doi: doi:10.4018/978-1-4666-4422-9.ch035

Miscione, G. (2011). Telemedicine and Development: Situating Information Technologies in the Amazon. [IJSKD]. *International Journal of Sociotechnology and Knowledge Development, 3*(4), 15–26. doi:10.4018/jskd.2011100102

Miwa, N., & Wang, Y. (2011). Online Interaction Between On-Campus and Distance Students: Learners' Perspectives. [IJCALLT]. *International Journal of Computer-Assisted Language Learning and Teaching, 1*(3), 54–69. doi:10.4018/ijcallt.2011070104

Moore, M. J., Nakano, T., Suda, T., & Enomoto, A. (2013). Social Interactions and Automated Detection Tools in Cyberbullying. In L. Caviglione, M. Coccoli, & A. Merlo (Eds.), *Social Network Engineering for Secure Web Data and Services* (pp. 67–87). Hershey, PA: Information Science Reference. doi:10.4018/978-1-4666-3926-3.ch004

Morueta, R. T., Gómez, J. I., & Gómez, Á. H. (2012). B-Learning at Universities in Andalusia (Spain): From Traditional to Student-Centred Learning. [IJTHI]. *International Journal of Technology and Human Interaction, 8*(2), 56–76. doi:10.4018/jthi.2012040104

Mosindi, O., & Sice, P. (2011). An Exploratory Theoretical Framework for Understanding Information Behaviour. [IJTHI]. *International Journal of Technology and Human Interaction, 7*(2), 1–8. doi:10.4018/jthi.2011040101

Mott, M. S., & Williams-Black, T. H. (2014). Media-Enhanced Writing Instruction and Assessment. In J. Keengwe, G. Onchwari, & D. Hucks (Eds.), *Literacy Enrichment and Technology Integration in Pre-Service Teacher Education* (pp. 1–16). Hershey, PA: Information Science Reference.

Mulvey, F., & Heubner, M. (2014). Eye Movements and Attention. In Assistive Technologies: Concepts, Methodologies, Tools, and Applications (pp. 1030-1054). Hershey, PA: Information Science Reference. doi: doi:10.4018/978-1-4666-4422-9.ch053

Muro, B. F., & Delgado, E. C. (2014). RACEM Game for PC for Use as Rehabilitation Therapy for Children with Psychomotor Disability and Results of its Application. In Assistive Technologies: Concepts, Methodologies, Tools, and Applications (pp. 740-757). Hershey, PA: Information Science Reference. doi: doi:10.4018/978-1-4666-4422-9.ch037

Muwanguzi, S., & Lin, L. (2014). Coping with Accessibility and Usability Challenges of Online Technologies by Blind Students in Higher Education. In Assistive Technologies: Concepts, Methodologies, Tools, and Applications (pp. 1227-1244). Hershey, PA: Information Science Reference. doi: doi:10.4018/978-1-4666-4422-9.ch064

Najjar, M., Courtemanche, F., Hamam, H., Dion, A., Bauchet, J., & Mayers, A. (2014). DeepKøver: An Adaptive Intelligent Assistance System for Monitoring Impaired People in Smart Homes. In Assistive Technologies: Concepts, Methodologies, Tools, and Applications (pp. 634-661). Hershey, PA: Information Science Reference. doi: doi:10.4018/978-1-4666-4422-9.ch031

Nap, H. H., & Diaz-Orueta, U. (2014). Rehabilitation Gaming. In J. Bishop (Ed.), *Gamification for Human Factors Integration: Social, Education, and Psychological Issues* (pp. 122–147). Hershey, PA: Information Science Reference. doi:10.4018/978-1-4666-5071-8.ch008

Neves, J., & Pinheiro, L. D. (2012). Cyberbullying: A Sociological Approach. In R. Luppicini (Ed.), *Ethical Impact of Technological Advancements and Applications in Society* (pp. 132–142). Hershey, PA: Information Science Reference. doi:10.4018/978-1-4666-1773-5.ch010

Nguyen, P. T. (2012). Peer Feedback on Second Language Writing through Blogs: The Case of a Vietnamese EFL Classroom. [IJCALLT]. *International Journal of Computer-Assisted Language Learning and Teaching*, *2*(1), 13–23. doi:10.4018/ijcallt.2012010102

Ninaus, M., Witte, M., Kober, S. E., Friedrich, E. V., Kurzmann, J., & Hartsuiker, E. et al. (2014). Neurofeedback and Serious Games. In T. Connolly, T. Hainey, E. Boyle, G. Baxter, & P. Moreno-Ger (Eds.), *Psychology, Pedagogy, and Assessment in Serious Games* (pp. 82–110). Hershey, PA: Information Science Reference.

Olla, V. (2014). An Enquiry into the use of Technology and Student Voice in Citizenship Education in the K-12 Classroom. In K-12 Education: Concepts, Methodologies, Tools, and Applications (pp. 892-913). Hershey, PA: Information Science Reference. doi: doi:10.4018/978-1-4666-4502-8.ch053

Orange, E. (2013). Understanding the Human-Machine Interface in a Time of Change. In R. Luppicini (Ed.), *Handbook of Research on Technoself: Identity in a Technological Society* (pp. 703–719). Hershey, PA: Information Science Reference. doi:10.4018/978-1-4666-4607-0.ch082

Palmer, D., Warren, I., & Miller, P. (2014). ID Scanners and Überveillance in the Night-Time Economy: Crime Prevention or Invasion of Privacy? In M. Michael, & K. Michael (Eds.), *Uberveillance and the Social Implications of Microchip Implants: Emerging Technologies* (pp. 208–225). Hershey, PA: Information Science Reference.

Papadopoulos, F., Dautenhahn, K., & Ho, W. C. (2013). Behavioral Analysis of Human-Human Remote Social Interaction Mediated by an Interactive Robot in a Cooperative Game Scenario. In R. Luppicini (Ed.), *Handbook of Research on Technoself: Identity in a Technological Society* (pp. 637–665). Hershey, PA: Information Science Reference.

Patel, K. K., & Vij, S. K. (2014). Unconstrained Walking Plane to Virtual Environment for Non-Visual Spatial Learning. In Assistive Technologies: Concepts, Methodologies, Tools, and Applications (pp. 1580-1599). Hershey, PA: Information Science Reference. doi: doi:10.4018/978-1-4666-4422-9.ch083

Patrone, T. (2013). In Defense of the 'Human Prejudice'. [IJT]. *International Journal of Technoethics*, *4*(1), 26–38. doi:10.4018/jte.2013010103

Peevers, G., Williams, R., Douglas, G., & Jack, M. A. (2013). Usability Study of Fingerprint and Palmvein Biometric Technologies at the ATM. [IJTHI]. *International Journal of Technology and Human Interaction*, *9*(1), 78–95. doi:10.4018/jthi.2013010106

Pellas, N. (2014). Theoretical Foundations of a CSCL Script in Persistent Virtual Worlds According to the Contemporary Learning Theories and Models. In E. Nikoi, & K. Boateng (Eds.), *Collaborative Communication Processes and Decision Making in Organizations* (pp. 72–107). Hershey, PA: Business Science Reference.

Perakslis, C. (2014). Willingness to Adopt RFID Implants: Do Personality Factors Play a Role in the Acceptance of Uberveillance? In M. Michael, & K. Michael (Eds.), *Uberveillance and the Social Implications of Microchip Implants: Emerging Technologies* (pp. 144–168). Hershey, PA: Information Science Reference.

Pereira, G., Brisson, A., Dias, J., Carvalho, A., Dimas, J., & Mascarenhas, S. et al. (2014). Non-Player Characters and Artificial Intelligence. In T. Connolly, T. Hainey, E. Boyle, G. Baxter, & P. Moreno-Ger (Eds.), *Psychology, Pedagogy, and Assessment in Serious Games* (pp. 127–152). Hershey, PA: Information Science Reference.

Pérez Pérez, A., Callejas Carrión, Z., López-Cózar Delgado, R., & Griol Barres, D. (2014). On the Use of Speech Technologies to Achieve Inclusive Education for People with Intellectual Disabilities. In Assistive Technologies: Concepts, Methodologies, Tools, and Applications (pp. 1106-1117). Hershey, PA: Information Science Reference. doi: doi:10.4018/978-1-4666-4422-9.ch057

Peschl, M. F., & Fundneider, T. (2014). Theory U and Emergent Innovation: Presencing as a Method of Bringing Forth Profoundly New Knowledge and Realities. In O. Gunnlaugson, C. Baron, & M. Cayer (Eds.), *Perspectives on Theory U: Insights from the Field* (pp. 207–233). Hershey, PA: Business Science Reference.

Petrovic, N., Jeremic, V., Petrovic, D., & Cirovic, M. (2014). Modeling the Use of Facebook in Environmental Higher Education. In G. Mallia (Ed.), *The Social Classroom: Integrating Social Network Use in Education* (pp. 100–119). Hershey, PA: Information Science Reference.

Phua, C., Roy, P. C., Aloulou, H., Biswas, J., Tolstikov, A., Foo, V. S., et al. (2014). State-of-the-Art Assistive Technology for People with Dementia. In Assistive Technologies: Concepts, Methodologies, Tools, and Applications (pp. 1606-1625). Hershey, PA: Information Science Reference. doi: doi:10.4018/978-1-4666-4422-9.ch085

Potts, L. (2011). Balancing McLuhan With Williams: A Sociotechnical View of Technological Determinism. [IJSKD]. *International Journal of Sociotechnology and Knowledge Development, 3*(2), 53–57. doi:10.4018/jskd.2011040105

Potts, L. (2013). Balancing McLuhan With Williams: A Sociotechnical View of Technological Determinism. In J. Abdelnour-Nocera (Ed.), *Knowledge and Technological Development Effects on Organizational and Social Structures* (pp. 109–114). Hershey, PA: Information Science Reference.

Potts, L. (2014). Sociotechnical Uses of Social Web Tools during Disasters. In Crisis Management: Concepts, Methodologies, Tools and Applications (pp. 531-541). Hershey, PA: Information Science Reference. doi: doi:10.4018/978-1-4666-4707-7.ch024

Proença, R., Guerra, A., & Campos, P. (2013). A Gestural Recognition Interface for Intelligent Wheelchair Users. [IJSKD]. *International Journal of Sociotechnology and Knowledge Development, 5*(2), 63–81. doi:10.4018/jskd.2013040105

Quilici-Gonzalez, J. A., Kobayashi, G., Broens, M. C., & Gonzalez, M. E. (2012). Ubiquitous Computing: Any Ethical Implications? In R. Luppicini (Ed.), *Ethical Impact of Technological Advancements and Applications in Society* (pp. 47–59). Hershey, PA: Information Science Reference. doi:10.4018/978-1-4666-1773-5.ch005

Rambaree, K. (2014). Computer-Aided Deductive Critical Discourse Analysis of a Case Study from Mauritius with ATLAS-ti 6.2. In H. Lim, & F. Sudweeks (Eds.), *Innovative Methods and Technologies for Electronic Discourse Analysis* (pp. 346–368). Hershey, PA: Information Science Reference.

Ratan, R. (2013). Self-Presence, Explicated: Body, Emotion, and Identity Extension into the Virtual Self. In R. Luppicini (Ed.), *Handbook of Research on Technoself: Identity in a Technological Society* (pp. 322–336). Hershey, PA: Information Science Reference.

Rechy-Ramirez, E. J., & Hu, H. (2014). A Flexible Bio-Signal Based HMI for Hands-Free Control of an Electric Powered Wheelchair. [IJALR]. *International Journal of Artificial Life Research*, *4*(1), 59–76. doi:10.4018/ijalr.2014010105

Reiners, T., Wood, L. C., & Dron, J. (2014). From Chaos Towards Sense: A Learner-Centric Narrative Virtual Learning Space. In J. Bishop (Ed.), *Gamification for Human Factors Integration: Social, Education, and Psychological Issues* (pp. 242–258). Hershey, PA: Information Science Reference. doi:10.4018/978-1-4666-5071-8.ch015

Reinhardt, J., & Ryu, J. (2013). Using Social Network-Mediated Bridging Activities to Develop Socio-Pragmatic Awareness in Elementary Korean. [IJCALLT]. *International Journal of Computer-Assisted Language Learning and Teaching*, *3*(3), 18–33. doi:10.4018/ijcallt.2013070102

Revuelta, P., Jiménez, J., Sánchez, J. M., & Ruiz, B. (2014). Automatic Speech Recognition to Enhance Learning for Disabled Students. In Assistive Technologies: Concepts, Methodologies, Tools, and Applications (pp. 478-493). Hershey, PA: Information Science Reference. doi: doi:10.4018/978-1-4666-4422-9.ch023

Ribeiro, J. C., & Silva, T. (2013). Self, Self-Presentation, and the Use of Social Applications in Digital Environments. In R. Luppicini (Ed.), *Handbook of Research on Technoself: Identity in a Technological Society* (pp. 439–455). Hershey, PA: Information Science Reference.

Richet, J. (2013). From Young Hackers to Crackers. [IJTHI]. *International Journal of Technology and Human Interaction*, *9*(3), 53–62. doi:10.4018/jthi.2013070104

Rigas, D., & Almutairi, B. (2013). An Empirical Investigation into the Role of Avatars in Multimodal E-government Interfaces. [IJSKD]. *International Journal of Sociotechnology and Knowledge Development*, *5*(1), 14–22. doi:10.4018/jskd.2013010102

Rodríguez, W. R., Saz, O., & Lleida, E. (2014). Experiences Using a Free Tool for Voice Therapy based on Speech Technologies. In Assistive Technologies: Concepts, Methodologies, Tools, and Applications (pp. 508-523). Hershey, PA: Information Science Reference. doi: doi:10.4018/978-1-4666-4422-9.ch025

Rothblatt, M. (2013). Mindclone Technoselves: Multi-Substrate Legal Identities, Cyber-Psychology, and Biocyberethics. In R. Luppicini (Ed.), *Handbook of Research on Technoself: Identity in a Technological Society* (pp. 105–122). Hershey, PA: Information Science Reference.

Rowe, N. C. (2012). The Ethics of Cyberweapons in Warfare. In R. Luppicini (Ed.), *Ethical Impact of Technological Advancements and Applications in Society* (pp. 195–207). Hershey, PA: Information Science Reference. doi:10.4018/978-1-4666-1773-5.ch015

Russo, M. R. (2014). Emergency Management Professional Development: Linking Information Communication Technology and Social Communication Skills to Enhance a Sense of Community and Social Justice in the 21st Century. In Crisis Management: Concepts, Methodologies, Tools and Applications (pp. 651-665). Hershey, PA: Information Science Reference. doi: doi:10.4018/978-1-4666-4707-7.ch031

Sajeva, S. (2011). Towards a Conceptual Knowledge Management System Based on Systems Thinking and Sociotechnical Thinking. [IJSKD]. *International Journal of Sociotechnology and Knowledge Development*, *3*(3), 40–55. doi:10.4018/jskd.2011070103

Sajeva, S. (2013). Towards a Conceptual Knowledge Management System Based on Systems Thinking and Sociotechnical Thinking. In J. Abdelnour-Nocera (Ed.), *Knowledge and Technological Development Effects on Organizational and Social Structures* (pp. 115–130). Hershey, PA: Information Science Reference.

Saleeb, N., & Dafoulas, G. A. (2014). Assistive Technologies and Environmental Design Concepts for Blended Learning and Teaching for Disabilities within 3D Virtual Worlds and Learning Environments. In Assistive Technologies: Concepts, Methodologies, Tools, and Applications (pp. 1382-1404). Hershey, PA: Information Science Reference. doi: doi:10.4018/978-1-4666-4422-9.ch073

Salvini, P. (2012). Presence, Reciprocity and Robotic Mediations: The Case of Autonomous Social Robots. [IJT]. *International Journal of Technoethics*, *3*(2), 9–16. doi:10.4018/jte.2012040102

Samanta, I. (2013). The Impact of Virtual Community (Web 2.0) in the Economic, Social, and Political Environment of Traditional Society. In S. Saeed, M. Khan, & R. Ahmad (Eds.), *Business Strategies and Approaches for Effective Engineering Management* (pp. 262–274). Hershey, PA: Business Science Reference. doi:10.4018/978-1-4666-3658-3.ch016

Samanta, S. K., Woods, J., & Ghanbari, M. (2011). Automatic Language Translation: An Enhancement to the Mobile Messaging Services. [IJTHI]. *International Journal of Technology and Human Interaction*, *7*(1), 1–18. doi:10.4018/jthi.2011010101

Sarkar, N. I., Kuang, A. X., Nisar, K., & Amphawan, A. (2014). Hospital Environment Scenarios using WLAN over OPNET Simulation Tool. [IJICTHD]. *International Journal of Information Communication Technologies and Human Development*, *6*(1), 69–90. doi:10.4018/ijicthd.2014010104

Sarré, C. (2013). Technology-Mediated Tasks in English for Specific Purposes (ESP): Design, Implementation and Learner Perception. [IJCALLT]. *International Journal of Computer-Assisted Language Learning and Teaching*, *3*(2), 1–16. doi:10.4018/ijcallt.2013040101

Saykili, A., & Kumtepe, E. G. (2014). Facebook's Hidden Potential: Facebook as an Educational Support Tool in Foreign Language Education. In G. Mallia (Ed.), *The Social Classroom: Integrating Social Network Use in Education* (pp. 120–146). Hershey, PA: Information Science Reference.

Sayoud, H. (2011). Biometrics: An Overview on New Technologies and Ethic Problems. [IJT]. *International Journal of Technoethics*, *2*(1), 19–34. doi:10.4018/jte.2011010102

Scott, C. R., & Timmerman, C. E. (2014). Communicative Changes Associated with Repeated Use of Electronic Meeting Systems for Decision-Making Tasks. In E. Nikoi, & K. Boateng (Eds.), *Collaborative Communication Processes and Decision Making in Organizations* (pp. 1–24). Hershey, PA: Business Science Reference.

Scott, K. (2013). The Human-Robot Continuum of Self: Where the Other Ends and Another Begins. In R. Luppicini (Ed.), *Handbook of Research on Technoself: Identity in a Technological Society* (pp. 666–679). Hershey, PA: Information Science Reference.

Shasek, J. (2014). ExerLearning®: Movement, Fitness, Technology, and Learning. In J. Bishop (Ed.), *Gamification for Human Factors Integration: Social, Education, and Psychological Issues* (pp. 106–121). Hershey, PA: Information Science Reference. doi:10.4018/978-1-4666-5071-8.ch007

Shen, J., & Eder, L. B. (2011). An Examination of Factors Associated with User Acceptance of Social Shopping Websites. [IJTHI]. *International Journal of Technology and Human Interaction*, *7*(1), 19–36. doi:10.4018/jthi.2011010102

Shrestha, P. (2012). Teacher Professional Development Using Mobile Technologies in a Large-Scale Project: Lessons Learned from Bangladesh. [IJCALLT]. *International Journal of Computer-Assisted Language Learning and Teaching*, *2*(4), 34–49. doi:10.4018/ijcallt.2012100103

Silvana de Rosa, A., Fino, E., & Bocci, E. (2014). Addressing Healthcare On-Line Demand and Supply Relating to Mental Illness: Knowledge Sharing About Psychiatry and Psychoanalysis Through Social Networks in Italy and France. In A. Kapoor, & C. Kulshrestha (Eds.), *Dynamics of Competitive Advantage and Consumer Perception in Social Marketing* (pp. 16–55). Hershey, PA: Business Science Reference.

Smith, M., & Murray, J. (2014). Augmentative and Alternative Communication Devices: The Voices of Adult Users. In Assistive Technologies: Concepts, Methodologies, Tools, and Applications (pp. 991-1004). Hershey, PA: Information Science Reference. doi: doi:10.4018/978-1-4666-4422-9.ch050

Smith, P. A. (2013). Strengthening and Enriching Audit Practice: The Socio-Technical Relevance of "Decision Leaders". In J. Abdelnour-Nocera (Ed.), *Knowledge and Technological Development Effects on Organizational and Social Structures* (pp. 97–108). Hershey, PA: Information Science Reference.

So, J. C., & Lam, S. Y. (2014). Using Social Networks Communication Platform for Promoting Student-Initiated Holistic Development Among Students. [IJISSS]. *International Journal of Information Systems in the Service Sector*, *6*(1), 1–23. doi:10.4018/ijisss.2014010101

Söderström, S. (2014). Assistive ICT and Young Disabled Persons: Opportunities and Obstacles in Identity Negotiations. In Assistive Technologies: Concepts, Methodologies, Tools, and Applications (pp. 1084-1105). Hershey, PA: Information Science Reference. doi: doi:10.4018/978-1-4666-4422-9.ch056

Son, J., & Rossade, K. (2013). Finding Gems in Computer-Assisted Language Learning: Clues from GLoCALL 2011 and 2012 Papers. [IJCALLT]. *International Journal of Computer-Assisted Language Learning and Teaching*, *3*(4), 1–8. doi:10.4018/ijcallt.2013100101

Sone, Y. (2013). Robot Double: Hiroshi Ishiguro's Reflexive Machines. In R. Luppicini (Ed.), *Handbook of Research on Technoself: Identity in a Technological Society* (pp. 680–702). Hershey, PA: Information Science Reference.

Spillane, M. (2014). Assistive Technology: A Tool for Inclusion. In Assistive Technologies: Concepts, Methodologies, Tools, and Applications (pp. 1-11). Hershey, PA: Information Science Reference. doi: doi:10.4018/978-1-4666-4422-9.ch001

Stahl, B. C., Heersmink, R., Goujon, P., Flick, C., van den Hoven, J., & Wakunuma, K. et al. (2012). Identifying the Ethics of Emerging Information and Communication Technologies: An Essay on Issues, Concepts and Method. In R. Luppicini (Ed.), *Ethical Impact of Technological Advancements and Applications in Society* (pp. 61–79). Hershey, PA: Information Science Reference. doi:10.4018/978-1-4666-1773-5.ch006

Stern, S. E., & Grounds, B. E. (2011). Cellular Telephones and Social Interactions: Evidence of Interpersonal Surveillance. [IJT]. *International Journal of Technoethics*, *2*(1), 43–49. doi:10.4018/jte.2011010104

Stinson, J., & Gill, N. (2014). Internet-Based Chronic Disease Self-Management for Youth. In Assistive Technologies: Concepts, Methodologies, Tools, and Applications (pp. 224-245). Hershey, PA: Information Science Reference. doi: doi:10.4018/978-1-4666-4422-9.ch012

Stockwell, G. (2011). Online Approaches to Learning Vocabulary: Teacher-Centred or Learner-Centred? [IJCALLT]. *International Journal of Computer-Assisted Language Learning and Teaching*, *1*(1), 33–44. doi:10.4018/ijcallt.2011010103

Stradella, E. (2012). Personal Liability and Human Free Will in the Background of Emerging Neuroethical Issues: Some Remarks Arising From Recent Case Law. [IJT]. *International Journal of Technoethics*, *3*(2), 30–41. doi:10.4018/jte.2012040104

Stubbs, K., Casper, J., & Yanco, H. A. (2014). Designing Evaluations for K-12 Robotics Education Programs. In K-12 Education: Concepts, Methodologies, Tools, and Applications (pp. 1342-1364). Hershey, PA: Information Science Reference. doi: doi:10.4018/978-1-4666-4502-8.ch078

Suki, N. M., Ramayah, T., Ming, M. K., & Suki, N. M. (2011). Factors Enhancing Employed Job Seekers Intentions to Use Social Networking Sites as a Job Search Tool. [IJTHI]. *International Journal of Technology and Human Interaction*, *7*(2), 38–54. doi:10.4018/jthi.2011040105

Sweeney, P., & Moore, C. (2012). Mobile Apps for Learning Vocabulary: Categories, Evaluation and Design Criteria for Teachers and Developers. [IJCALLT]. *International Journal of Computer-Assisted Language Learning and Teaching*, *2*(4), 1–16. doi:10.4018/ijcallt.2012100101

Szeto, A. Y. (2014). Assistive Technology and Rehabilitation Engineering. In Assistive Technologies: Concepts, Methodologies, Tools, and Applications (pp. 277-331). Hershey, PA: Information Science Reference. doi: doi:10.4018/978-1-4666-4422-9.ch015

Tamim, R. (2014). Technology Integration in UAE Schools: Current Status and Way Forward. In K-12 Education: Concepts, Methodologies, Tools, and Applications (pp. 41-57). Hershey, PA: Information Science Reference. doi: doi:10.4018/978-1-4666-4502-8.ch004

Tan, R., Wang, S., Jiang, Y., Ishida, K., & Fujie, M. G. (2014). Motion Control of an Omni-Directional Walker for Walking Support. In Assistive Technologies: Concepts, Methodologies, Tools, and Applications (pp. 614-622). Hershey, PA: Information Science Reference. doi: doi:10.4018/978-1-4666-4422-9.ch029

Tankari, M. (2014). Cultural Orientation Differences and their Implications for Online Learning Satisfaction. In J. Keengwe, G. Schnellert, & K. Kungu (Eds.), *Cross-Cultural Online Learning in Higher Education and Corporate Training* (pp. 20–61). Hershey, PA: Information Science Reference. doi:10.4018/978-1-4666-5023-7.ch002

Tchangani, A. P. (2014). Bipolarity in Decision Analysis: A Way to Cope with Human Judgment. In A. Masegosa, P. Villacorta, C. Cruz-Corona, M. García-Cascales, M. Lamata, & J. Verdegay (Eds.), *Exploring Innovative and Successful Applications of Soft Computing* (pp. 216–244). Hershey, PA: Information Science Reference.

Tennyson, R. D. (2014). Computer Interventions for Children with Disabilities: Review of Research and Practice. In Assistive Technologies: Concepts, Methodologies, Tools, and Applications (pp. 841-864). Hershey, PA: Information Science Reference. doi: doi:10.4018/978-1-4666-4422-9.ch042

Terrell, S. S. (2011). Integrating Online Tools to Motivate Young English Language Learners to Practice English Outside the Classroom. [IJCALLT]. *International Journal of Computer-Assisted Language Learning and Teaching*, *1*(2), 16–24. doi:10.4018/ijcallt.2011040102

Tiwary, U. S., & Siddiqui, T. J. (2014). Working Together with Computers: Towards a General Framework for Collaborative Human Computer Interaction. In Assistive Technologies: Concepts, Methodologies, Tools, and Applications (pp. 141-162). Hershey, PA: Information Science Reference. doi:doi:10.4018/978-1-4666-4422-9.ch008

Tomas, J., Lloret, J., Bri, D., & Sendra, S. (2014). Sensors and their Application for Disabled and Elderly People. In Assistive Technologies: Concepts, Methodologies, Tools, and Applications (pp. 357-376). Hershey, PA: Information Science Reference. doi:doi:10.4018/978-1-4666-4422-9.ch018

Tomasi, A. (2013). A Run for your [Techno]Self. In R. Luppicini (Ed.), Handbook of Research on Technoself: Identity in a Technological Society (pp. 123-136). Hershey, PA: Information Science Reference. doi:doi:10.4018/978-1-4666-2211-1.ch007

Tootell, H., & Freeman, A. (2014). The Applicability of Gaming Elements to Early Childhood Education. In J. Bishop (Ed.), *Gamification for Human Factors Integration: Social, Education, and Psychological Issues* (pp. 225–241). Hershey, PA: Information Science Reference. doi:10.4018/978-1-4666-5071-8.ch014

Tsai, C. (2011). How Much Can Computers and Internet Help?: A Long-Term Study of Web-Mediated Problem-Based Learning and Self-Regulated Learning. [IJTHI]. *International Journal of Technology and Human Interaction*, *7*(1), 67–81. doi:10.4018/jthi.2011010105

Tsai, W. (2013). An Investigation on Undergraduate's Bio-Energy Engineering Education Program at the Taiwan Technical University. [IJTHI]. *International Journal of Technology and Human Interaction*, *8*(3), 46–53. doi:10.4018/jthi.2012070105

Tsiakis, T. (2013). Using Social Media as a Concept and Tool for Teaching Marketing Information Systems. In M. Pătruţ, & B. Pătruţ (Eds.), *Social Media in Higher Education: Teaching in Web 2.0* (pp. 24–44). Hershey, PA: Information Science Reference. doi:10.4018/978-1-4666-2970-7.ch002

Tu, C., McIsaac, M. S., Sujo-Montes, L. E., & Armfield, S. (2014). Building Mobile Social Presence for U-Learning. In F. Neto (Ed.), *Technology Platform Innovations and Forthcoming Trends in Ubiquitous Learning* (pp. 77–93). Hershey, PA: Information Science Reference.

Valeria, N., Lu, M. V., & Theng, L. B. (2014). Collaborative Virtual Learning for Assisting Children with Cerebral Palsy. In Assistive Technologies: Concepts, Methodologies, Tools, and Applications (pp. 786-810). Hershey, PA: Information Science Reference. doi:doi:10.4018/978-1-4666-4422-9.ch040

Van Leuven, N., Newton, D., Leuenberger, D. Z., & Esteves, T. (2014). Reaching Citizen 2.0: How Government Uses Social Media to Send Public Messages during Times of Calm and Times of Crisis. In Crisis Management: Concepts, Methodologies, Tools and Applications (pp. 839-857). Hershey, PA: Information Science Reference. doi:doi:10.4018/978-1-4666-4707-7.ch041

Vargas-Hernández, J. G. (2013). International Student Collaboration and Experiential Exercise Projects as a Professional, Inter-Personal and Inter-Institutional Networking Platform. [IJTEM]. *International Journal of Technology and Educational Marketing*, *3*(1), 28–47. doi:10.4018/ijtem.2013010103

Velicu, A., & Marinescu, V. (2013). Usage of Social Media by Children and Teenagers: Results of EU KIDS Online II. In M. Pătruţ, & B. Pătruţ (Eds.), *Social Media in Higher Education: Teaching in Web 2.0* (pp. 144–178). Hershey, PA: Information Science Reference. doi:10.4018/978-1-4666-2970-7.ch008

Vidaurre, C., Kübler, A., Tangermann, M., Müller, K., & Millán, J. D. (2014). Brain-Computer Interfaces and Visual Activity. In Assistive Technologies: Concepts, Methodologies, Tools, and Applications (pp. 1549-1570). Hershey, PA: Information Science Reference. doi: doi:10.4018/978-1-4666-4422-9.ch081

Viswanathan, R. (2012). Augmenting the Use of Mobile Devices in Language Classrooms. [IJCALLT]. *International Journal of Computer-Assisted Language Learning and Teaching*, *2*(2), 45–60. doi:10.4018/ijcallt.2012040104

Wallgren, L. G., & Hanse, J. J. (2012). A Two-Wave Study of the Impact of Job Characteristics and Motivators on Perceived Stress among Information Technology (IT) Consultants. [IJTHI]. *International Journal of Technology and Human Interaction*, *8*(4), 75–91. doi:10.4018/jthi.2012100105

Wang, H. (2014). A Guide to Assistive Technology for Teachers in Special Education. In Assistive Technologies: Concepts, Methodologies, Tools, and Applications (pp. 12-25). Hershey, PA: Information Science Reference. doi: doi:10.4018/978-1-4666-4422-9.ch002

Wang, S., Ku, C., & Chu, C. (2013). Sustainable Campus Project: Potential for Energy Conservation and Carbon Reduction Education in Taiwan. [IJTHI]. *International Journal of Technology and Human Interaction*, *8*(3), 19–30. doi:10.4018/jthi.2012070103

Wang, Y., & Tian, J. (2013). Negotiation of Meaning in Multimodal Tandem Learning via Desktop Videoconferencing. [IJCALLT]. *International Journal of Computer-Assisted Language Learning and Teaching*, *3*(2), 41–55. doi:10.4018/ijcallt.2013040103

Wareham, C. (2011). On the Moral Equality of Artificial Agents. [IJT]. *International Journal of Technoethics*, *2*(1), 35–42. doi:10.4018/jte.2011010103

Warwick, K., & Gasson, M. N. (2014). Practical Experimentation with Human Implants. In M. Michael, & K. Michael (Eds.), *Uberveillance and the Social Implications of Microchip Implants: Emerging Technologies* (pp. 64–132). Hershey, PA: Information Science Reference.

Welch, K. C., Lahiri, U., Sarkar, N., Warren, Z., Stone, W., & Liu, C. (2014). Affect-Sensitive Computing and Autism. In Assistive Technologies: Concepts, Methodologies, Tools, and Applications (pp. 865-883). Hershey, PA: Information Science Reference. doi: doi:10.4018/978-1-4666-4422-9.ch043

Wessels, B., Dittrich, Y., Ekelin, A., & Eriksén, S. (2014). Creating Synergies between Participatory Design of E-Services and Collaborative Planning. In Assistive Technologies: Concepts, Methodologies, Tools, and Applications (pp. 163-179). Hershey, PA: Information Science Reference. doi: doi:10.4018/978-1-4666-4422-9.ch009

White, E. L. (2014). Technology-Based Literacy Approach for English Language Learners. In K-12 Education: Concepts, Methodologies, Tools, and Applications (pp. 723-740). Hershey, PA: Information Science Reference. doi: doi:10.4018/978-1-4666-4502-8.ch042

Whyte, K. P., List, M., Stone, J. V., Grooms, D., Gasteyer, S., & Thompson, P. B. et al. (2014). Uberveillance, Standards, and Anticipation: A Case Study on Nanobiosensors in U.S. Cattle. In M. Michael, & K. Michael (Eds.), *Uberveillance and the Social Implications of Microchip Implants: Emerging Technologies* (pp. 260–279). Hershey, PA: Information Science Reference.

Wilson, S., & Haslam, N. (2013). Reasoning about Human Enhancement: Towards a Folk Psychological Model of Human Nature and Human Identity. In R. Luppicini (Ed.), *Handbook of Research on Technoself: Identity in a Technological Society* (pp. 175–188). Hershey, PA: Information Science Reference.

Woodhead, R. (2012). What is Technology? [IJSKD]. *International Journal of Sociotechnology and Knowledge Development, 4*(2), 1–13. doi:10.4018/jskd.2012040101

Woodley, C., & Dorrington, P. (2014). Facebook and the Societal Aspects of Formal Learning: Optional, Peripheral, or Essential. In G. Mallia (Ed.), *The Social Classroom: Integrating Social Network Use in Education* (pp. 269–291). Hershey, PA: Information Science Reference.

Yamazaki, T. (2014). Assistive Technologies in Smart Homes. In Assistive Technologies: Concepts, Methodologies, Tools, and Applications (pp. 663-678). Hershey, PA: Information Science Reference. doi: doi:10.4018/978-1-4666-4422-9.ch032

Yan, Z., Chen, Q., & Yu, C. (2013). The Science of Cell Phone Use: Its Past, Present, and Future. [IJCBPL]. *International Journal of Cyber Behavior, Psychology and Learning, 3*(1), 7–18. doi:10.4018/ijcbpl.2013010102

Yang, Y., Wang, X., & Li, L. (2013). Use Mobile Devices to Wirelessly Operate Computers. [IJTHI]. *International Journal of Technology and Human Interaction, 9*(1), 64–77. doi:10.4018/jthi.2013010105

Yartey, F. N., & Ha, L. (2013). Like, Share, Recommend: Smartphones as a Self-Broadcast and Self-Promotion Medium of College Students. [IJTHI]. *International Journal of Technology and Human Interaction, 9*(4), 20–40. doi:10.4018/ijthi.2013100102

Yaseen, S. G., & Al Omoush, K. S. (2013). Investigating the Engage in Electronic Societies via Facebook in the Arab World. [IJTHI]. *International Journal of Technology and Human Interaction, 9*(2), 20–38. doi:10.4018/jthi.2013040102

Yeo, B. (2012). Sustainable Economic Development and the Influence of Information Technologies: Dynamics of Knowledge Society Transformation. [IJSKD]. *International Journal of Sociotechnology and Knowledge Development, 4*(3), 54–55. doi:10.4018/jskd.2012070105

Yu, L., & Ureña, C. (2014). A Review of Current Approaches of Brain Computer Interfaces. In Assistive Technologies: Concepts, Methodologies, Tools, and Applications (pp. 1516-1534). Hershey, PA: Information Science Reference. doi: doi:10.4018/978-1-4666-4422-9.ch079

Zelenkauskaite, A. (2014). Analyzing Blending Social and Mass Media Audiences through the Lens of Computer-Mediated Discourse. In H. Lim, & F. Sudweeks (Eds.), *Innovative Methods and Technologies for Electronic Discourse Analysis* (pp. 304–326). Hershey, PA: Information Science Reference.

Compilation of References

Abbagnano, N. (1990). *Ricordi di un filosofo: A cura di Marcello Staglieno*. Milano: Rizzoli.

Abou-Tair, D., & Berlik, S. (2006). An ontology-based approach for managing and maintaining privacy in information systems. *Lecture Notes in Computer Science*, *4275*, 983–994. doi:10.1007/11914853_63

Ackerman, M. S., & Cranor, L. (1999). Privacy critics: UI components to safeguard users' privacy. In *Extended Abstracts of CHI* (pp. 258-259). New York: ACM Press.

Ackoff, R. L. (1989). From Data to Wisdom. *Journal of Applies Systems Analysis*, *16*(4), 3–9.

Agre, P. E. (1997). Introduction. In P. E. Agre, & M. Rotenberg (Eds.), *Technology and privacy: The new landscape* (pp. 1–28). Cambridge, MA: The MIT Press.

Alien Tort Claims Act (2000) 28 U.S.C. § 1350.

Allen, C., Wallach, W., & Smit, I. (2006). Why Machine Ethics? *IEEE Intelligent Systems*, *21*(4), 12–17. doi:10.1109/MIS.2006.83

Anderson, M., Anderson, S., & Armen, C. (2005). Towards Machine Ethics: Implementing Two Action-Based Ethical Theories. In *Proceeding of AAAI 2005 Fall Symposiom on Machine Ethics*. AAAI Press.

Ardia, D. S. (2010). Free speech savior or shield for scoundrels: An empirical study of intermediary immunity under section 230 of the communications decency act. *Loyola of Los Angeles Law Review*, *43*(2), 373–505.

Arnaudo, L. (2010). Diritto cognitivo: Prolegomeni a una ricerca. *Politica del Diritto, 1*.

Ashford, W. (2010, June 9). Data Protection Act is out of kilter with EU law, warns privacy lawyer. *Computer Weekly*. Retrieved from http://www.computerweekly.com/Articles/2010/06/09/241513/Data-Protection-Act-is-out-of-kilter-with-EU-law-warns-privacy.htm

Asimov, I. (1983). Asimov on Science fiction. London: Granada.

Assange, J. et al. (2012). *Cyberpunks: Freedom and the future of the internet*. Academic Press.

Baarda, R., & Rocci, L. (2012). The Use and Abuse of Digital Democracy. *International Journal of Technoethics*, *3*, 50–69.

Baase, S. (2008). *Gift of fire: Social, legal, and ethical issues for computing and the internet* (3rd ed.). Cambridge, MA: Pearson.

Baeva, L. V. (2012). Anthropogenesis and Dynamics of Values under Conditions of Information Technology. *International Journal of Technoethics*, *3*(3), 33–50.

Baeva, L. V. (2012). Existential Axiology. *Culture: International Journal of Philosophy of Culture and Axiology*, *l*(9), 73–85.

Baier, K. (1967). Concept of Value. *The Journal of Value Inquiry*, *1*(1), 1–11.

Bailey, J. A. (1979). On intrinsic value. *Philosophia*, *9*(1), 1–8. doi:10.1007/BF02379981

Bailie, J. L., & Jortberg, M. A. (2009). Online learner authentication: Verifying the identity of online learners. *MERLOT Journal of Online Learning and Teaching*, *5*(2).

Bainbridge, D. (2008). *Introduction to information technology law*. London, UK: Pearson.

Baker, S. (2006). The changing face of current affairs programmes in New Zealand, the United States and Britain 1984-2004. *Communication Journal of New Zealand, 7*, 1–22.

Baudrillard, J. (1981). Requiem for the Media. In *For a Critique of the Political Economy of the Sign.* St. Louis, MO: Telos Press.

Baudrillard, J. (1983). *Simulations, Semiotext(e), Inc.* Columbia University.

Baudrillard, J. (1994). *Simulacra and Simulation: The Body in Theory: Histories of Cultural Materialism.* University of Michigan Press.

Baumgartner, J., & Morris, J. (2010). MyFaceTube politics: Social networking web sites and political engagement of young adults. *Social Science Computer Review, 28*(1), 24–44. doi:10.1177/0894439309334325

Baum, M. A. (2003). Soft news and political knowledge: Evidence of absence or absence of evidence. *Political Communication, 20*, 173–190. doi:10.1080/10584600390211181

Bazerman, M. H., & Tenbrunsel, A. E. (2011, April). Ethical breakdowns. *Harvard Business Review, 89*(4), 58–65. PMID:21510519

Beardsley, M. C. (1965). Intrinsic value. *Philosophy and Phenomenological Research, 26*(1), 1–17. doi:10.2307/2105465

Bedau, M. A. (1998). Philosophical Content and Method of Artificial Life. In *The Digital Phoenix: How Computers are Changing Philosophy* (pp. 135–152). Oxford, UK: Blackwell Publishers.

Bennett, W. The personalization of politics: Political identity, social media, and changing patterns of participation. *The Annals of the American Academy of Political and Social Science, 644*(1), 20–39. doi:10.1177/0002716212451428

Bentham, J. (2007). *An introduction to the principles of morals and legislation.* Mineola, NY: Dover Publications.

Bernardi, A. (2009). Soft law e diritto penale: antinomie, convergenze, intersezioni. In *Soft law e hard law nelle società postmoderne.* Turin, Italy: Giappichelli.

Bernstein, A. J. (2013, May 4). Where have all the jobs gone? *The New York Times,* p. A21.

Bertot, J. C., Jaeger, P. T., & Hansen, D. (2012). The impact of policies of government social media usage: Issues, challenges, and recommendations. *Government Information Quarterly, 29*(1), 30–40. doi:10.1016/j.giq.2011.04.004

Bibeau, G. (2011). What is human in humans? Responses from biology, anthropology, and philosophy. *The Journal of Medicine and Philosophy, 36*, 354–363. doi:10.1093/jmp/jhr025 PMID:21859676

Bird, S. J. (2013). Security and Privacy: Why Privacy Matters. *Science and Engineering Ethics, 19*(3), 669–671. doi:10.1007/s11948-013-9458-z PMID:23893336

Birsch, D., & Fielder, J. H. (1994). *The Ford Pinto case: A study in applied ethics, business, and technology.* Albany, NY: University of New York Press.

Boczkowski, P. J., & Peer, L. (2011). The choice gap: The divergent online news preferences of journalists and consumers. *The Journal of Communication, 61*(5), 857–876. doi:10.1111/j.1460-2466.2011.01582.x

Boella, L. (2008). *Neuroetica: La morale prima della morale.* Milan, Italy: Raffaello Cortina.

Boorse, C. (1975). On the distinction between disease and illness. *Philosophy & Public Affairs, 5*(1), 49–68.

Boorse, C. (1977). Health as a theoretical concept. *Philosophy of Science, 44*(4), 542–573. doi:10.1086/288768

Borenstein, J., & Pearson, Y. (2010). Robot caregivers: Harbingers of expanded freedom for all? *Ethics and Information Technology, 12*, 277–288. doi:10.1007/s10676-010-9236-4

Borning, A., Friedman, B., & Kahn, P. (2004). Designing for human values in an urban simulation system: Value sensitive design and participatory design. In *Proceedings of Eighth Biennal Participatory Design Conference* (pp. 64-67). Toronto, Canada: ACM Press.

Bottalico, B. (2011). *European Center for Law, Science and New Technologies.* Retrieved from http://www.unipv.lawtech.eu

Bottalico, B. (2011). Neuroscience and law in a nutshell. *Diritti Comparati.* Retrieved from http://www.dirittícomparati.it

Boyd, C. (2008). Profile: Gary McKinnon. *BBC News*. Retrieved from http://news.bbc.co.uk/1/hi/technology/4715612.stm

Breazeal, C. (2002). *Desining socialble robots*. The MIT Press.

Breuker, J., Casanovas, P., Klein, M. C. A., & Francesconi, E. (Eds.). (2008). *Law, ontologies and the semantic web: Channelling the legal informationflood*. Amsterdam, The Netherlands: IOS Press.

Brey, P. (1999). The Ethics of Representation and Action in Virtual Reality. *Ethics and Information Technology*, *1*(1), 5–14.

Brin, D. (1998). *The transparent society*. Reading, MA: Addison-Wesley.

Brody, J. E. (2011, June 7). Law on end-of-life care rankles doctors. *The New York Times*, p. D7.

Brouwer, E. (2009). The EU passenger name record (PNR) system and human rights: Transferring passenger data or passenger freedom?. *CEPS Working Document, 320.*

Brues, A. M., & Snow, C. C. (1965). Physical anthropology. *Biennial Review of Anthropology*, *4*, 1–39.

Brumbaugh, R. S. (1972). Changes of Value Order and Choices in Time. In *Value and Valuation: Axiological Studies in Honor of Robert S. Hartman*. The University of Tennessee Press.

Brumfiel, G. (2013). Replaceable you. *Smithsonian*, *44*(5), 68–76.

Bryan, J. (1986). *High-technology medicine: Benefits and burdens*. Oxford, UK: Oxford University Press.

Buchen, L. (2012). Arrested Development. *Nature*, *484*, 304–306. doi:10.1038/484304a PMID:22517146

Bucy, E. P. (2003). Media credibility reconsidered: Synergy effects between on-air and online news. *Journalism & Mass Communication Quarterly*, *80*, 247–264. doi:10.1177/107769900308000202

Bunge, M. (1977). Toward a technoethics. *The Monist*, *60*, 96–107. doi:10.5840/monist197760134

Burgan, M. (1996). Teaching the subject: Developmental identities in teaching. In *Ethical dimensions of college and university teaching: Understanding and honoring the special relationship between teachers and students*. San Francisco, CA: Jossey-Bass Publishers.

Buruma, I. (2010, January 30). Battling the information barbarians. *The Wall Street Journal*, p. W1-2.

Bynum, T. W. (1998). Global Information Ethics and the Information Revolution. In *The Digital Phoenix: How Computers are Changing Philosophy* (pp. 274–291). Oxford, UK: Blackwell Publishers.

Bynum, T. W. (2000). The Foundation of Computer Ethics. *Computers & Society*, *30*(2), 6–13. doi:10.1145/572230.572231

Byrne, M. (2008). When in Rome: Aiding and abetting in Wang Xiaoning v. Yahoo. *Brooklyn Journal of International Law*, *34*, 151.

Campbell, A. V. (2009). *The body in bioethics*. New York, NY: Routledge-Cavendish.

Camporesi, S., & Bottalico, B. (2011). Can we finally 'see pain'? Brian imaging techniques and implications for the law. *Journal of Consciousness Studies*, *18*(9-10), 257–276.

Camps, F. (1981). Warning an auto company about an unsafe design. In A. F. Westin (Ed.), *Whistle blowing: Loyalty and dissent in the corporation* (pp. 19–129). New York: McGraw-Hill Book Company.

Canil, T., & Amin, Z. (2002). Neuroimaging of emotion and personality: Scientific evidence and ethical considerations. *Brain and Cognition*, *50*(3), 414–431. doi:10.1016/S0278-2626(02)00517-1 PMID:12480487

CAPER. (2013). *Consolidated Review Report (EU Project 261712)*. CAPER.

Capurro, R. (2006). Towards an Ontological Foundation of Information Ethics. *Ethics and Information Technology*, *8*(4), 175–186.

Casanovas, P., Pagallo, U., Sartor, G., & Ajani, G. (2010). *AI approaches to the complexity of legal systems: Complex systems, the semantic web, ontologies, argumentation, and dialogue*. Dordrecht, The Netherlands: Springer.

Casellas, N., Nieto, J.-E., Meroño, A., Roig, A., Torralba, S., Reyes, M., & Casanovas, P. (2010). Ontological semantics for data privacy dompliance: The NEURONA project. In *Proceedings of AAAI Spring Symposium: Intelligent Information Privacy Management*. AAAI.

Castells, M. (1997). *The End of the Millennium, the Information Age: Economy, Society and Culture* (Vol. 13). Cambridge: MA, Oxford, UK: Blackwell.

Cavalieri, P., & Singer, P. (1993). *The Great Ape Project: Equality Beyond Humanity*. London, UK: Fourth Estate.

Cavalier, R. (Ed.). (2005). *The Impact of Internet on Our Moral Life*. Albany, NY: State University of New York Press.

Cavoukian, A. (2010). Privacy by design: the definitive workshop. *Identity in the Information Society*, *3*(2), 247–251. doi:10.1007/s12394-010-0062-y

Cerqui, D., & Arras, K. O. (2003). Human Beings and Robots: Towards a Symbiosis? A 2000 People Survey. In *Proceedings of International Conference on Socio Political Informatics and Cybernetics* (PISTA'03). Orlando, FL: PISTA.

Chang, C. L. (2011). The Effect of an Information Ethics Course on the Information Ethics Values of Students–A Chinese guanxi culture perspective. *Computers in Human Behavior*, *27*, 2028–2038.

Cheating Statistics. (2001). *Caveon test security*. Retrieved from http://www.caveon.com/resources/cheating-statistics/

Chen, K. (2006, January 6). Microsoft defends censoring a dissident's blog in China. *The Wall Street Journal*, p. A9.

Childress, A. R., Mozley, D., McElgin, W., Fitzgerald, J., Reivich, M., & O'Brien, C. P. (1999). Limbic activation during cue-induced cocaine craving. *The American Journal of Psychiatry*, *156*(1), 11–18. PMID:9892292

Choza, J. (1988). *Manual de antropologia filosofica*. Madrid, Spain: Rialp.

Christians, C. G. (2005). Ethical theory in communications research. *Journalism Studies*, *6*(1), 3–14. doi:10.1080/1461670052000328168

Christians, C. G. (2007). Utilitarianism in media ethics and its discontents. *Journal of Mass Media Ethics*, *22*, 113–131. doi:10.1080/08900520701315640

Christians, C. G. (2008). Media ethics in education. *Journalism & Communication Monographs*, *9*(4), 180–221.

Christians, C. G., & Lambeth, E. B. (1996). The status of ethics instruction in communication departments. *Communication Education*, *45*(3), 236–243. doi:10.1080/03634529609379052

Chua, A. Y. K., Goh, D. H., & Ang, R. P. Web 2.0: Applications in government web sites. *Online Information Review*, *36*(2), 175–195. doi:10.1108/14684521211229020

Churchland, P. M. (1998). The Neural Representation of the Social World. In *The Digital Phoenix: How Computers are Changing Philosophy* (pp. 153–170). Oxford, UK: Blackwell Publishers.

Cockton, G. (2004). From quality in Use to Value in the World. In *Proceedings of the CHI'04 Extended Abstracts on Human factors in Computing Systems* (CHI'04). New York: ACM Press.

Code of Ethics for the Information Society proposed by the Intergovernmental Council of the Information for All Programme (IFAP). (2011). *General Conference for consideration UNESCO, 36th session 2011*. Retrieved from http://unesdoc.unesco.org/images/0021/002126/212696e.pdf

Cohen, P. (2011, June 19). Genetic basis for crime: A new look. *The New York Times*.

Cohen, C. J. (2012). Obama, neoliberalism, and the 2012 election. *Souls: A Critical Journal of Black Politics. Cultura e Scuola*, *14*(1-2), 19–27.

Cohn, J. (2013). The robot will see you now. *Atlantic (Boston, Mass.)*, *311*(2), 58–67.

Coleman, S. (1999). Cutting out the middle man: From virtual representation to direct deliberation. In B. N. Hague, & B. Loader (Eds.), *Digital democracy: Discourse and decision making in the information age* (pp. 195–210). New York, NY: Routledge.

Comeau, S. (1997). Getting high on gambling. *McGill Reporter*. Retrieved from http://reporter-archive.mcgill.ca/Rep/r3004/gambling.html

Conboy, M. (2006). *Tabloid Britain: Constructing a community through language*. New York, NY: Routledge.

Consumer Focus. (2010). *Outdated copyright law confuses consumers*. Retrieved from http://www.consumerfocus.org.uk/news/outdated-copyright-law-confuses-consumers

Cooper, T. (1990). Comparative international media ethics. *Journal of Mass Media Ethics*, *5*(1), 3–14. doi:10.1207/s15327728jmme0501_1

Copeland, R. (1990). The presence of mediation. *The Drama Review*, *34*(3).

Costera Meijer, I. (2013). When news hurts. *Journalism Studies*, *14*(1), 13–28. doi:10.1080/1461670X.2012.662398

Council for International Organizations of Medical Sciences/World Health Organization (CIOMS/WHO). (1993). *International ethical guidelines for biomedical research involving human subjects*. Geneva, Switzerland: CIOMS/WHO.

Cowan, R. S. (1997). *A social history of American technology*. New York: Oxford University Press.

Crisp, R. (1997). *Mill on Utilitarianism*. London, UK: Routledge.

Crowell, C. R., Narvaez, D., & Gomberg, A. (2005). Moral psychology and information ethics: Psychological distance and the components of moral behavior in a digital world. In *Information ethics: Privacy and intellectual property*. Hershey, PA: Information Science Publishing.

Cullen, F. T. et al. (1987). *Corporate crime under attack: The Ford Pinto case and beyond*. Cincinnati, OH: Anderson.

Cutas, D. E. (2008). Life extension, overpopulation and the right to life: Against lethal ethics. *Journal of Medical Ethics*, *34*, e7. doi:10.1136/jme.2007.023622 PMID:18757626

Cutter, C., & Smith, M. (2008, April). Gambling addiction and problem gambling, signs, symptoms and treatment. *helpguide.org*. Retrieved from http://www.helpguide.org/mental/gambling_addiction.htm

Cybercrime. (2012). *Britannica online encyclopedia*. Retrieved, October 15, 2012, from www.britannica.com/EBchecked/topic/130595/cybercrime

Dahlberg, L. (2011). Reconstructing digital democracy: An outline of four 'positions'. *New Media & Society*, *13*, 855–866. doi:10.1177/1461444810389569

Danielson, P. (1998). How Computers Extend Artificial Morality. In *The Digital Phoenix: How Computers are Changing Philosophy* (pp. 292–307). Oxford, UK: Blackwell Publishers.

Dario, P., Laschi, C., & Guglielmelli, E. (1999). Design and experiments on a personal robotic assistant. *Advanced Robotics*, *13*(2).

Davis, J. K. (2004). Collective suttee: Is it unjust to develop life extension if it will not be possible to provide it to everyone? *Annals of the New York Academy of Sciences*, *1019*, 535–541. doi:10.1196/annals.1297.099 PMID:15247081

Davis, K. (2012). Brain Trials: Neuroscience is Taking a Stand in the Courtroom. *ABA Journal*.

Davis, R. (1999). *The web of politics: The internet's impact on the American political system*. New York: Oxford University Press.

Dawson, M. C. (2012). Beyond 2012. *Souls: A Critical Journal of Black Politics. Cultura e Scuola*, *14*(1-2), 117–122.

Day, L. A. (2006). *Ethics in media communications: Cases and controversies*. Belmont, CA: Thomson Wadsworth.

De Andres Argente, T. (2002). *Homo cybersapiens: La inteligencia artificial y la humana*. Pamplona, Spain: EUNSA.

De Angeli, A., & Johnson, G. I. (2004). Emotional Intelligence in Interactive Systems. In *Design and Emotion* (pp. 262–266). London: Taylor & Francis.

De Cataldo Neuburger, L. (2010). Aspetti psicologici nella formazione della prova: Dall'ordalia alle neuroscienze. *Diritto Penale e Processo, 5*.

De Grey, A., & Rae, M. (2007). *Ending aging: The rejuvenation breakthroughs that could reverse human aging in our lifetime*. New York, NY: St. Martin's Press.

de Mooij, M. (2014). Mass media, journalism, society, and culture. In M. de Mooij (Ed.), *Human and mediated communication around the world* (pp. 309–353). Cham, Switzerland: Springer International Publishing. doi:10.1007/978-3-319-01249-0_10

Dean, J. (2005, December 16). As Google pushes into China, it faces clashes with censors. *The Wall Street Journal*, p. A12.

Decew, J. (2013). Privacy. *The Stanford Encyclopedia of Philosophy*. Retrieved from http://plato.stanford.edu/archives/fall2013/entries/privacy/

Decew, J. (1997). *In pursuit of privacy: Law, ethics, and the rise of technology*. Ithaca, NY: Cornell University Press.

Decker, M. (2007). Can Humans Be Replaced by Autonomous Robots? Ethical Reflections in the Framework of an Interdisciplinary Technology Assessment. In *Proceedings of ICRA'07 Workshop on RoboEthics*. Rome, Italy: ICRA.

DeGeorge, R. T. (1990). *Business ethics* (3rd ed.). New York: Macmillan.

Deibert, R. J., Palfrey, J. G., Rohozinski, R., & Zittrain, J. (2008). *Access denied: The practice and policy of global internet filtering*. Cambridge, MA: The MIT Press.

Deplaces, U. (2011). Technological Enhancements of the Human Body: A Conceptual Framework. *Acta Philosiphica*, *20*, 53–72.

Dines, M. (2013). The conversation: Responses and reverberations. *Atlantic (Boston, Mass.)*, *311*(4), 10–11.

Douglas, T. (2008). Moral enhancement. *Journal of Applied Philosophy*, *25*, 228–245. doi:10.1111/j.1468-5930.2008.00412.x PMID:19132138

Duff, A. (2008). The Normative Crisis of the Information Society. *Cyberpsychology. Journal of Psychosocial Research on Cyberspace*, *2*(1).

Duffy, M. J., & Freeman, C. P. (2011). *Unnamed sources: A utilitarian exploration of their justification and guidelines for limited use*. Retrieved from http://digitalarchive.gsu.edu/cgi/viewcontent.cgi?filename=0&article=1012&context=communication_facpub&type=additional

Eastwood, L. (2008). Google faces the Chinese internet market and the Global Online Freedom Act. *Minnesota Journal of Law, Science, &. Technology (Elmsford, N.Y.)*, *9*, 287.

Economic Times. (2013, August 7). *78 govt websites hacked till June this year: Milind Deora*. Retrieved from http://articles.economictimes.indiatimes.com/2013-08-07/news/41167675_1_cyber-attacks-national-cyber-security-policy-websites

Efrati, A., & Chao, L. (2012, January 12). Google softens China stance. *The Wall Street Journal*, pp. B1-B2.

Eid, M. (2008). *Interweavement: International media ethics and rational decision-making*. Boston, MA: Pearson.

Eligon, J. (2013, August 20). Deaf student, denied interpreter by medical school, draws focus of advocates. *The New York Times*, p. A11.

Elliot, C. (2011). Enhancement technologies and the modern self. *The Journal of Medicine and Philosophy*, *36*, 364–374. doi:10.1093/jmp/jhr031 PMID:21903906

Elliott, D. (2007). Getting Mill right. *Journal of Mass Media Ethics*, *22*, 100–112. doi:10.1080/08900520701315806

Erickson, S. K. (2012). The limits of neurolaw. *Houston Journal of Health Law & Policy*, 11.

Ess, C. (2006). Universal Information Ethics? Ethical Pluralism and Social Justice. In E. Rooksby, & J. Weckert (Eds.), *Information technology and Social Justice* (pp. 69–92). Hershey, PA: Idea Publishing. doi:10.4018/978-1-59140-968-7.ch004

Ess, C., & Thorseth, M. (2010). Global information and computer ethics. In L. Floridi (Ed.), *The Cambridge handbook of information and computer ethics* (pp. 163–180). Cambridge, UK: Cambridge University Press. doi:10.1017/CBO9780511845239.011

Esser, F. (1999). Tabloidization' of news. *European Journal of Communication*, *14*, 291–324. doi:10.1177/0267323199014003001

Ey, H. (1998). *Conscience*. Bucharest: Editura Ştiinţifică.

Fabro, C. (1955). *L'anima*. Roma, Italy: Studium.

Farah, M. (2002). Emerging ethical issues in neuroscience. *Nature Neuroscience*, *5*(11), 1123–1129. doi:10.1038/nn1102-1123 PMID:12404006

Farrelly, C. (2010). Equality and the duty to retard human ageing. *Bioethics*, *24*(8), 384–394. doi:10.1111/j.1467-8519.2008.00712.x PMID:19222442

Fecht, S. (2013). Where the Amish and technology meet. *Popular Mechanics*, *190*(2), 17.

Feresin, E. (2009). Lighter sentence for murderer with bad genes. *Nature*. doi:10.1038/news.2009.1050

Filartiga v. Pena-Irala 630 F. 2d 876, 2d Cir. (1980).

Finnis, J. (1980). *Natural law and natural rights*. Oxford, UK: Oxford University Press.

Flanagan, M., Howe, D. C., & Nissenbaum, M. (2008). Embodying values in technology: Theory and practice. In J. van den Hoven, & J. Weckert (Eds.), *Information technology and moral philosophy* (pp. 322–353). New York: Cambridge University Press.

Flanagin, A. J., & Metzer, M. J. (2000). Perceptions of Internet information credibility. *Journalism & Mass Communication Quarterly*, *77*, 515–540. doi:10.1177/107769900007700304

Floridi, L. (1999). Information Ethics: On the Theoretical Foundation of Computer Ethics. *Ethics and Information Technology*, *1*(1), 37–56. doi:10.1023/A:1010018611096

Floridi, L. (2006). Four challenges for a theory of informational privacy. *Ethics and Information Technology*, *8*(3), 109–119. doi:10.1007/s10676-006-9121-3

Floridi, L. (2011). The informational nature of personal identity. *Minds and Machines*, *21*(4), 549–566. doi:10.1007/s11023-011-9259-6

Floridi, L. (2014). *The fourth revolution – The impact of information and communication technologies on our Lives*. Oxford, UK: Oxford University Press.

Fong, T., Nourbakhsh, I., & Dautenhahn, K. (2003). A survey of socially interactive robots. *Robotics and Autonomous Systems*, 42.

Foot, K. A., Schneider, S. M., & Dougherty, M. (2007). Online structure for political action in the 2004 U.S. congressional electoral web sphere. In R. Kluver, N.W. Jankowski, A. Foot, & S.M. Schneider. The internet and national elections: A comparative study of web campaigning (pp. 92-104). New York (NY).

Foot, P. (1979, June). *Moral relativism*. Paper presented for Lindley Lecture, Department of Philosophy, University of Kansas. Lawrence, KS.

Foot, P. (2001). *Natural Goodness*. Oxford, UK: Blackwell.

Forza, A. (2010). Le neuroscienze entrano nel processo penale. *Rivista Penale, 1*.

Foster, J. (2001). A Brief Defense of the Cartesian View. In *Soul, Body and Survival* (pp. 15–29). Ithaca, NY: Cornell University Press.

Fountain, J. E. (2002). Toward a theory of federal bureaucracy in the 21st century. In E. C. Kamarck, & J. S. Nye (Eds.), *Governance.com: Democracy in the information age* (pp. 117–140). Washington, D.C.: Brookings Institution Press.

FoxNews. (2010, April 15). 7,500 Online Shoppers Unknowingly Sold Their Souls. *Fox News*. Retrieved from http://www.foxnews.com/scitech/2010/04/15/online-shoppers-unknowingly-sold-souls/

Fox, R. C., & Swazey, J. P. (2002). *The courage to fail: A social view of organ transplants and dialysis*. Piscataway, NJ: Transaction Publishers.

Frankena, W. K. (1963). *Ethics*. Englewood Cliffs, NJ: Prentice-Hall.

Frankl, V. E. (1963). *Man's Search for Meaning: An Introduction to Logotherapy* (I. Lasch, Trans.). New York: Washington Square Press.

Frankl, V. E. (1997). *Man's Search for Ultimate Meaning*. New York: Peruses Book Publishing.

Freeman, L. A., & Peace, A. G. (2005). Revisiting Mason: The last 18 years and onward. In *Information ethics: Privacy and intellectual property*. Hershey, PA: Information Science Publishing.

Freitas, R. (2001). Some limits to global ecophagy by biovorous nanoreplicators, with public policy recommendations. *Foresight Institute*. Retrieved July 26, 2012, from http://www.foresight.org/nano/Ecophagy.html

Friedel, S. D. (2007). *A culture of improvement: Technology and the Western millennium*. Cambridge, MA: The MIT Press.

Friedman, B., Howe, D. C., & Felten, E. (2002). Informed consent in the Mozilla browser: Implementing value-sensitive design. In *Proceedings of 35th Annual Hawaii International Conference on System Sciences*. IEEE Computer Society.

Friedman, B., Kahn, P. H., Jr., & Borning, A. (2006). Value sensitive design and information systems. In Human-computer interaction in management information systems: Foundations (pp. 348-372). New York: Armonk.

Friedman, B. (1986). Value-sensitive design. *Interaction, 3*(6), 17–23.

Friedman, B. (1997). *Human Values and the Design of Computer Technology*. CSLI Publications.

Friedman, B., & Kahn, P. H. Jr. (2003). Human values, ethics, and design. In J. Jacko, & A. Sears (Eds.), *The human-computer interaction handbook* (pp. 1177–1201). Mahwah, NJ: Lawrence Erlbaum Associates.

Frondizi, R. (1971). *What is Value? An Introduction to Axiology by Risiery Frondizi*. La Salle.

Fukuyama, F. (2002). *Our posthuman future: Consequences of the biotechnology revolution*. New York, NY: Picador.

Gabbay, D., Thagard, P., & Woods, J. (2008). *Philosophy of Information*. North-Holland.

Galatino, N. (2005). Koerper e Leib: tra determinismo biologico e determinismo culturale. In *La sfida del post-umano: Verso nuovi modelli di esistenza?* (pp. 49–65). Roma, Italy: Studium.

Gale, K., & Bunton, K. (2005). Assessing the impact of ethics instruction on advertising and public relations graduates. *Journalism & Mass Communication Educator, 60*(3), 272–285. doi:10.1177/107769580506000306

Galimberti, U. (2000). *Psichè e tecnè*. Milano, Italy: Feltrinelli.

Galvan, J. M. (2008). Creation and casuality: The case in Christian theological anthropology. In M. Negrotti (Ed.), Yearbook of the artificial: Nature, culture & technology: Natural chance, artificial chance (vol. 5, pp. 129.141). Bern, Switzerland: Peter Lang Verlag.

Galvan, J. M. (2004). On technoethics. *IEEE Robotics and Automation Society Magazine, 10*, 58–63.

Galvan, J. M. (2011). La tecnoetica e le speranze umane. In *Scienza, Tecnologia e Valori Morali: Quale futuro?* (pp. 175–185). Roma, Italy: Armando Editore.

Gambleaware. (n.d.). Responsible gambling. *Gambleaware*. Retrieved from http://gambleaware.co.uk/responsible-gambling

Gans, H. J. (2009). Can popularization help the news media. In B. Zeilzer (Ed.), *The changing faces of journalism tabloidization, technology, and truthiness* (pp. 17–28). New York, NY: Routledge.

GAO. (2002). Internet Gambling: An Overview of the Issues. *United States General Accounting Office*. Retrieved from http://www.gao.gov/new.items/d0389.pdf

Garfinkel, S., & Spafford, G. (1997). *Web security and commerce*. Sebastopol, CA: O'Reilly.

Garland, B., & Glimcher, P. W. (2006). Cognitive neuroscience and the law. *Current Opinion in Neurobiology, 16*, 130–134. doi:10.1016/j.conb.2006.03.011 PMID:16563731

Garrett, J. (2004). *A Simple and Usable (Although Incomplete) Ethical Theory Based on the Ethics of W. D. Ross*. Western Kentucky University. Retrieved from http://www.wku.edu/~jan.garrett/ethics/rossethc.htm

Garvy, H. (2007). *Rebels with a cause: A collective memoir of the hopes, rebellions and repression of the 1960s*. Los Gatos, California: Shire Press.

Gaziano, C., & McGrath, K. (1986). Measuring the concept of credibility. *The Journalism Quarterly, 63*, 451–462. doi:10.1177/107769908606300301

Geach, P. T. (1956). Good and evil. *Analysis, 17*(2), 33–42. doi:10.1093/analys/17.2.33

Gearhart, D. (2000). *Ethics in distance education: Are distance educators and administrators following ethical practices?* Unpublished Manuscript.

Gearhart, D. (2005). *The ethical use of technology and the Internet in research and learning*. Paper presented at the Center of Excellence in Computer Information Systems 2005 Spring Symposium. New York, NY.

Gems, D. (2003). Is more life always better? The new biology of aging and the meaning of life. *The Hastings Center Report, 33*(4), 31–39. doi:10.2307/3528378 PMID:12971059

George, R. P. (2007). Natural law. *The American Journal of Jurisprudence, 52*, 55. doi:10.1093/ajj/52.1.55

Gibson, W. (1984). *Neuromancer*. New York: Ace Books.

Giddens, A. (1990). *The Consequences of Modernity*. Cambridge, MA: Polity.

Gilbert, R. L., Murphy, N. A., & Ávalos, C. M. (2011). Communication Patterns and Satisfaction Levels in Three-Dimensional Versus Real-Life Intimate Relationships. *Cyberpsychology, Behavior, and Social Networking, 14*(10), 585–589. PMID:21381970

Gingras, Y. (2005). Éloge de l'homo techno-logicus. Saint-Laurent, Canada: Les Ed.s Fides.

Ginsburg, J. (2003). From having copies to experiencing works: The development of an access right in US copyright law. *Journal of the Copyright Society of the USA, 50*, 113–131.

Gladney, G. A., Shaprio, I., & Castaldo, J. (2007). Online editors rate web news quality criteria. *Newspaper Research Journal, 28*, 55–69.

Glannon, W. (2002). Identity, prudential concern, and extended lives. *Bioethics, 16*(3), 266–283. doi:10.1111/1467-8519.00285 PMID:12211249

Glimne, D. (n.d.). Gambling. *Encyclopaedia Britannica*. Retrieved from http://www.britannica.com

Golan, G. J., & Kiousis, S. K. (2010). Religion, media credibility, and support for democracy in the Arab world. *Journal of Media and Religion, 9*, 84–98. doi:10.1080/15348421003738793

Goldsmith, J. (1998). Against cyberanarchy. *The University of Chicago Law Review. University of Chicago. Law School, 65*(4), 1199–1250. doi:10.2307/1600262

Goldsmith, J., & Wu, T. (2008). *Who controls the internet?* Oxford, UK: Oxford University Press.

Goldstein, M. B. (2011). *Living with the (rest of the) new program integrity regulations in an online environment*. Paper presented at the 23rd Annual WCET Conference. Denver, CO.

Goodenough, O. R., & Tucker, M. (2010). Law and Cognitive Neuroscience. *Annu. Rev. Law Soc. Sci., 6*, 61–92. doi:10.1146/annurev.lawsocsci.093008.131523

Goodwin, D. (2013, November 14). Google fails to gain search market share. *Search Engine Watch*. Retrieved December 28, 2013, from http://searchenginewatch.com/article/2307115/Google-Fails-to-Gain-Search-Market-Share

Google in China Editorial. (2006, January 30). *The Wall Street Journal*, p. A18.

Gordon, A. D. (1999). Media codes of ethics are useful and necessary to the mass media and to society. In A. D. Gordon, & J. M. Kittross (Eds.), *Controversies in media ethics* (pp. 61–68). New York: Longman.

Greene, J. D., & Cohen, J. D. (2004). For the law, neuroscience changes nothing and everything. *Philosophical Transactions of the Royal Society of London. Series B, Biological Sciences, 359*, 1775–1778. doi:10.1098/rstb.2004.1546 PMID:15590618

Greenwald, G., MacAskill, E., & Poitras, L. (2013). Edward Snowden: The whistleblower behind the NSA surveillance revelations. *The Guardian*, p. 9.

Gregory, R. (2000). *Future of Mind-Makers*. Bucharest: Editura Ştiinţifică.

Griffin, J. (1986). *Well-being. Its meaning, measurement and moral importance*. New York, NY: Clarendon Press.

Griffin, J. (1997). Incommensurability: What's the problem? In R. Chang (Ed.), *Incommensurability, incomparability, and practical reason* (pp. 35–51). Cambridge, MA: Harvard University Press.

Grisez, G. (1991). *Difficult moral questions.* Chicago: Franciscan Herald Press.

Ground Zero Summit 2013, Asia's Largest Information Security Conference, Launched. (2013, August 9). Retrieved August 18, 2013, from http://www.prnewswire.co.in/: http://www.prnewswire.co.in/news-releases/ground-zero-summit-2013-asias-largest-information-security-conference-launched-218973741.html

Guardini, R. (1994). *Ethik: Vorlesungen an der Universität München, 1950-1962.* Padenborn, Germany: Matthias-Gruenewald Verlag.

Gyanfinder. (2012). *Ethics in online education.* Retrieved from http://www.gyanfinder.com/blogs/1/42/ethics-in-online-education

H2 Gambling Capital. (2010). United States Internet Gambling: Job Creation – Executive Summary. *H2 Gambling Capital.* Retrieved from www.safeandsecureig.org/news/InternetGamblingStudy_04_15_2010.pdf

Haddon, L. (2004). *Information and Communication Technologies in Everyday Life: A Concise Introduction and Research Guide.* Berg.

Haftendorn, H. (1991). The security puzzle: Theory building and discipline building in international security. *International Studies Quarterly, 35*(1), 3–17. doi:10.2307/2600386

Hainz, T. (2012b). Aging as a disease. In M. G. Weiss & H. Greif (Ed.), *Ethics – society – politics: Papers of the 35th International Wittgenstein Symposium* (pp. 109-111). Kirchberg: Austrian Ludwig Wittgenstein Society.

Hainz, T. (2012a). Value lexicality and human enhancement. *International Journal of Technoethics, 3*(4), 54–65. doi:10.4018/jte.2012100105

Hanson, G. (2002). Learning journalism ethics: The classroom versus the real world. *Journal of Mass Media Ethics, 17*(3), 235–247. doi:10.1207/S15327728JMME1703_05

Hansson, S. O. (2001). *The Structure of Value and Norms.* New York: Cambridge University Press.

Hardin, G. (1968). The tragedy of the commons. *Science, 162,* 1243–1248. doi:10.1126/science.162.3859.1243 PMID:5699198

Hargreaves, I. (2011). *Digital Opportunity: A Review of Intellectual Property and Growth.* Retrieved from http://www.ipo.gov.uk/ipreview-finalreport.pdf

Harrington, S. (2008). Popular news in the 21st century. *Journalism, 9,* 266–284. doi:10.1177/1464884907089008

Harris, J. (2007). *Enhancing evolution: The ethical case for making better people.* Princeton, NJ: Princeton University Press.

Hart, H. L. (1983). *Essays in jurisprudence and philosophy.* Oxford, UK: Oxford University Press. doi:10.1093/acprof:oso/9780198253884.001.0001

Hartman, R. S. (1973). Formal Axiology and the Measurement of Values. In *Value Theory in Philosophy and Social Science.* New York: Gordon and Breach.

Hausmanninger, T. (2007). Allowing for difference: Some preliminary remarks concerning intercultural information ethics. In *Localizing the Internet: Ethical issues in intercultural perspective* (pp. 39–56). Munich: Wilhelm Fink Verlag.

Hayflick, L. (2004). Anti-aging is an oxymoron. *Journal of Gerontology, 59A*(6), 573–578.

Heathwood, C. (2010). Welfare. In J. Skorupski (Ed.), *The Routledge companion to ethics* (pp. 645–655). Abingdon, UK: Routledge.

Heidegger, M. (1927). *Sein und Zeit.* Frankfurt am Main: Vittorio Klostermann Verlag.

Heidegger, M. (2000). *Being and Time* (J. Macquarrie, & E. Robinson, Trans.). London: Blackwell Publishing Ltd.

Heide, T. (2001). Copyright in the EU and the U.S.: What access right? *European Intellectual Property Review, 23*(8), 469–477.

Heisler, W., Westfall, F., & Kitahara, R. (2012). Technological approaches to maintaining academic integrity in management education. In *Handbook of research on teaching ethics in business and management education.* Hershey, PA: IGI Global.

Heller, P. B. (1992). Ethical considerations in the international distribution of new contraceptive technology: The case of Norplant and RU 486. In J. R. Wilcox (Ed.), *The internationalization of American business: Ethical issues and cases* (pp. 57–75). New York: McGraw-Hill.

Heller, P. B. (2012a). Technoethics. *International Journal of Technoethics*, *3*(1), 14–27. doi:10.4018/jte.2012010102

Heller, P. B. (2012b). *Technology and Society Reader: Case Studies* (rev. ed.). Lanham, MD: University Press of America.

Hemingway, C. J., & Gough, T. G. (2000). The Value of Information Systems Teaching and Research in the Knowledge Society. *Informing Science*, *4*(3), 167–184.

Herrscher, R. (2002). A universal code of journalism ethics: Problems, limitations, and proposals. *Journal of Mass Media Ethics*, *17*(4), 277–289. doi:10.1207/S15327728JMME1704_03

Herzfeld, N. (2009). *Technology and religion: Remaining human in a co-created world.* West Conshohocken, PA: Templeton Press.

Hildt, E. (2009). Living longer: Age retardation and autonomy. *Medicine, Health Care, and Philosophy*, *12*, 179–185. doi:10.1007/s11019-008-9162-y PMID:18668344

Hillis, W. D. (2001). *The Pattern on the Stone: The Simple Ideas that Make Computers Work.* Bucharest: Editura Humanitas.

Holtzman, D. H. (2006). *Privacy lost: How technology is endangering your privacy*. New York: Jossey-Bass.

Hong, S., & Nadler, D. (2012). Which candidates do the public discuss online in an election campaign?: The use of social media by 2012 presidential candidates and its impact on candidate salience. *Government Information Quarterly*, *29*(4), 455–461. doi:10.1016/j.giq.2012.06.004

Horrobin, S. (2006). Immortality, human nature, the value of life and the value of life extension. *Bioethics*, *20*(6), 279–292. doi:10.1111/j.1467-8519.2006.00506.x

Hovland, C. I., & Weiss, W. (1951). The influence of source credibility on communication effectiveness. *Public Opinion Quarterly*, *15*, 635–650. doi:10.1086/266350

iGambling Business. (n.d.). Rosy future for online casino industry. *iGambling Business*. Retrieved from http://www.igamingbusiness.com/content/rosy-future-online-casino-industry

India Infoline News Service. (2013, August 10). *10% of all cyber crimes globally happen in India: Study*. Retrieved December 12, 2013, from http://www.indiainfoline.com/Markets/News/10-percent-of-all-cyber-crimes-globally-happen-in-India-Study/5754085645

International Journal of Technoethics. (n.d.). Retrieved from http://www.igi-global.com/journal/international-journal-technoethics-ijt/1156

Internet Crime Complaint Center (IC3). (2012). *Internet Crime Report*. Retrieved from https://www.ic3.gov/media/annualreport/2012_IC3Report.pdf

Irrgang, B. (2005). *Posthumanes Menschsein?* Weisbaden, Germany: Franz Steiner Verlag.

Jarfinkel, S. (2000). *Database nation: The death of privacy in the 21st century*. Sebastopol, CA: O'Reilly.

Jennett, B. (1986). *High-technology medicine: Burdens and benefits* (p. 174). Oxford: Oxford Universty Press.

Jewkes, Y. (Ed.). (2003). *Dot.cons: Crime, deviance and identity on the internet*. Cullompton, UK: Willan.

Johansson, S. (2008). Gossip, sport and pretty girls. *Journalism Practice*, *2*, 402–413. doi:10.1080/17512780802281131

Johnson, D. (1985). *Computer ethics*. Englewood Cliffs, NJ: Prentice-Hall.

Johnson, T. J., & Kaye, B. K. (2010). Still cruising and believing? An analysis of online credibility across three presidential campaigns. *The American Behavioral Scientist*, *54*, 57–77. doi:10.1177/0002764210376311

Johnson, T. J., Kaye, B. K., Bichard, S. L., & Wong, W. J. (2007). Every blog has its day: Politically-interested internet users' perceptions of blog credibility. *Journal of Computer-Mediated Communication*, *13*, 100–122. doi:10.1111/j.1083-6101.2007.00388.x

Jonas, H. (1985). *The imperative of responsibility: In search of an ethics for the technological age*. The University of Chicago Press.

Joy, B. (2000). Why the future doesn't need us. *Wired*. Retrieved July 26, 2012, from http://www. wired.com/wired/archive/8.04/joy_pr.html

Kadic v. Karadzic 70 F.3d 232, 2d Cir. (1995).

Kahn, P. H. Jr, Ishiguro, Friedman, Kanda, Freier, Severson, & Miller. (2007). What is a Human? Toward Psychological Benchmarks in the Field of Human-Robot Interaction. *Interaction Studies: Social Behaviour and Communication in Biological and Artificial Systems*, *8*(3), 363–390. doi:10.1075/is.8.3.04kah

Kallenberg, B. J. (2001). *Ethics and Grammar: Changing the Postmodern Subject*. Notre Dame Press.

Kaspersky. (2012, October 10). *Kaspersky Security Bulletin2012*. Retrieved from http://www.securelist.com/en/analysis/204792255/Kaspersky_Security_Bulletin_2012_The_overall_statistics_for_2012

Katyal, N. (2002). Architecture as crime control. *The Yale Law Journal*, *111*(5), 1039–1139. doi:10.2307/797618

Katyal, N. (2003). Digital architecture as crime control. *The Yale Law Journal*, *112*(6), 101–129.

Kaufman, D., Stasson, M. F., & Hart, J. W. (1999). Are the tabloids always wrong or is that just what we think? Need for cognition and perceptions of articles in print media. *Journal of Applied Social Psychology*, *29*, 1984–1997. doi:10.1111/j.1559-1816.1999.tb00160.x

Kelly, K. (2010). *What technology wants* (p. 194). New York: Penguin Books.

Kennedy, M. (2012, December 14). Gary McKinnon will Face No Charges in UK. *The Guardian*. Retrieved from http://www.theguardian.com/world/2012/dec/14/gary-mckinnon-no-uk-charges

Kerdellant, C., & Crezillon, G. (2006). *Children of the Processor: How Internet and Video Games Form the Adults of Tomorrow* (A. Luschanova, Trans.). U-Faktoriya.

Kevorkian, J. Obituary (2011), *The Economist*, 399(8737), 88.

Kevorkian, J. Obituary. (2011). *The Economist*, *399*(8737), 88.

Khosbin, L. S., & Khosbin, S. (2007). Imaging the mind, minding the image: An historical introduction to brain imaging and the law. *American Journal of Law & Medicine*, *33*, 171–192. PMID:17910156

Kim, K. S., & Pasadeos, Y. (2007). Study of partisan news readers reveals hostile media perceptions of balanced stories. *Newspaper Research Journal*, *28*, 99–106.

Kiobel v. Royal Dutch Petroleum Co. 133 U.S. 1659 (2013).

Klaming, L. (2011). The influence of neuroscientific evidence on legal decision-making: the effect of presentation mode. In *Technologies on the stand: Legal and ethical questions in neuroscience and robotics*. Njimegen, The Netherlands: WLP.

Kraybill, D. B., Johnson-Weiner, K. M., & Nolt, S. M. (2013). *The Amish*. Baltimore, MD: The Johns Hopkins University Press.

Kurtz, H. (1993). *Media circus: The trouble with America's newspapers*. New York, NY: Times Books.

Lain Entralgo, P. (1991). *El cuerpo humano: Una teoria actual*. Madrid, Spain: Espasa.

Lain Entralgo, P. (1995). *Alma, cuerpo, persona*. Barcelona, Spain: Circulo de Lectores.

Laruelle, F. (1990). *Théorie des identitées, fractalité generalisée et philosophie artificielle*. Paris: P. U. F.

Lebedev, M. A., & Nicolelis, M. A. L. (2006). Brain-machines interfaces: Past, present and future. *Trends in Neurosciences*, *29*(9), 536–546. doi:10.1016/j.tins.2006.07.004

Lee, B., & Padgett, G. (2000). Evaluating the effectiveness of a mass media ethics course. *Journalism and Mass Communication Educator*, *55*(2), 27–39. doi:10.1177/107769580005500204

Lee, T. (2010). Why they don't trust the media: An examination of factors predicting trust. *The American Behavioral Scientist*, *54*, 8–21. doi:10.1177/0002764210376308

Leslie, L. Z. (2000). *Mass communication ethics: Decision making in postmodern culture*. Boston, MA: Houghton Mifflin Company.

Lessig, L. (1999). *Code and other laws of cyberspace*. New York: Basic Books.

Letters to the editor, Dr. machine will see you now.... (2011, March 2). *The New York Times*, p. A24.

Letters to the editor, The special doctor-patient relationship (2011, April 26). *The New York Times*, p. A24.

Letters to the editor, Treat the patient not the CT scan (2011, February 28). *The New York Times*, p. WK10.

Levinas, E. (1979). The Contemporary Criticism of the Idea of Value and the Prospects for Humanism. In *Value and Values in Evolution*. New York: Gordon and Breach.

Lewis, E. (2003). *Gambling and Islam: Clashing and Co-existing*. Brigham Young University. Retrieved from http://www.math.byu.edu/~jarvis/gambling/student-papers/eric-lewis.pdf

Lilleker, D. G., & Jackson, N. A. (2010). Towards a more participatory style of election campaigning: The impact of web 2.0 on the UK 2010 general election. *Policy & Internet*, *2*(3), 69–98. doi:10.2202/1944-2866.1064

Linders, D. (2012). From e-government to we-government: Defining a typology for citizen coproduction in the age of social media. *Government Information Quarterly*, *29*(4), 446–454. doi:10.1016/j.giq.2012.06.003

Lindlof, T. R., & Taylor, B. C. (2002). *Qualitative communication research methods*. Thousand Oaks, California: Sage.

Lioukadis, G., Lioudakisa, G., Koutsoloukasa, E., Tselikasa, N., Kapellakia, S., & Prezerakosa, G. et al. (2007). A middleware architecture for privacy protection. *The International Journal of Computer and Telecommunications Networking*, *51*(16), 4679–4696.

Lloyd, I. (2008). *Information technology law*. Oxford, UK: Oxford University Press.

Logan, R. A. (1985-1986). USA Today's innovations and their impact on journalism ethics. *Journal of Mass Media Ethics*, *1*, 74-87.

Lombardo, T. J. (1987). *The reciprocity of Perceiver and Environment: The evolution of James J. Gibson's ecological psychology*. Hillsdale, NJ: L. Erlbaum Associates.

LoSchiavo, F. M., & Shatz, M. A. (2011). The impact of an honor code on cheating in online courses. *MERLOT Journal of Online Learning and Teaching*, *7*(2).

Luppicini, R. (2008). Introducing technoethics. In *Handbook of research on technoethics* (pp. 1–18). Hershey, PA: Idea Group. doi:10.4018/978-1-60566-022-6.ch001

Luppicini, R. (2009). Technoethical inquiry: From technological systems to society. *Global Media Journal*, *2*(1), 5–21.

Luppicini, R. (2010). *Technoethics and the evolving knowledge society*. Hershey, PA: Idea Group. doi:10.4018/978-1-60566-952-6

Luppicini, R. (2013). *Handbook of research on technoself: Identity in a technological society* (Vol. 1-2). Hershey, PA: IGI Global.

Luppicini, R. (Ed.). (2012b). *Ethical impact of technological advancements and applications in society*. Hershey, PA: Idea Group Publishing. doi:10.4018/978-1-4666-1773-5

Luppicini, R., & Adell, R. (Eds.). (2008). *Handbook of research on technoethics*. Hershey, PA: Idea Group. doi:10.4018/978-1-60566-022-6

MacBride, S. (1980). *Many voices, one world: Towards a new more just and more efficient world information and communication order*. Paris: UNESCO.

Maccarone, E. M. (2010). Ethical responsibilities to subjects and documentary filmmaking. *Journal of Mass Media Ethics*, *25*, 192–206. doi:10.1080/08900523.2010.497025

Mackey, T. (2003). An ethical assessment of anti-aging medicine. *Journal of Anti-aging Medicine*, *6*(3), 187–204. doi:10.1089/109454503322733045 PMID:14987433

MacKinnon, R. (2010, March 24). Testimony for the congressional-executive commission on China. *Congressional Hearing on Google and Internet Control in China*. Retrieved January 2, 2014 from http://rconversation.blogs.com/MacKinnonCECC_Mar24.pdf

MacKinnon, R. (2012). *Consent of the networked*. New York: Basic Books.

Malina, A. (1999). Perspectives on citizen democratization and alienation in the virtual public sphere. In B. N. Hague, & B. Loader (Eds.), *Digital democracy: Discourse and decision making in the Information Age* (pp. 23–38). New York, NY: Routledge.

Maner, W. (1996). Unique ethical problems in information technology. *Science and Engineering Ethics*, *2*, 137–154. doi:10.1007/BF02583549

Mann, S., Nolan, J., & Wellman, B. (2003). Sousveillance: Inventing and using wearable computing devices for data collection in surveillance environments. *Surveillance & Society*, *1*(3), 331–355.

Marino, D., & Tamburrini, G. (2006). Learning robots and human responsibility. *Int. Rev. Inform. Ethics, 6.*

Markel, M. (2001). *Ethics in technical communication: A critique and synthesis*. Westport, CT: Ablex Publishing.

Maslow, A. (1966). *The Psychology of Science: A Reconnaissance*. New York: Harper & Row.

Mason, R. O. (1986). Four ethical issues of the information age. *Management Information Systems Quarterly*, *10*(1). doi:10.2307/248873

Masuda, Y. (1983). *The Information Society as Post-Industrial Society*. Washington, DC: Academic Press.

Matthias, A. (2004). The responsibility gap: Ascribing responsibility for the actions of learning automata. *Ethics and Information Technology*, 6.

McCroskey, J. C., & Teven, J. J. (1999). Goodwill: A reexamination of the construct and its measurement. *Communication Monographs*, *66*, 90–103. doi:10.1080/03637759909376464

McKenzie, R. (2006). *Comparing media from around the world*. Boston, MA: Pearson Education.

McKeown, T. (1979). *The role of medicine: Dream, mirage, or nemesis?* Princeton, NJ: Princeton University Press.

McKinnon (Appellant) *v* Government of the United States of America (Respondents) and Another (2008) UKHL 59. House of Lords.

McLachlan, S., & Golding, P. (2000). Tabloidization in the British press: A quantitative investigation into changes in British Newspapers. In C. Sparks, & J. Tulloch (Eds.), *Tabloid tales: Global debates over media standards* (pp. 76–90). Lanham, MD: Rowman and Littlefield.

McLaren, B. M., & Ashley, K. D. (2000). Assessing Relevance with Extensionally Defined Principles and Cases. In *Proceedings of the 17th National Conference of Artificial Intelligence*. AAAI Press.

McLhuan, M. (1994). *Understanding Media: The Extensions of Man*. The MIT Press.

McLuhan, M. (1951). *Mechanical bride: Folklore of industrial man*. New York: Vanguard Press.

McMillan, J., & Buhle, P. (2003). *The new left revisited*. Philadelphia, PA: Temple University Press.

McNabb, L., & Olmstead, A. (2009). Communities of integrity in online courses: Faculty member beliefs and strategies. *MERLOT Journal of Online Learning and Teaching, 5*(2).

Meilaender, G. C. (1995). *Body, soul, and bioethics*. Notre Dame, IN: University of Notre Dame Press.

Melican, D. B., & Dixon, T. L. (2008). News on the net: Credibility, selective exposure, and racial prejudice. *Communication Research*, *35*, 151–168. doi:10.1177/0093650207313157

Merleau-Ponty, M. (1968). The Intertwining - The chiasm. In *The Visible and the Invisible*. Evanston, IL: Northwestern University Press.

Mill, J. S. (2009). Utilitarianism. R. Crisp (Ed.), J.S. Mill Utilitarianism. New York, NY: Oxford.

Miller, B. (2001). The concept of security: Should it be redefined? *The Journal of Strategic Studies*, *24*(2), 13–42. doi:10.1080/01402390108565553

Miller, D. W., Urbaczewski, A., & Salisbury, W. D. (2005). Does public access imply ubiquitous or immediate? Issues surrounding public documents online. In *Information ethics: Privacy and intellectual property*. Hershey, PA: Information Science Publishing.

Miller, R. A. (2002). Extending life: Scientific prospects and political obstacles. *The Milbank Quarterly*, *80*(1), 155–174. doi:10.1111/1468-0009.00006 PMID:11933792

Millgram, E. (2005). Incommensurability and practical reasoning. In E. Millgram (Ed.), *Ethics done right: Practical reasoning as a foundation for moral theory* (pp. 273–294). Cambridge, UK: Cambridge University Press. doi:10.1017/CBO9780511610615.011

Mill, J. S. (1871). *Utilitarianism*. London: Longmans, Green, Reader, and Dyer.

Mintu-Wimsatt, A., Kernek, C., Lozada, H. R. (2010). Netiquette: Make it part of your syllabus. *MERLOT Journal of Online Learning and Teaching, 6*(1).

Mitcham, C. (1995). Ethics into design. In R. Buchanan, & V. Margolis (Eds.), *Discovering design* (pp. 173–179). Chicago: University of Chicago Press.

Mitre, H., González-Tablas, A., Ramos, B., & Ribagorda, A. (2006). A legal ontology to support privacy preservation in location-based services. *Lecture Notes in Computer Science, 4278*, 1755–1764. doi:10.1007/11915072_82

Moor, J. (1985). What is computer ethics? *Metaphilosophy, 16*(4), 266–275. doi:10.1111/j.1467-9973.1985.tb00173.x

Morely, D., & Robins, K. (2013). *Space of identity: Global media, electronic landscapes and cultural boundaries*. London: Routledge.

Morse, S. J. (2007). The non-problem of free will in forensic psychiatry and psychology. *Behavioral Sciences & the Law, 25*, 203–220. doi:10.1002/bsl.744 PMID:17393403

Morse, S. J. (2011). NeuroLaw exuberance: A plea for neuromodesty. In *Technologies on the stand: Legal and ethical questions in neuroscience and robotics*. Njimegen, The Netherlands: WLP.

Nam, T. (2012). Suggesting frameworks of citizen-sourcing via Government 2.0. *Government Information Quarterly, 29*(1), 12–20. doi:10.1016/j.giq.2011.07.005

National Crime Records Bureau. (2012). *Crime in India 2012 Compendium*. New Delhi: NCRB, Government of India. Retrieved from http://ncrb.gov.in/CD-CII2012/Compendium2012.pdf

Neher, W. W., & Sandin, P. J. (2007). *Communicating ethically: Character, duties, consequences, and relationships*. Boston, MA: Pearson Education.

Net-Security. (2010). Botnets drive the rise of ransomware. *Net Security*. Retrieved from http://www.net-security.org/secworld.php?id=9095

New York Times. (2013, June 2). The banality of don't be evil. *The New York Times*, p. SR4.

News, B. B. C. (2005, September 7). Yahoo helped jail China writer. *BBC Online*. Retrieved July 31, 2010 from http://www.news.bbc.co.uk/l/hi/world/4221538.stm

News, B. B. C. (2010, April 4). Jailed Dubai kissing pair lose appeal over conviction. *BBC News*. Retrieved from http://news.bbc.co.uk/1/hi/uk/8602449.stm

Nice, L. (2007). Tabloidization and the teen market. *Journalism Studies, 8*, 117–136. doi:10.1080/14616700601056882

Niculescu, C., & Pană, L. (2010). Architecture of a Multi-Framework Set for Collaborative Knowledge Generation. In *Proceedings of the 11th European Conference on Knowledge Management* (pp.73-75). Academic Publishing Limited.

Nielsen, L. W. (2011). The concept of nature and the enhancement technologies debate. In J. Savulescu, R. ter Meulen, & G. Kahane (Eds.), *Enhancing human capacities* (pp. 19–33). Malden, MA: Wiley-Blackwell. doi:10.1002/9781444393552.ch2

Nissenbaum, H. (2004). Privacy as contextual integrity. *Washington Law Review (Seattle, Wash.), 79*(1), 119–158.

Norton. (2013). *Norton Symantec report on Cyber Crime, 2013*. Retrieved from http://now-static.norton.com/now/en/pu/images/Promotions/2012/cybercrimeReport/2012_Norton_Cybercrime_Report_Master_FINAL_050912.pdf

Odobleja, S. (1982). *Consonantal psychology*. Bucharest: Editura Ştiinţifică si Enciclopedică.

Olt, M. R. (2009). Seven strategies for plagiarism-proofing discussion threads in online courses. *MERLOT Journal of Online Learning and Teaching, 5*(2).

Olt, M. R. (2002). Ethics and distance education: Strategies for minimizing academic dishonesty in online assessment. *Online Journal of Distance Learning Administration, 5*(3).

Online Etymology Dictionary. (n.d.). Retrieved from http://www.etymonline.com

Ott, M., & Pozzi, F. (2011). Towards a New Era for Cultural Heritage Education: Discussing the Role of ICT. *Computers in Human Behavior*, *27*, 1365–1371.

Ovide, S. (2013, August 5). Free speech a test for Twitter. *The Wall Street Journal*, pp. B1-2.

Owen, J., & Shen, F. (2011). Law and Neuroscience in the United States. In *International Neurolaw – A Comparative Analysis* (pp. 349–380). Academic Press.

Pagallo, U., & Durante, M. (Forthcoming). *Legal memories and the right to be forgotten.* Paper presented at the Workshop at the European University in Fiesole. Fiesole, Italy.

Pagallo, U. (2008). *La tutela della privacy negli USA e in Europa: Modelli giuridici a confront.* Milano: Giuffrè.

Pagallo, U. (2010). As law goes by: topology, ontology, evolution. In P. Casanovas et al. (Eds.), *AI approaches to the complexity of legal systems* (pp. 12–26). Dordrecht, The Netherlands: Springer. doi:10.1007/978-3-642-16524-5_2

Pagallo, U. (2011). ISPs & rowdy web sites before the law: Should we change today's safe harbour clauses? *Philosophy and Technology*, *24*(4), 419–436. doi:10.1007/s13347-011-0031-x

Pagallo, U. (2012). On the principle of privacy by design and its limits: Technology, ethics, and the rule of Law. In S. Gutwirth, R. Leenes, P. De Hert, & Y. Poullet (Eds.), *European Data Protection: In Good Health?* (pp. 331–346). Dordrecht, The Netherlands: Springer. doi:10.1007/978-94-007-2903-2_16

Pagallo, U. (2013). Robots in the cloud with privacy: A new threat to data protection? *Computer Law & Security Report*, *29*(5), 501–508. doi:10.1016/j.clsr.2013.07.012

Pagallo, U., & Bassi, E. (2011). The future of EU working parties' the future of privacy and the principle of privacy by design. In M. Bottis (Ed.), *An Information Law for the 21st Century* (pp. 286–309). Athens, Greece: Nomiki Bibliothiki Group.

Pagallo, U., & Durante, M. (2009). Three roads to P2P systems and their impact on business practices and ethics. *Journal of Business Ethics*, *90*(4), 551–564. doi:10.1007/s10551-010-0606-y

Palmirani, M., Pagallo, U., Sartor, G., & Casanovas, P. (2012). *AI approaches to the complexity of legal systems: Models and ethical challenges for legal systems, legal language and legal ontologies, argumentation and software agents.* Dordrecht, The Netherlands: Springer.

Pană, L. (2003a). The Intelligent Environment as an Answer to Complexity. In *Proceedings of the IUAES Congress: XVICAES2K3 Humankind/Nature Interaction: Past, Present and Future*, (Vol. 2, p. 1198). International Union of Anthropological and Ethnological Sciences.

Pană, L. (2005a, March). *From Virtue Ethics to Virtual Ethics.* Paper presented at the Interdisciplinary Research Group of the Romanian Committee for History and Philosophy of Science and Technology of the Romanian Academy. Bucharest, Romania.

Pană, L. (2005b). Moral Intelligence for Artificial and Human Agents, Machine Ethics. In *Papers from the AAAI Fall Symposium Series.* AAAI Press.

Pană, L. (2006a). Knowledge Management and Intellectual Techniques – Intellectual Invention and Its Forms. In R. Trapl (Ed.), *Cybernetics and Systems: Proceedings of the Eighteenth European Meeting on Cybernetics and Systems Research*, (vol. 2, pp. 422-427). Austrian Society for Cybernetic Studies.

Pană, L. (2006b, June). *Values Inventing and Motives Computing: Elements of Artificial Ethics for Cognitive and Operative Moral Agents.* Paper presented at the 4th European Computing and Philosophy Conference. Trondheim, Norway.

Pană, L. (2007, June). *From Information Flows and Nets to Knowledge Groups and Works.* Paper presented at the 5th European Computing and Philosophy Conference. Enschede, The Netherlands.

Pană, L. (2008c). The Preferential Sense as a Source of Natural and Artificial Evolution. In *Proceedings of the 14th International Congress of Cybernetics and Systems of WOSC* (pp. 984-993). Wroclaw, Poland: Wroclaw University of Technology and the World Organization of Systems and Cybernetics.

Pană, L. (1988). Cognition and action centered values in human conduct structuring. *Revue Roumaine des Sciences Sociales. Série de Philosophie et de Logique*, *32*(1-2), 11–18.

Pană, L. (2000a). Information machine, information action, information thinking and information man. In L. Pană (Ed.), *Philosophy of Technical Culture* (pp. 430–452). Bucharest: Editura Tehnică.

Pană, L. (2000b). Moral Culture. In L. Pană (Ed.), *Philosophy of Technical Culture* (pp. 104–112). Bucharest: Editura Tehnică.

Pană, L. (2004a). Artificial Philosophy. In *Philosophy of Information and Information Technology*. Bucharest: Politehnica Press.

Pană, L. (2004b). Modeling some Evolutions of Value Systems from the Perspective of Technical Culture. In L. Pană (Ed.), *Evolutions in Value Systems under the Influence of Technical Culture*. Bucharest: Politehnica Press.

Pană, L. (2005c). Philosophy of Artificial and Artificial Philosophy. *Academica, 15*(34), 32–34.

Pană, L. (2008a). An Integrative Model of Brain, Mind, Cognition and Conscience. *Noema, 7*(1), 120–137.

Pană, L. (2008b). Moral Intelligence: Elements of Artificial Ethics for Cognitive and Moral Agents. *Noésis (Pau), 33*(1), 39–51.

Parent, W. (1983). Privacy, morality and the law. *Philosophy & Public Affairs, 12*, 269–288.

Parker-Pope, T. (2011, June 14). Digital flirting: Easy to get caught. *The New York Times,* p. D5.

Park, H. M., & Perry, J. L. (2008). Does internet use really facilitate civic engagement? Empirical evidence from the American national election studies. In K. Yang, & E. Bergrud (Eds.), *Civic engagement in a network society* (pp. 237–269). Charlotte, North Carolina: Information Age Publishing.

Park, H. M., & Perry, J. L. (2009). Do campaign web sites really matter in electoral civic engagement? Empirical evidence from the 2004 and 2006 post-election internet tracking survey. In C. Panagopoulos (Ed.), *Politicking online: The transformation of election campaign communications* (pp. 101–123). New Brunswick, NJ: Rutgers University Press.

Parorobots. (n.d.). Retrieved from http://www.parorobots.com/

Pasick, A. (2007). FBI checks gambling in Second Life virtual world. *The Reuters*. Retrieved from http://www.reuters.com/article/technologyNews/idUSHUN43981820070405

Patrick, Keith, & George. (2011). Robot ethics: Mapping the issues for a mechanized world. *Artificial Intelligence, 175*, 5–6.

Patterson, T. E. (2000). *Doing well and doing good: How soft news are shrinking the news audience and weakening democracy*. Cambridge, MA: Harvard University Press.

Peck, L. A. (2006). A ''fool satisfied''? Journalists and Mill's principle of utility. *Journalism and Mass Communication Educator, 61*, 205–213. doi:10.1177/107769580606100207

Pellegrini, S., & Pietrini, P. (2010). Siamo davvero liberi? Il comportamento tra geni e cervello. *Sistemi Intelligenti, 22*, 281–293.

Pellegrini, S., & Pietrini, P. (2010). Verso un'etica. molecolare? *Giornale Italiano di Psicologia, 37*, 841–846.

Perez, O. (2013). Open government, technological innovation, and the politics of democratic disillusionment: (E-)democracy from Socrates to Obama. *I/S: A Journal of Law and Policy for the Information Society, 9*(1), 61-138.

Perlmutter, D. D., & Schoen, M. (2007). If I break a rule, what do I do, fire myself? Ethics codes of independent blogs. *Journal of Mass Media Ethics, 22*, 37–48. doi:10.1080/08900520701315269

Peters, C. (2011). Emotion aside or emotional side? Crafting an 'experience of involvement' in the news. *Journalism, 12*, 297–316. doi:10.1177/1464884910388224

Phillips, A. (2013). Transparency and the news ethics of journalism. In B. Franklin (Ed.), *The future of journalism* (pp. 289–298). London: Routledge.

Pietrini, P., & Bambini, V. (2009). Homo ferox: The contribution of functional brain studies to understanding the neural bases of aggressive and criminal behavior. *International Journal of Law and Psychiatry, 32*, 259–265. doi:10.1016/j.ijlp.2009.04.005 PMID:19477522

Pijnenburg, M. A. M., & Leget, C. (2007). Who wants to live forever? Three arguments against extending the human lifespan. *Journal of Medical Ethics*, *33*(10), 585–587. doi:10.1136/jme.2006.017822 PMID:17906056

Plaisance, P. L. (2007). An assessment of media ethics education: Course content and the values and ethical ideologies of media ethics students. *Journalism & Mass Communication Educator*, *61*(4), 378–396. doi:10.1177/107769580606100404

Plato, . (1935). *The republic*. Cambridge, MA: Harvard University Press.

Polo, L. (1993). *Presente y futuro del hombre*. Madrid, Spain: Rialp.

Polo, L. (1996). *Etica: Hacia una version moderna de los temas clasicos*. Madrid, Spain: Union Editorial.

Porto, M. (2007). TV news and political change in Brazil: The impact of democratization on TV Globo's journalism. *Journalism*, *8*, 263–284. doi:10.1177/1464884907078656

Post, D. G. (2002). Against against cyberspace. *Berkeley Technology Law Journal*, *17*(4), 1365–1383.

Postman, N. (2013). Five things we need to know about technological change. In *Technologies, social media, and society* (18th ed., pp. 3–6). New York: McGraw-Hill.

President's Council on Bioethics. (2003). *Beyond therapy: Biotechnology and the pursuit of happiness*. Washington, DC: Author.

Price, J. (2006). Gambling - Internet Gambling. *The Ethics and Religious Library Commission*. Retrieved from http://erlc.com/article/gambling-internet-gambling

Prior, M. (2003). Any good news in soft news? The impact of soft news preference on political knowledge. *Political Communication*, *20*, 149–171. doi:10.1080/10584600390211172

Puech, M. (2008). *Homo sapiens technologicus: Philosophie de la technologie contemporaine, philosophie de la sagesse contemporaine*. Paris: Éditions Le Pommier.

Putnam, R. D. (1995). Bowling alone: America's declining social capital. *Journal of Democracy*, *6*(1), 65–78. doi:10.1353/jod.1995.0002

Quiring, O. (2009). What do users associate with 'interactivity'? A qualitative study on user schemata. *New Media & Society*, *11*(6), 899–920. doi:10.1177/1461444809336511

Raban, D. R. (2009). Self-Presentation and the Value of Information in Q&A Websites. *Journal of the American Society for Information Science and Technology*, *60*(12), 2465–2473.

Rao, S., & Johal, N. S. (2006). Ethics and news making in the changing Indian mediascape. *Journal of Mass Media Ethics*, *21*, 286–303. doi:10.1207/s15327728jmme2104_5

Rauch, J. (2013). How not to die. *Atlantic (Boston, Mass.)*, *311*(4), 64–69.

Rawls, J. (1971). *A theory of justice*. Cambridge, MA: Harvard University Press.

Rawls, J. (2001). *The law of peoples*. Cambridge, MA: Harvard University Press.

Reed, C. (2012). *Making laws for cyberspace*. Oxford, UK: Oxford University Press.

Rees, M. (2003). *Our final hour: A scientist's warning: How terror, error, and environmental disaster threaten humankind's future in this century—On earth and beyond*. New York, NY: Basic Book.

Reeves, B., & Nass, C. (1998). *The media equation: How people treat computers, television, and new media like real people and places*. CSLI Publications.

Reis, R. (2000). Teaching media ethics in a multicultural setting. *Journal of Mass Media Ethics*, *15*(3), 194–205. doi:10.1207/S15327728JMME1503-5

Reno v. ACLU. 521 U.S. 844. (1997).

RIAA. (n.d.). Piracy: Online and On The Street. *Recording Industry Association of America (RIAA) homepage*. Retrieved from http://www.riaa.com/physicalpiracy.php

Richerson, R., & Boyd, R. (2005). *Not by genes alone: How culture transformed human evolution*. Chicago: University of Chicago Press.

Richmond, S. (2010, April 17). Gamestation Collects Customers' Souls in April Fools Gag. *Telegraph*. Retrieved from http://blogs.telegraph.co.uk/technology/shanerichmond/100004946/gamestation-collects-customers-souls-in-april-fools-gag/

Roberts, C. J., & Shalin, H. (2009). Issues of academic integrity: an online course for students Addressing academic dishonesty. *MERLOT Journal of Online Learning and Teaching, 5*(2).

Robins, K. (1996). *Into the Image: Culture and politics in the field of vision.* London: Routledge. doi:10.4324/9780203440223

Rogerson, S. (1998). *Social Values in the Information Society.* Retrieved from http://dehn.slu.edu/courses/fall06/493/rogerson.pdf

Ronchi, A. M. (2009). *E-Culture.* New York: Springer-Verlag, LLC.

Rooney, D. (2000). Thirty years of competition in British tabloid press: *The Mirror* and *The Sun* 1968-1998. In C. Sparks, & J. Tulloch (Eds.), *Tabloid tales: Global debates about media standards* (pp. 91–109). Lanham, MD: Rowman & Littlefield.

Ross, W. D. (1930). *The Right and the Good.* Oxford, UK: Oxford University Press.

Rowe, D. (2011). Obituary for the newspaper? Tracking the tabloid. *Journalism, 12*, 449–466. doi:10.1177/1464884910388232

Russell, J. (2006). Europe: Responsible gambling: A safer bet? *Ethical Corporation.* Retrieved from http://www.ethicalcorp.com/content.asp?ContentID=4291

SACS COC. (2012). *The principles of accreditation: Foundations for quality enhancement.* Decatur, GA: Southern Association of Colleges and Schools Commission on Colleges.

Saha, P. (2005). Gambling with responsibilities. *Ethical Corporation.* Retrieved from http://www.ethicalcorp.com/content.asp?ContentID=3774

Sakamoto, D., & Hiroshi. (2009). Geminoid: Remote-Controlled Android System for Studying Human Presence. *Kansei Engineering International, 8*(1).

Salvini, P. (2006). Presence: A Network of Reciprocal Relations. In *Proceedings of PRESENCE 2006: The 9th Annual International Workshop on Presence.* PRESENCE.

Sandel, M. J. (2009). The case against perfection: What's wrong with designer children, bionic athletes, and genetic engineering. In J. Savulescu, & N. Bostrom (Eds.), *Human enhancement* (pp. 71–89). New York, NY: Oxford University Press.

Santosuosso, A. (2008, December 19). Il dilemma del diritto di fronte alle neuroscienze. In Atti del convegno Le neuroscienze e il diritto. Academic Press.

Santosuosso, A. (Ed.). (2009). *Le neuroscienze e il diritto.* Como-Pavia, Italy: Ibis.

Santosuosso, A. (Ed.). (2011). *Diritto, scienza, nuove tecnologie.* Padua, Italy: Cedam.

Savulescu, J., Sandberg, A., & Kahane, G. (2011). Well-being and enhancement. In J. Savulescu, R. ter Meulen, & G. Kahane (Eds.), *Enhancing human capacities* (pp. 3–18). Malden, MA: Wiley-Blackwell. doi:10.1002/9781444393552

Schaefer, G. O., Kahane, G., & Savulescu, J. (2013). Autonomy and enhancement. *Neuroethics.* doi: doi:10.1007/s12152-013-9189-5

Schaler, J. A. (Ed.). (2009). *Peter Singer Under Fire: The Moral Iconoclast Faces His Critics.* Chicago, IL: Open Court.

Scherer, M. (2012, November). Exclusive: Obama's 2012 digital fundraising outperformed 2008. *Time Magazine.* Retrieved from http://swampland.time.com/2012/11/15/exclusive-obamas-2012-digital-fundraising-outperformed-2008/

Schmidt, C. T. A., & Kraemer, F. (2006). Robots, Dennett and the Autonomous: A Terminological Investigation. *Minds and Machines, 16*(1). doi:10.1007/s11023-006-9014-6

Schmidt, E., & Cohen, J. (2013). *The new digital age.* New York: Knopf.

Schramme, T. (2008). Should we prevent non-therapeutic mutilation and extreme body modification? *Bioethics, 22*(1), 8–15. PMID:18154584

Schultz, R. A. (2006). *Contemporary issues in ethics and information technology.* Hershey, PA: IRM Press.

Schwartz, S. H. (1994). Are There Universal Aspects in the Structure and Contents of Human Values? *The Journal of Social Issues*, *50*(4), 19–46.

Schwartz, S. H., & Bilsky, W. (1987). Toward a Psychological Structure of Human Values. *Journal of Personality and Social Psychology*, *53*, 550–562.

Scott, D. K., & Gobetz, R. H. (1992). Hard News/Soft News Content of the National Broadcast Networks, 1972-1987. *The Journalism Quarterly*, *69*(2), 406–412. doi:10.1177/107769909206900214

Selwyn, N., & Robson, K. (1998). Using e-mail as a research tool. *Social Research Update 21*. Retrieved from http://sru.soc.surrey.ac.uk/SRU21.html

Shelton, K. (2011). A review of paradigms for evaluating the quality of online education programs. *The Online Journal for Distance Learning Administrators*, *4*(1).

Shetty, A. (2013, July 2). *Sibal says 2013 National Cyber Security Policy will be a 'real task' to operate*. Retrieved August 18, 2013, from http://tech.firstpost.com/news-analysis/sibal-says-2013-national-cyber-security-policy-will-be-a-real-task-to-operate-101759.html

Shibata, T., Kawaguchi, Y., & Wada, K. (2010). *Investigation on people living with Paro at home*. Paper presented at 2010 IEEE-RO-MAN. Viareggio, Italy.

Shneiderman, N. (2000). Universal usability. *Communications of the ACM*, *43*(3), 84–91. doi:10.1145/332833.332843

Shoemaker, P. J., & Cohen, A. A. (2006). *News around the world: Practitioners, content and the public*. Oxford, UK: Routledge.

Signore, M. (2006). *Lo sguardo della responsabilità: Politica, economia e tecnica per un antropocentrismo relazionale*. Roma, Italy: Studium.

Simon, H. A. (1996). *The sciences of the artificial*. Cambridge, MA: The MIT Press.

Singer, P. (1991). Research into aging: should it be guided by the interests of present individuals, future individuals, or the species? In F. C. Ludwig (Ed.), *Life span extension: Consequences and open questions* (pp. 132–145). New York, NY: Springer.

Skolnicki, Z., & Arciszewski, T. (2003). Intelligent Agents in Design. In *Proceedings of 2003 ASME International Design Engineering Technical Conferences & The Computers and Information in Engineering Conference*. Mason University.

Sloman, A. (1990). Motives, Mechanisms, Emotions. In M. Boden (Ed.), *The Philosophy of Artificial Intelligence* (pp. 231–247). New York: Oxford University Press.

Smith Nash, S. (2006). *Ethics and mobile learning: Should we worry?* Retrieved from http://elearnqueen.blogspot.com/2006/09/ethics-and-mobile-learning-should-we.html

Spaemann, R. (1991). *Moralische Grundbegriffe*. München, Germany: C.H. Beck Verlag.

Sparks, C., & Tulloch, J. (2000). *Tabloid tales: Global debates over media standards*. New York, NY: Rowman & Littlefield Publishers, Inc.

Sparrow, R. (2002). The March of the Robot Dogs. *Ethics and Information Technology*, *4*(4), 305–318. doi:10.1023/A:1021386708994

Spence, E. H., & Quinn, A. (2008). Information ethics as a guide for new media. *Journal of Mass Media Ethics*, *23*(4), 264–279. doi:10.1080/08900520802490889

Spinello, R. A. (2003). *Case studies in information technology ethics* (2nd ed.). Upper Saddle River, NJ: Prentice Hall.

Spinello, R. A. (2014). *Cyberethics: Morality and law in cyberspace* (5th ed.). Sudbury, MA: Jones & Bartlett.

Spinello, R. A., & Tavani, H. (2004). *Readings in cyberethics* (2nd ed.). Sudbury, MA: Jones & Bartlett.

Stein, A. R. (1998). The unexceptional problem of jurisdiction in cyberspace. *International Lawyer*, *32*, 1167–1194.

Stradella, E. (2009). Recenti tendenze del diritto penale simbolico. In *Il diritto penale nella giurisprudenza costituzionale*. Turin, Italy: Giappichelli.

Strike, K. A., & Ternasky, P. L. (1991). *Ethics for professional in education: Perspectives for preparation and practice*. New York, NY: Teachers College Press, Columbia University.

Strobel, L. P. (1980). *Reckless homicide? Ford's Pinto trial*. South Bend, IN: And Books.

Sykes, C. (1999). *The end of privacy: The attack on personal rights at home, at work, on-line, and in court*. New York: St. Martin's Griffin.

Talbot, D. (2008, September/October). How Obama really did it: Social technology helped bring him to the brink of the presidency. *Technology Review*, *111*(5), 78–83.

Tavani, H. T. (2004). *Ethics and technology: Ethical issues in an age of information and communication technology*. Hoboken, NJ: John Wiley & Sons.

Tavani, H. T. (2007). Philosophical theories of privacy: Implications for an adequate online privacy policy. *Metaphilosophy*, *38*(1), 1–22. doi:10.1111/j.1467-9973.2006.00474.x

Tavani, H. T., & Moor, J. H. (2001). Privacy protection, control of information, and privacy-enhancing technologies. *Computers & Society*, *31*(1), 6–11. doi:10.1145/572277.572278

Tenner, E. (1997). *Why things bite back: Technology and the revenge of unintended consequences*. New York: Random House Vintage.

Thompson, C. (2006, April 23). China's Google problem. *New York Times Magazine*, 36-41, 73-76.

Torralba Roselló, F. (2005). *Qeé es la dignidad humana?* Barcelona, Spain: Herder.

Towner, T. L., & Duleo, D. A. (2012). New media and political marketing in the United States: 2012 and beyond. *Journal of Political Marketing*, *11*(1-2), 95–119. doi:10.1080/15377857.2012.642748

Trammell, K., Porter, L., Chung, D., & Kim, E. (2006). *Credibility and the uses of blogs among professionals in the communication industry*. Paper presented to the Credibility divide Communication Technology Division at the 2006 Association for Education in Journalism and Mass Communication. San Francisco, CA.

Trouble at the Lab. (2013). *The Economist, 409*(8858), 26-30.

Tsfati, Y. (2010). Online news exposure and trust in the mainstream media: Exploring possible associations. *The American Behavioral Scientist*, *54*, 22–42. doi:10.1177/0002764210376309

Tulloch, J. (2000). The eternal recurrence of new journalism. In C. Sparks, & J. Tulloch (Eds.), *Tabloid tales: Global debates about media standards* (pp. 13–46). Lanham, MD: Rowman & Littlefield.

Tumpey, T. M., Basler, C. F., Aguilar, P. V., Zeng, H., Solórzano, A., & Swayne, D. E. (2005). Character- ization of the reconstructed 1918 Spanish influenza pandemic virus. *Science*, *310*, 77–80. doi:10.1126/science.1119392 PMID:16210530

Turkle, S. (2007). Authenticity in the age of digital companions. *Interaction Studies: Social Behaviour and Communication in Biological and Artificial Systems*, *8*(3). doi:10.1075/is.8.3.11tur

Turkle, S. O. (2011). *Alone together: Why we expect more from technology and less from each other*. New York: Basic Books.

Turner, R., & Eden, A. H. (2008). The Philosophy of Computer Science. *Journal of Applied Logic*, *6*, 459–626.

Ukrinform. (2012, June). *Ukraine among 20 countries with highest percentage of computer attacks*. Retrieved August 12, 2013, from http://www.ukrinform.ua/eng/news/ukraine_among_20_countries_with_highest_percentage_of_computer_attacks_305071

Ullman, R. (1983). Redefining security. *International Security*, *8*(1), 129–150. doi:10.2307/2538489

United Nations Charter. (2007). The Universal Declaration of Human Rights. In R. A. Spinello (Ed.), *Moral philosophy for managers* (5th ed., pp. 293–297). New York: McGraw-Hill.

United Nations Office on Drugs and Crime. (2013). *Comprehensive Study on Cybercrime Draft*. United Nations Office on Drugs and Crime, Vienna. Retrieved from http://www.unodc.org/documents/organized-crime/UNODC_CCPCJ_EG.4_2013/CYBERCRIME_STUDY_210213.pdf

United States of America v. David Bermingham, et al., 18 U.S.C. §§ 1343 & 2. CR-02-00597.

United States of America v. Gary McKinnon. 18 U.S.C. §§ 1030(a)(5)(A) & 2 Indictment.

United States White House. (2010). *The Obama administration's commitment to an open government: A status report*. Retrieved from http://www.whitehouse.gov/sites/default/files/opengov_report.pdf

Unsworth, K., & Townes, A. (2012). Social media and E-Government: A case study assessing Twitter use in the implementation of the open government directive. *Proceedings of the American Society for Information Science and Technology*, *49*(1), 1–3. doi:10.1002/meet.14504901298

Uribe, R., & Gunter, B. (2007). Are 'sensational' news stories more likely to trigger viewers' emotions than non-sensational news stories? *European Journal of Communication*, *22*, 207–228. doi:10.1177/0267323107076770

Van Nedervelde, P. (2006). Lifeboat foundation security preserver. *Lifeboat Foundation*. Retrieved July 26, 2012, from http://lifeboat.com/ex/security.preserver

van Schewick, B. (2010). *Internet architecture and innovation*. Cambridge, MA: The MIT Press.

Vartapetiance, A., & Gillam, L. (2010). DNA dataveillance: Protecting the innocent? *Journal of Information. Communication and Ethics in Society*, *8*(3), 270–288. doi:10.1108/14779961011071079

Veltman, K. H. (2004). Towards a Semantic Web for Culture. *Journal of Digital Information*, *4*(4).

Vernon, G. M. (1973). Values, Value Definitions, and Symbolic Interaction. In *Value Theory in Philosophy and Social Science*. New York: Gordon and Breach.

Verplanken, B., & Holland, R. W. (2002). Motivated Decision Making: Effects of Activation and Self-Centrality of Values on Choices and Behavior. *Journal of Personality and Social Psychology*, *82*(3), 434–447. PMID:11902626

Veruggio, G., & Operto, F. (2008). Roboethics: Social and Ethical Implications of Robotics. In *Springer Handbook of Robotics*. Springer-Verlag. doi:10.1007/978-3-540-30301-5_65

Vincent, R. J. (1988). *Human rights and international relations*. New York: Cambridge University Press.

Virilio, P. (1997). *Open Sky*. London: Verso.

Virilio, P. (2000). *From Modernism to Hypermodernism and Beyond*. Sage Publications.

Volti, R. (2014). *Society and technological change* (7th ed.). New York: Worth Publishers.

Ward, S. J. A. (2007). Utility and impartiality: Being impartial in a partial world. *Journal of Mass Media Ethics*, *22*, 151–167. doi:10.1080/08900520701315913

Warneke, B., Last, M., Liebowitz, B., & Pister, K. S. J. (2001). Smart dust: Communicating with a cubic-millimeter computer. *Computer*, *34*, 44–51. doi:10.1109/2.895117

Warwick, K. (2003). Cyborg morals, cyborg values, cyborg ethics. *Ethics and Information Technology*, *5*(3), 131–137. doi:10.1023/B:ETIN.0000006870.65865.cf

Waters, R. (2010, March 24). Realism lies behind decision to quit. *Financial Times*, p. B1.

Wax, M. L., & Cassell, J. (1979). *Federal regulations ethical issues and social research*. Boulder, CO: Westview Press, Inc.

Weber, T. (2007, January 25). Criminals 'may overwhelm the web'. *BBC News*. Retrieved from http://news.bbc.co.uk/1/hi/business/6298641.stm

Wenger, A., & Wollenmann, R. (Eds.). (2007). *Bioterrorism: Confronting a complex threat*. Boulder, CO: Lynne Rienner.

Wennberg, J. E., Fisher, E. S., Goodman, D. C., & Skinner, E. S. (2008). *Tracking the care of patients with severe chronic illness*. NH, Lebanon: The Dartmouth Institute for Health Policy and Clinical Practice.

Wessberg, J., Stambaugh, C. R., Kralik, J. D., Beck, P. D., Laubach, M., & Chapin, J. K. (2000). Real time prediction of hand trajectory by ensembles of cortical neurons in primate. *Nature*, *408*, 361–365. doi:10.1038/35042582 PMID:11099043

West, C. K. (2001). *Techno-human MESH: The growing power of information technologies*. Westport, CT: Quorum Books.

Westerwick, A. (2013). Effects of sponsorship, web site design, and google ranking on the credibility of online information. *Journal of Computer-Mediated Communication, 18*(2), 80–97. doi:10.1111/jcc4.12006

Westin, A. (1967). *Privacy and freedom.* New York, NY: Atheneum.

Whitbeck, C. (1996). Ethics as design: doing justice to moral problems. *The Hastings Center Report, 26*(3), 9–16. doi:10.2307/3527925 PMID:8736668

Wiener, N. (1954). *The Human use of Human Beings – Cybernetics and society.* Da Capo Press.

Wierbicki, A. P., & Nakamori, Y. (2005). *Creative Space: Models of Creative Processes for the Knowledge Civilization Age.* Springer. doi:10.1007/b137889

Williams, C. B., & Gordon, J. A. (2003). Dean Meetups: Using the Internet for grassroots organizing. Retrieved from http://www.deanvolunteers.org/Survey/Dean_Meetup_Analysis.htm.

Williams, C. B., & Gultai, G. J. (2009). The political impact of Facebook: Evidence from the 2006 elections and the 2008 nomination contest. In C. Panagopoulos (Ed.), *Politicking online: The transformation of election campaign communications* (pp. 249–271). New Brunswick, NJ: Rutgers University Press.

Williams, J. (2012). Change you can believe in, You better not believe it. *Critical Sociology, 38*(5), 747–769. doi:10.1177/0896920511426804

Wojtyła, K. (1982). *Persona e Atto: Città del Vaticano.* Vatican State, Italy: Libreria Editrice Vaticana.

Wolpe, P. R., Foster, K. R., & Langleben, D. D. (2005). Emerging neurotechnologies for lie-detection: Promises and perils. *The American Journal of Bioethics, 5*(2), 40–48. doi:10.1080/15265160590923367 PMID:16036657

WordNet. (n.d.). Retrieved from http://wordnetweb.princeton.edu/perl/webwn

Wright, D. K. (1996). Communication ethics. In M. B. Salwen, & D. W. Stacks (Eds.), *An integrated approach to communication theory and research* (pp. 519–535). Lawrence Erlbaum Associates Publishers.

Yeung, K. (2007). Towards an understanding of regulation by design. In R. Brownsword, & K. Yeung (Eds.), *Regulating technologies: Legal futures, regulatory frames and technological fixes* (pp. 79–108). London, UK: Hart Publishing.

Young, A. (2006). *Gambling or Gaming, Entertainment or Exploitation?* Ethical Investment Advisory Group of the Church of England. Retrieved from http://www.cofe.anglican.org/info/ethical/policystatements/gambling.pdf

Zhou, L., Ding, L., & Finin, T. (2011). How is the Semantic Web evolving? A Dynamic Social Network Perspective. *Computers in Human Behavior, 27,* 1294–1302.

Zittrain, J. (2003). Internet points of control. *Boston College Law Review. Boston College. Law School, 44,* 653.

Zittrain, J. (2007). Perfect enforcement on tomorrow's internet. In R. Brownsword, & K. Yeung (Eds.), *Regulating technologies: legal futures, regulatory frames and technological fixes* (pp. 125–156). London, UK: Hart Publishing.

Zittrain, J. (2008). *The future of the internet and how to stop it.* New Haven, CT: Yale University Press.

About the Contributors

Rocci Luppicini is an Associate Professor in the Department of Communication and affiliate of the Institute for Science, Society, and Policy (ISSP) at the University of Ottawa (Canada) and acts as the editor-in-chief for the *International Journal of Technoethics*. He is a leading expert in Technology Studies (TS) and technoethics. He has published over 25 peer reviewed articles and has authored and edited several books, including *Online Learning Communities in Education* (IAP, 2007), the *Handbook of Conversation Design for Instructional Applications* (IGI, 2008), *Trends in Canadian Educational Technology and Distance Education* (VSM, 2008), the *Handbook of Research on Technoethics: Volume Is & II* (with R. Adell) (IGI, 2008, 2009), *Technoethics and the Evolving Knowledge Society: Ethical Issues in Technological Design, Research, Development, and Innovation* (2010), *Cases on Digital Technologies in Higher Education: Issues and Challenges* (with A. Haghi) (IGI, 2010), *Education for a Digital World: Present Realities and Future Possibilities* (AAP, in press). His most recent edited work, the *Handbook of Research on Technoself: Identity in a Technological Society: Volumes I & II* (IGI, 2013) provides the first comprehensive reference work in the English language on human enhancement and identity within an evolving technological society.

***

Rachel Baarda graduated from Brock University in 2008 with an Honours BA in English Language and Literature and a minor in Psychology. In 2012, she graduated from the University of Ottawa with a MA in Communication. She became interested in the subject of digital media and political participation after taking Communications courses on new media and reading news reports about Barack Obama's unprecedented Internet campaign. As a result, her Master's—"Case Study of my.barackobama.com: Promoting Participatory Democracy?"—explored the ways in which Barack Obama's campaign social networking site influenced political participation.

Liudmila Baeva is Dean of the Department of Social Communication, Head of the Chair of Philosophy of Astrakhan State University (Astrakhan, Russia). She is an expert of the Federal scientist research consulting Center (Moscow) for planning studies for 2014-2020 in the thematic area "Humanitarian problems of innovative development." She is an expert of the Russian Foundation for Humanities (Moscow) and the author of over 160 scientific articles and 6 monographs. A specialist in the field of axiology, philosophical anthropology, the study of the problems of the information society, member of the Philosophical Society of the Russian Philosophical Society, the Russian Political Science Association, a member of the editorial board of the international journal *Socioloska Luca: Journal of Social*

Anthropology, Social Demography, and Social Psychology (Montenegro), and a member of the editorial board of the international journal *The Caspian Region: Economics, Politics, and Culture* (Russian), she is a Visiting Professor of Philosophy at Moscow State Technical University (2005), Hainan University, China (2007), and South Kazakhstan State University (2011). She is a Member of the World Congress of Philosophy in Istanbul, Seoul, and Athens.

Erica Bailey is a Doctoral student in the Mass Communication program at Penn State University, where she is also a University Fellow. She received her BA from Otterbein University in Public Relations and Philosophy and her MA in Communication from Virginia Tech. She has taught undergraduate public speaking courses as well as assisted in teaching public relations and media classes. Her research interests include the ways in which technology affects journalists' ethical decisions and the psychological aspects of entertainment media selection and enjoyment.

Angela Di Carlo (Verbania, 1984) is Post-doc Research Fellow in Constitutional Law at Scuola Superiore Sant'Anna, Pisa, where she obtained the PhD in Individual and Legal Protections in the area of "Constitutional Protection of the Individual and of Social Groups." Her main research interests are in the field of Public Law, specifically sources of law, case law of the Italian Constitutional Court, fundamental rights' protection, and social rights – in particular the right to health. She is currently a member of the research Team of Scuola Superiore Sant'Anna working on the European RoboLaw Project, funded by the European Commission under the Seventh Framework Programme. Since 2010, she has collaborated in research and consulting activities at the Department of Legal and Legislative Affairs of the Italian Government.

Mahmoud Eid is an Associate Professor at the Department of Communication, University of Ottawa, Canada. Dr. Eid is the author of *Interweavement: International Media Ethics and Rational Decision-Making* (2008), co-author of *Mission Invisible: Race, Religion, and News at the Dawn of the 9/11 Era* (2014), editor of *Exchanging Terrorism Oxygen for Media Airwaves: The Age of Terroredia* (2014) and *Research Methods in Communication* (2011), and co-editor of *Basics in Communication and Media Studies* (2012) and *The Right to Communicate: Historical Hopes, Global Debates and Future Premises* (2009). Dr. Eid is the Editor of the *Global Media Journal -- Canadian Edition*, serves on the editorial boards of several academic journals and as an organizing committee member for various international conferences, has contributed several book chapters and journal articles, and has presented numerous papers at global conferences. His research interests focus on international communication, media ethics, media representations, decision-making, crisis management, conflict resolution, terrorism, Islam, Arab culture, Middle East politics, research methods, and the political economy of communication.

Jose M. Galvan worked as surgeon in the "San Carlos" Clinic Hospital of the School of Medicine of the Complutense University of Madrid, in the II Chair of Surgery. He was ordained a Catholic priest in 1981. From 1986, he was Professor of Dogmatic Theology in the School of Theology of the Pontifical University of the Holy Cross and of Anthropology in the "Apollinare Higher Institute of Religious Sciences" (ISSRA); at present, he is Professor of Moral Theology in the University examination. His areas of interest are Trinity, Divine Attributes, Pneumatology, Theological Anthropology, Art and Aesthetics, and Technoethics. From 1988 to 1992, he was a member of the Council of ISSRA. From 1995 to 2001, he was vice-director of the Department of Fundamental and Dogmatic Theology of the University of the Holy Cross. He was President of the Scientific Committees of the II International Symposium "The Justification in Christ" of the School of Theology of the University of the Holy Cross (1996) and of the International Congress of Theology "Christ in Human Pilgrimage" in occasion of the International Meeting of the Jubilee of the Universities in Rome (2000). He was a Founding Academician of the "Roman Academy of the Arts." Since 1998, he has given several courses on Aesthetics. He collaborates with the Department of Bio-Engineering of the Scuola Superiore Sant'Anna of Pisa, with the Ethicbot Project of the European Robotics Network (EURON) and with the Techno-Ethical Committee of the IEEE-RAS.

Deb Gearhart is the Vice Provost for E-Learning and Strategic Partnerships at Ohio University. Dr. Gearhart has worked in the field of distance education for 27 years. Previously, Dr. Gearhart served as director of eTROY at Troy University for 5 years and served as the founding Director of E-Education Services at Dakota State University in Madison, South Dakota for 11 years. Before joining Dakota State, she spent 10 years with the Department of Distance Education at Penn State, now Penn State World Campus. Dr. Gearhart has earned a BA in Sociology from Indiana University of Pennsylvania. She earned a MEd in Adult Education with a distance education emphasis and an MPA in Public Administration, both from Penn State. Deb completed her PhD program in Education, with a certificate in distance education, from Capella University.

Lee Gillam is a Senior Lecturer in the Department of Computing at the University of Surrey and Chartered IT Professional Fellow of the British Computer Society (FBCS CITP). His research interests, and teaching activities, cover Cloud Computing, Information Retrieval and Information Extraction, and Ontology Learning with over 80 publications. He is the founding editor-in-chief of the Springer *Journal of Cloud Computing Advances, Systems, and Applications* (JoCCASA), an editor of a Springer book on Cloud Computing, and PI on a current TSB Project with Jaguar Land Rover investigating protection of Intellectual Property through supply chains and mechanisms that help to assure trust in collaboration through the Cloud.

Tobias Hainz holds positions as a postdoctoral researcher at Hannover Medical School, Institute for History, Ethics, and Philosophy of Medicine, as well as Leibniz University Hannover, Faculty of Law. In July 2013, he completed his dissertation on "Radical Life Extension: An Ethical Analysis" at Heinrich-Heine-University Düsseldorf, where he was a member of the graduate school "Age(ing): Cultural Concepts and Practical Realisations." From April to June 2012, he was a recognised student at the Oxford Uehiro Centre for Practical Ethics, University of Oxford. His research focuses on the ethics of human enhancement, especially life extension technologies, the philosophy of transhumanism, and theoretical problems of public involvement in the context of biomedical research and innovation. Further research interests include value theory and its practical applications, the regulation of emerging and future technologies, as well as metaphysical and ontological problems of social entities, especially with regard to moral philosophy.

Peter Heller, after attending French and English schools and Oxford and Cambridge Universities Joint Board School Certificate (honors), the author received his BA in English (magna cum laude) from New York University. At the graduate level, for his MA and PhD degrees, he concentrated on Middle

East Area Studies, also at New York University. His work experience includes employment at The British Council in Cairo, Egypt, and a stint as an associate textbook editor at the Pitman Publishing Corporation in New York. His academic record includes a faculty position in the Department of Government and Politics at Manhattan College in Riverdale, New York, where he has taught a course in Technology and Society for a number of years. He held adjunct positions at other New York institutions. His relevant publications include *Technology Transfer and Human Values: Concepts, Applications Cases* (University Press of America, 1985), *Technology and Society Reader: Case Studies* (rev. ed.), also published by University Press of America, 2012, as well as "Frankenstein's Monster: The Downsides of Technology" appearing in the *International Journal of Technology, Knowledge, and Society* (Spring 2010), based on a paper delivered in Berlin, Germany, in that year.

Jenn Burleson Mackay is assistant professor of multimedia journalism at Virginia Tech. Her primary research interests are journalistic ethics and the influence of technology on journalism. She has studied the various factors that constrain the ethical choices of journalists, the use of mobile technology by journalists, the influence of blogging on journalism, and the influence that new technology had on the 2008 Obama campaign. She enjoys teaching communication ethics, multimedia reporting, and classes that delve into the influence that the media has on society. She has been published in the *Journal of Mass Media Ethics, Journalism Practice, Journalism & Mass Communication Quarterly, Newspaper Research Journal*, the *International Journal of Technoethics*, and several other journals. She has contributed chapters to several books and is co-editor of the book *Media Bias: Finding It, Fixing It.*

Saurabh Mittal is a core faculty member in the area of Information Technology and Online Marketing and serving as Area Chairperson (IT). Prof. Mittal has done PhD, MPhil Computer Sc, and MCA. He has around nine years of experience in industry and academia. He has been actively organising international conferences, workshops, and seminars on information technology and corporate social responsibility. His experience in academic industry spans more than six years. Dr. Mittal has many publications in national and international journals of repute. Dr. Mittal has written and edited books in the domain of Information Technology and Social Responsibility.

Ugo Pagallo is Professor of Jurisprudence at the Department of Law, University of Turin, since 2000, faculty at the Center for Transnational Legal Studies (CTLS) in London, and faculty fellow at the NEXA Center for Internet and Society at the Politecnico of Turin. Member of the European RPAS Steering Group (2011-2012), and the Group of Experts for the Onlife Initiative set up by the European Commission (2012-2013), he is chief editor of the Digitalica series published by Giappichelli in Turin and co-editor of the AICOL series by Springer. Author of nine monographs and numerous essays in scholarly journals, his main interests are AI and law, network and legal theory, robotics, and information technology law (especially data protection law, copyright, and online security). He currently is a member of the Ethical Committee of the CAPER project, supported by the European Commission through the Seventh Framework Programme for Research and Technological Development.

Laura Pană graduated from the Faculty of Philosophy, the University of Bucharest, and continued her dissertation paper in "Philosophy of Action and Philosophy of Conscience" with a doctoral thesis in "Philosophy of Action and Sociology of Knowledge." She first taught Philosophy of Science and Social

Philosophy, as well as Ethics. She was hired at the Polytechnic University in Bucharest, where she still teaches today. In this context, she wrote five books in the field of the Philosophy of Technical Culture (two of which she coordinated). After the computational turn in philosophy, she dealt with the Philosophy of Computing, the Philosophy of Artificial Intelligence, the Philosophy of Virtual Reality—the Philosophy of the Artificial, in general—using skills from the field of psychology, such as the Cognitive Psychology and Psychology of Intelligence and Creativity. She participated in five editions of the European Conference on Computing and Philosophy and attended a Congress and four Conferences on Cybernetics and Systems, and this summer, the 23rd World Congress of Philosophy. Her recent book, illustrative of the possibilism she adopted in her studies, is called *Possibility, Infinity, Predictability* (2009) and follows other books such as *Social and Technological Prognosis*, which is the sixth in the series of books dedicated to the Philosophy of Technology, and which also forms a handbook for a Masters of Studies course. Other studies such as Intellectics and Inventics or Social Invention and Change Management capitalize on personal inquiries in Sociocybernetics. Her interdisciplinary research in the philosophy of technical and human intelligent systems and their social interaction is called Sociomatics. Artificial Ethics can be viewed as an aspect of researching and implementing sociomatic systems.

Ashu Singh is Assistant Professor of Human Resources at Asia-Pacific Institute of Management, New Delhi. She has more than 18 years of teaching experience and has taught MBA, PGDM, MCom, BCom, and BBA scholars. She holds PhD in Commerce from Chaudhary Charan Singh University, Meerut. She has written articles on Management, Disaster Management, and many social issues, which have been published in newspapers and magazines. She has a vast teaching experience in the areas of HR, Marketing, General Management, and Economics. Her areas of interest are Personality, Development and Growth, and Recruitment and Training.

Pericle Slavini graduated in Foreign Languages and Literatures from the University of Pisa in 2000 with a thesis on "Theatre and Technologies." In 2005, he completed a Master of Research degree in Theatre Studies at Lancaster University (UK), where he delved into the telepresence artworks of E. Kac, R. Ascott, and P. Sermon. In 2008, he received his PhD in Biorobotics Science and Engineering from IMT Lucca Institute for Advanced Studies (Italy) and Scuola Superiore Sant'Anna (Italy). In his PhD thesis, he focused on the ethical, legal, and social implications of robot design and deployment. He is currently research fellow at the BioRobotics Institute of Scuola Superiore Sant'Anna, Pisa, Italy. His main research interests are in the fields of Human-Robot Interaction (i.e. aesthetic design, human factor, and social acceptance of robots), and Science and Technologies Studies (i.e. ethical, legal, and social implications of robotics research and applications). He is also involved in activities concerning the use of robots in education and art.

Richard A. Spinello is an Associate Research Professor in the Carroll School of Management at Boston College, where he teaches courses on ethics, social issues in management, corporate strategy, and globalization. Prior to joining the faculty of Boston College, he worked as a consultant and product manager in the software industry. He has written and edited ten books on applied ethics and related areas, including *CyberEthics: Morality and Law in Cyberspace* and *A Defense of Intellectual Property Rights*. He has also written numerous articles and scholarly papers on ethics and management that have appeared in journals such as *Business Ethics Quarterly, The Journal of Business Ethics*, and *Ethics and Information Technology*.

Elettra Stradella (Genoa, 1980), PhD, is Assistant Professor in Comparative Public Law at the University of Pisa. She's the author of more than 70 scientific publications in the field of constitutional rights and freedoms, public institutions, and sources of law, among them a book on *Freedom of Expression and its Limits: Between Theories and Practices* (Giappichelli, Turin, 2008), a book on *The Election of the President of the Republic: Ideas from Europe, Perspectives for Italy* (Pisa University Press, 2013), and she recently edited with Erica Palmerini, in the field of the project RoboLaw (FP7), the book *Law and Technologies: The Challenge of Regulating Technological Development* (Pisa University Press, RoboLaw Series, 2013).

Anna Vartapetiance is presently a PhD (ABD) student in the Department of Computing at the University of Surrey, and has approaching 20 publications relating to Deception Detection and Professional Ethics. She is a Committee Member of British Computer Society's ICT Ethics Specialist Group. She is a Member of British Computer Society (MBCS), Member of International Federation of Information Processing (IFIP) Special Interest Group 9.2.2 Framework on Ethics of Computing (SIG 9.2.2), and Guest Member of International Federation of Information Processing (IFIP) Working Group on Social Accountability and Computing (WG 9.2).

Mark Walker is an Associate Professor in the Philosophy Department, where he occupies the Richard L. Hedden Endowed Chair in Advanced Philosophical Studies. Mark's PhD is from the Australian National University. He previously taught at McMaster University in the Department of Philosophy and in the Arts and Science Program. He serves on the editorial board of the *Journal of Evolution and Technology* and on the board of directors of the Institute for Ethics and Emerging Technologies. Dr. Walker's teaching and research interests include ethics, epistemology, philosophy of law, philosophy of religion, and philosophy of science. His current primary research interest is in ethical issues arising out of emerging technologies (e.g., genetic engineering, advanced pharmacology, artificial intelligence research, and nanotechnology).

Index

A

C

D

E

F

G

H

I

J

L

M

N

O

P

Q

R

S

T

U

V